Methods in Molecular Biology

Series Editor
John M. Walker
School of Life and Medical Sciences
University of Hertfordshire
Hatfield, Hertfordshire, AL10 9AB, UK

For further volumes:
http://www.springer.com/series/7651

Photorespiration

Methods and Protocols

Edited by

Alisdair R. Fernie

Max Planck Institute of Molecular Plant Physiology, Potsdam-Golm, Germany

Hermann Bauwe

Plant Physiology Department, University of Rostock, Rostock, Germany

Andreas P.M. Weber

Institute of Plant Biochemistry, Cluster of Excellence on Plant Sciences (CEPLAS), Heinrich-Heine-University, Düsseldorf, Germany

Humana Press

Editors
Alisdair R. Fernie
Max Planck Institute of Molecular Plant Physiology
Potsdam-Golm, Germany

Hermann Bauwe
Plant Physiology Department
University of Rostock
Rostock, Germany

Andreas P.M. Weber
Institute of Plant Biochemistry, Cluster of Excellence on Plant Sciences (CEPLAS)
Heinrich-Heine-University
Düsseldorf, Germany

ISSN 1064-3745 ISSN 1940-6029 (electronic)
Methods in Molecular Biology
ISBN 978-1-4939-7224-1 ISBN 978-1-4939-7225-8 (eBook)
DOI 10.1007/978-1-4939-7225-8

Library of Congress Control Number: 2017947304

Printed on acid-free paper

This Humana Press imprint is published by Springer Nature
The registered company is Springer Science+Business Media LLC
The registered company address is: 233 Spring Street, New York, NY 10013, U.S.A.

Preface

Photorespiration renders oxygenic photosynthesis possible by acting as a metabolic repair mechanism by which the toxic by-product 2-phosphoglycolate is scavenged and recycled. However, it also leads to high losses of newly assimilated CO_2 from most land plants. The latter fact historically led to photorespiration being classified as wasteful; however, cumulative evidence has suggested that it is rather an important ancillary component of both photosynthesis and the global carbon cycle. Indeed photorespiration is one of the major flux bearing pathways in the terrestrial biosphere, which is multiply linked to other major pathways of central metabolism in land plants. While the study of photorespiratory mutants was initiated some 40 years ago, recent technological advances have been made in the full spectrum of approaches which have been used to analyze photorespiration including multi-parallel approaches to quantify transcripts, enzymes, proteins, and their modifications, photorespiratory metabolites that have followed the adoption of systems biology approaches as well as the exploitation of t-DNA knockout mutants, development of ever sophisticated gas exchange measurements, and cell biological analysis of the evolution of photorespiration. While these methods are becoming commonly adopted there are a number of hurdles which need to be met in their establishment. This volume is therefore intended to present a comprehensive overview of contemporary methods to analyze photorespiration in higher plants.

Potsdam-Golm, Germany
Rostock, Germany
Düsseldorf, Germany

Alisdair R. Fernie
Hermann Bauwe
Andreas P.M. Weber

Acknowledgements

The origins of this book lie within the activities of the DFG project PROMICS-Photorespiration: Origin and Metabolic Integration in Interacting Compartments (http://www.promics.uni-rostock.de) wherein a total of 12 groups based in 5 different German cities collectively collaborated on improving methods and understanding of the role of photorespiration. As such the content for the majority of the chapters has therefore been derived from the project itself as well as contributions from experts in complementary fields, many of whom were invited to one of the two international conferences organized under the auspices of the grant. We would therefore like to thank the Deutsche Forschungsgemeinschaft and the 12 project partners and 10 associate partners for financially supporting the Promics project (DFG FOR1186) and thus for making this book possible.

Contents

Contributors

LUCAS DE ÁVILA SILVA • *Departmento de Biologia Vegetal, Universidade Federal de Viçosa, Viçosa, MG, Brazil*

DOUG K. ALLEN • *Donald Danforth Plant Science Center, St. Louis, MO, USA; United States Department of Agriculture, Agricultural Research Service, St. Louis, MO, USA*

ARMAND D. ANOMAN • *Departament de Biologia Vegetal, Facultat de Farmàcia, Universitat de València, Burjassot, Spain; Estructura de Recerca Interdisciplinar en Biotecnologia i Biomedicina (ERI BIOTECMED), Universitat de València, Burjassot, Spain*

WAGNER L. ARAÚJO • *Departmento de Biologia Vegetal, Universidade Federal de Viçosa, Viçosa, MG, Brazil*

STÉPHANIE ARRIVAULT • *Max Planck Institute of Molecular Plant Physiology, Potsdam, Germany*

HERMANN BAUWE • *Plant Physiology Department, University of Rostock, Rostock, Germany*

RUBEN M. BENSTEIN • *Botanical Institute II, University of Cologne, Cologne, Germany*

PAUL J. BOERSEMA • *Proteomics and Signal Transduction, Max-Planck Institute of Biochemistry, Martinsried, Germany; Department of Biology, Institute of Biochemistry, ETH Zurich, Zurich, Switzerland*

FRANK VAN BREUSEGEM • *Department of Plant Systems Biology, VIB, Ghent, Belgium; Department of Plant Biotechnology and Bioinformatics, Ghent University, Ghent, Belgium*

FLORIAN A. BUSCH • *Research School of Biology, The Australian National University, Canberra, ACT, Australia*

JÉRÉMY COUTURIER • *Faculté des Sciences et Technologies, UMR 1136 Interactions Arbres/Microorganismes, Université de Lorraine/INRA, Vandoeuvre-lès-Nancy, France*

ROSS M. DEANS • *Research School of Biology, The Australian National University, Canberra, ACT, Australia*

MARTIN K.M. ENGQVIST • *Department of Biology and Biological Engineering, Division of Systems and Synthetic Biology, Chalmers University of Technology, Göteborg, Sweden*

ALISDAIR R. FERNIE • *Max-Planck-Institute of Molecular Plant Physiology, Potsdam-Golm, Germany*

IRIS FINKEMEIER • *Plant Proteomics, Max-Planck Institute for Plant Breeding Research, Köln, Germany; Institute of Plant Biology and Biotechnology, University of Münster, Münster, Germany*

MARÍA FLORES-TORNERO • *Departament de Biologia Vegetal, Facultat de Farmàcia, Universitat de València, Burjassot, Spain; Estructura de Recerca Interdisciplinar en Biotecnologia i Biomedicina (ERI BIOTECMED), Universitat de València, Burjassot, Spain*

MARKUS HARTL • *Plant Proteomics, Max-Planck Institute for Plant Breeding Research, Köln, Germany; Mass Spectrometry Facility, Max F. Perutz Laboratories (MFPL), University of Vienna, Vienna Biocenter (VBC), Vienna, Austria*

MEISHA-MARIKA HOLLOWAY-PHILLIPS • *Research School of Biology, The Australian National University, Canberra, ACT, Australia*

FABIAN HOSP • *Proteomics and Signal Transduction, Max-Planck Institute of Biochemistry, Martinsried, Germany*

LARA J. JAZMIN • *Department of Chemical and Biomolecular Engineering, Vanderbilt University, Nashville, TN, USA*
ANIKA KÜKEN • *Systems Biology and Mathematical Modeling Group, Max Planck Institute of Molecular Plant Physiology, Potsdam-Golm, Germany*
OLIVIER KEECH • *Department of Plant Physiology, Umeå Plant Science Centre, Umeå University, Umeå, Sweden*
PAVEL I. KERCHEV • *Department of Plant Systems Biology, VIB, Ghent, Belgium; Department of Plant Biotechnology and Bioinformatics, Ghent University, Ghent, Belgium*
ROXANA KHOSHRAVESH • *Department of Ecology and Evolutionary Biology, University of Toronto, Toronto, ON, Canada*
STEPHAN KRUEGER • *Botanical Institute II, University of Cologne, Cologne, Germany*
FRANZISKA KUHNERT • *Institute of Plant Biochemistry, Cluster of Excellence on Plant Sciences (CEPLAS), Heinrich-Heine-University, Düsseldorf, Germany*
INES LASSOWSKAT • *Plant Proteomics, Max-Planck Institute for Plant Breeding Research, Köln, Germany*
VANESSA LUNDSGAARD-NIELSEN • *Department of Ecology and Evolutionary Biology, University of Toronto, Toronto, ON, Canada*
FANGFANG MA • *Donald Danforth Plant Science Center, St. Louis, MO, USA*
MATTHIAS MANN • *Proteomics and Signal Transduction, Max-Planck Institute of Biochemistry, Martinsried, Germany*
VERONICA G. MAURINO • *Institute of Developmental and Molecular Biology of Plants, Plant Molecular Physiology and Biotechnology Group, Cluster of Excellence on Plant Sciences (CEPLAS), Heinrich-Heine-Universität, Düsseldorf, Germany*
AMNA MHAMDI • *Department of Plant Systems Biology, VIB, Ghent, Belgium; Department of Plant Biotechnology and Bioinformatics, Ghent University, Ghent, Belgium*
KATHARINA MODDE • *Plant Physiology Department, University of Rostock, Rostock, Germany*
ZORAN NIKOLOSKI • *Systems Biology and Mathematical Modeling Group, Max Planck Institute of Molecular Plant Physiology, Potsdam-Golm, Germany*
GRAHAM NOCTOR • *Institut des Sciences des Plantes de Paris-Saclay, Unité Mixte de Recherche 8618 Centre National de la Recherche Scientifique, Université de Paris-Sud, Orsay cedex, France; Unité Mixte de Recherche 9213/Unité Mixte de Recherche 1403, Université Paris-Sud, Centre National de la Recherche Scientifique, Institut National de la Recherche Agronomique, Université d'Evry, Université Paris-Diderot, Sorbonne Paris-Cité, Orsay, France*
ADRIANO NUNES-NESI • *Departmento de Biologia Vegetal, Universidade Federal de Viçosa, Viçosa, MG, Brazil*
TOSHIHIRO OBATA • *Department of Biochemistry, Center for Plant Science Innovation, University of Nebraska-Lincoln, Lincoln NE, USA*
FRANKLIN MAGNUM DE OLIVEIRA SILVA • *Departmento de Biologia Vegetal, Universidade Federal de Viçosa, Viçosa, MG, Brazil*
MARC-SVEN ROELL • *Institute of Plant Biochemistry, Cluster of Excellence on Plant Sciences (CEPLAS), Heinrich-Heine-University, Düsseldorf, Germany*
ROC ROS • *Departament de Biologia Vegetal, Facultat de Farmàcia, Universitat de València, Burjassot, Spain; Estructura de Recerca Interdisciplinar en Biotecnologia i Biomedicina (ERI BIOTECMED), Universitat de València, Burjassot, Spain*
NICOLAS ROUHIER • *Faculté des Sciences et Technologies, UMR 1136 Interactions Arbres/Microorganismes, Université de Lorraine/INRA, Vandoeuvre-lès-Nancy, France*

TAMMY L. SAGE • *Department of Ecology and Evolutionary Biology, University of Toronto, Toronto, ON, Canada*

LEONARDO PEREZ DE SOUZA • *Max-Planck-Institute of Molecular Plant Physiology, Potsdam-Golm, Germany*

STEFANIE SULTMANIS • *Department of Ecology and Evolutionary Biology, University of Toronto, Toronto, ON, Canada*

LEE J. SWEETLOVE • *Department of Plant Sciences, University of Oxford, Oxford, UK*

MAREK SZECÓWKA • *Max-Planck-Institute of Molecular Plant Physiology, Potsdam-Golm, Germany*

STEFAN TIMM • *Plant Physiology Department, University of Rostock, Rostock, Germany*

TAKAYUKI TOHGE • *Max-Planck-Institute of Molecular Plant Physiology, Potsdam-Golm, Germany*

ANDREAS P.M. WEBER • *Institute of Plant Biochemistry, Cluster of Excellence on Plant Sciences (CEPLAS), Heinrich-Heine-University, Düsseldorf, Germany*

PATRICK WILLEMS • *Department of Plant Systems Biology, VIB, Ghent, Belgium; Department of Plant Biotechnology and Bioinformatics, Ghent University, Ghent, Belgium; Medical Biotechnology Center, VIB, Ghent, Belgium; Department of Biochemistry, Ghent University, Ghent, Belgium*

SABINE WULFERT • *Botanical Institute II, University of Cologne, Cologne, Germany*

YI XIAO • *Plant Systems Biology Group, Institute of Plant Physiology and Ecology, Chinese Academy of Sciences, Shanghai, China; University of Chinese Academy of Sciences, Beijing, China*

JAMEY D. YOUNG • *Department of Chemical and Biomolecular Engineering, Vanderbilt University, Nashville, TN, USA*

SHIRIN ZAMANI-NOUR • *Institute of Plant Biochemistry, Cluster of Excellence on Plant Sciences (CEPLAS), Heinrich-Heine-University, Düsseldorf, Germany*

FLAVIEN ZANNINI • *Faculté des Sciences et Technologies, UMR 1136 Interactions Arbres/Microorganismes, Université de Lorraine/INRA, Vandoeuvre-lès-Nancy, France*

HONGLONG ZHAO • *Plant Systems Biology Group, Institute of Plant Physiology and Ecology, Chinese Academy of Sciences, Shanghai, China; University of Chinese Academy of Sciences, Beijing, China*

XIN-GUANG ZHU • *Plant Systems Biology Group, Institute of Plant Physiology and Ecology, Chinese Academy of Sciences, Shanghai, China; University of Chinese Academy of Sciences, Beijing, China; Plant Systems Biology Groups, CAS Key Laboratory for Computational Biology, CAS-MPG Partner Institute for Computational Biology, CAS, Shanghai, China*

AGUSTIN ZSÖGÖN • *Departmento de Biologia Vegetal, Universidade Federal de Viçosa, Viçosa, MG, Brazil*

Chapter 1

Estimation of Photorespiratory Fluxes by Gas Exchange

Florian A. Busch, Ross M. Deans, and Meisha-Marika Holloway-Phillips

Abstract

Photorespiratory fluxes can be easily estimated by photosynthetic gas exchange using an infrared gas analyzer and applying the Farquhar, von Caemmerer, and Berry (Farquhar et al. Planta 149:78–90, 1980) photosynthesis model. For a more direct measurement of photorespiratory CO_2 release from glycine decarboxylation, infrared gas analysis can be coupled to membrane-inlet mass spectrometry, capable of separating the total CO_2 concentration into its $^{12}CO_2$ and $^{13}CO_2$ components in a continuous online fashion. This chapter discusses how to calculate rates of photorespiration from Rubisco kinetics and describes in detail a method for measuring the CO_2 release from glycine decarboxylation using $^{13}CO_2$.

Key words Photorespiration, Photosynthetic gas exchange, Membrane-inlet mass spectrometry, Carbon isotopes, Photosynthesis model, Rubisco kinetics

1 Introduction

The first step in photosynthetic carbon fixation occurs when the enzyme ribulose 1,5-bisphosphate carboxylase/oxygenase (Rubisco) catalyzes the addition of a CO_2 molecule to ribulose 1,5-bisphosphate (RuBP), forming two molecules of 3-phosphoglycerate (3-PGA)—a process referred to as carboxylation. However, as the name of Rubisco suggests, this important enzyme also catalyzes the addition of O_2 to RuBP by oxygenation. The oxygenation of RuBP results in the formation of one molecule of 3-PGA, along with one molecule of 2-phosphoglycolate (2-PG), which is toxic to the plant and must be quickly metabolized. Metabolism of 2-PG in the photorespiratory C_2 cycle is often seen as wasteful as it results in a net release of CO_2 - a process known as photorespiration.

There are several methods for estimating photorespiration by gas exchange [1, 2]. Here we first discuss the estimation of photorespiration via the FvCB model and then describe a method of directly measuring photorespiratory CO_2 release by glycine decarboxylation.

Alisdair R. Fernie et al. (eds.), *Photorespiration: Methods and Protocols*, Methods in Molecular Biology, vol. 1653, DOI 10.1007/978-1-4939-7225-8_1, © Springer Science+Business Media LLC 2017

The simplest method involves inferring the rate of oxygenation from the FvCB model [3]. Net carbon assimilation (A) represents the balance between the rate of gain of CO_2 through carboxylation (V_c) and the rate of loss of CO_2 from oxygenation (V_o) and mitochondrial respiration in the light (R_d) and can be expressed as

$$A = V_c - 0.5V_o - R_d \tag{1}$$

Here, a factor of 0.5 assumes a fully closed photorespiratory C_2 cycle, which equates to the release of 0.5 molecules of CO_2 per oxygenation reaction. However, deviation from 0.5 can occur, with higher values reported in Arabidopsis mutants of the photorespiratory cycle [4] and lower values observed in sunflower plants [5]. V_c and V_o can be related through the CO_2 compensation point in the absence of mitochondrial respiration (Γ^*) and the CO_2 concentration within the chloroplast (C_c) by

$$\frac{V_o}{V_c} = \frac{2\Gamma^*}{C_c} \tag{2}$$

Combining Eqs. 1 and 2 and expressing the result in terms of the photorespiratory flux of CO_2 (R_{PR}) gives

$$R_{PR} = 0.5V_o = \frac{A + R_d}{\frac{C_c}{\Gamma^*} - 1} \tag{3}$$

In inferring photorespiration from Eq. 3, A and R_d can be obtained as a routine measurement via gas exchange. However, good estimates of C_c and Γ^* are required. C_c is calculated from the intercellular CO_2 concentration of the leaf (C_i, determined during gas exchange) and the conductance to CO_2 between the intercellular spaces and the chloroplast, known as mesophyll conductance (g_m), by

$$C_c = C_i - \frac{A}{g_m} \tag{4}$$

Average values of mesophyll conductance are on the order of 0.3 mol $CO_2/m^2/s/bar$ [6] but vary greatly by species [7] or with growth condition [8]. Good estimates are therefore required, or g_m has to be measured concurrently with gas exchange by either chlorophyll fluorescence or stable isotope techniques [9, 10]. Only as a crude estimation should C_i be used.

The value of Γ^* depends on both Rubisco kinetic parameters and the ambient O_2 concentration, and varies with temperature [3]. Ideally Γ^* should be determined in the species under investigation or a literature value for that species used; however, values of Γ^* are difficult to measure accurately and have only been determined in a few species. Because of this, often the literature value of another species is used. This may cause issues if the kinetic properties of Rubisco are not appropriate for the plant under study [11], or have been altered by natural mutation or transgenic approaches [12].

Inferring photorespiration from the FvCB model is convenient as all that it requires is simple gas exchange, but it relies heavily on estimates of Rubisco kinetics and g_m. Here we describe an alternative method that directly measures photorespiration as the CO_2 release from glycine decarboxylation [1, 13]. This is achieved through combined measurement of gas exchange with online membrane-inlet mass spectrometry (MIMS). In this method, switching from natural composition CO_2 (99% $^{12}CO_2$, 1% $^{13}CO_2$) to a high [CO_2] mixture enriched in $^{13}CO_2$ (1% $^{12}CO_2$, 99% $^{13}CO_2$) allows the measurement of the $^{12}CO_2$ efflux from R_{PR} and R_d by MIMS. Switching to high [$^{13}CO_2$] suppresses photorespiration and causes the photorespiratory flux of CO_2 to decay over time. The decay transient for photorespiration is determined from the photorespiratory $^{12}CO_2$ release into the background gas containing $^{13}CO_2$, but being largely devoid of $^{12}CO_2$. Extrapolation of the transient back to the time of switching provides a measure of the sum of photorespiration and mitochondrial respiration prior to switching, while the final steady-state flux reflects the rate of mitochondrial respiration. From this the photorespiratory flux of CO_2 can be determined (*see* Fig. 1).

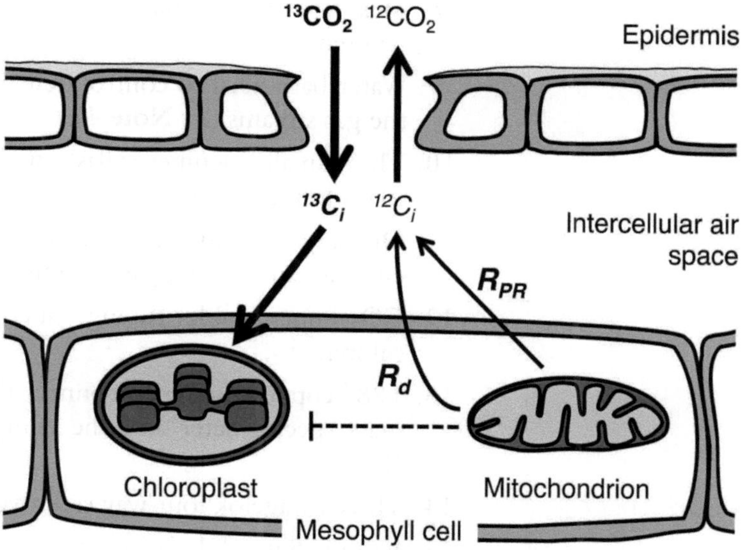

Fig. 1 Replacing $^{12}CO_2$ with $^{13}CO_2$ in the air surrounding the leaf allows us to transiently separate the gross fluxes of CO_2 into and out of the leaf. Newly assimilated CO_2 will be solely $^{13}CO_2$, while released CO_2 only consists of $^{12}CO_2$. The high concentration of $^{13}CO_2$ (*bold arrows*) prevents any $^{12}CO_2$ released from photorespiration and respiration to be reassimilated (*dashed line*). The measured efflux of $^{12}CO_2$ therefore represents the sum of CO_2 released from photorespiration and mitochondrial respiration

2 Materials

Here we describe the materials used to measure photorespiration according to the method of Busch et al. [13]. Other commercial instruments or custom-built devices of similar capability may also be used.

1. Open path infrared gas analyzer (IRGA; e.g., LI-6400XT, Li-Cor, Lincoln, NE, USA).
2. Membrane-inlet isotope ratio mass spectrometer (MIMS; e.g., Delta V Plus, Thermo Scientific, West Palm Beach, FL, USA).
3. Gas-permeable elastic polycarbonate-silicone membrane (e.g., MEM 213, General Electric, Waterford, NY, USA).
4. Tall, 1 L dewar filled with dry ice and topped up with ethanol (*see* **Note 1**).
5. Tank of compressed $^{12}CO_2$ ("normal" CO_2, contains approx. 1% $^{13}CO_2$ impurities), to be connected to the IRGA.
6. Tank of 10,000 μmol/mol $^{13}CO_2$ in air (*see* **Note 2**).
7. Mass flow controllers (e.g., Aalborg, Orangeburg, NY, USA): 5 L/min calibrated for nitrogen and 1 L/min calibrated for oxygen used to mix CO_2 free air, and 5 L/min calibrated for air to control the flow of premixed 10,000 μmol/mol $^{13}CO_2$ in air (*see* **Note 3**).
8. Mass flow meter, calibrated for air in the range of gas flows to be used.
9. Water bath used to control dew point and hence humidity in the gas streams (*see* **Note 4**).
10. Three small Erlenmeyer flasks with rubber stoppers and tubes for entering and exiting gas streams, used as humidifiers.
11. One large Erlenmeyer flask with rubber stopper and tubes for entering and exiting gas streams, used as a mixing flask.
12. CO_2-impermeable tubing, such as 1/4″ Bev-A-Line IV tubing.
13. 1/8″ copper tubing to connect the membrane holder to the mass spectrometer and the compressed $^{12}CO_2$ tank to the IRGA.
14. Three Swagelok four-way crossover ball valves 1/4″.
15. Various Swagelok fittings to connect individual components with each other via the tubing.

3 Methods

3.1 Experimental Setup

A schematic diagram of the experimental setup is outlined in Fig. 2.

1. Connect the tank of compressed $^{12}CO_2$ to the IRGA with copper tubing and use the regulator to set the CO_2 inlet pressure recommended by the manufacturer of the IRGA.
2. Connect the tanks of N_2 and O_2 to their respective mass flow controllers and combine gas streams with a T-fitting. The gas stream is directed into a mixing flask to ensure a homogenous gas mix. Between the mixing flask and the IRGA an additional T-fitting has to be inserted to vent any excess flow. The excess flow of the CO_2-free gas mix is directed to a four-way valve (Valve 1 via Valve 3) between the leaf chamber and the membrane inlet and vented to the atmosphere (*see* **Note 5**).

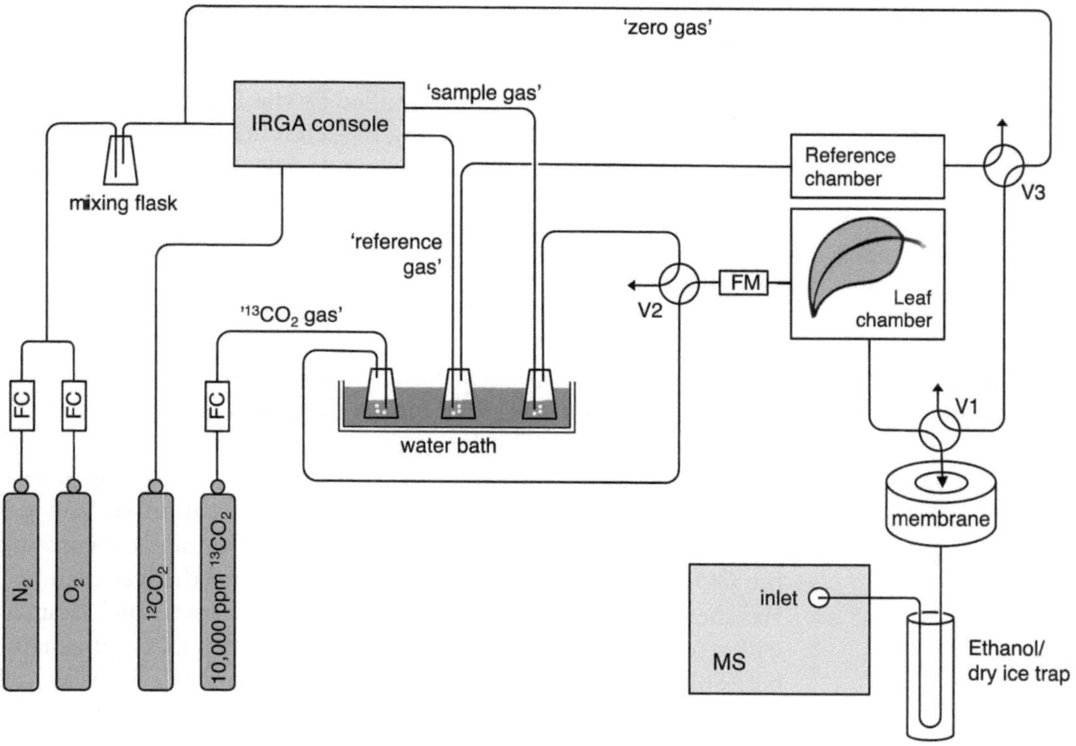

Fig. 2 Schematic diagram of the gas exchange system used to measure photorespiratory fluxes. The IRGA is supplied with CO_2-free air at the desired O_2 concentration. A tank of $^{12}CO_2$ is used as the source to set the chosen CO_2 concentration by the IRGA console, under which photorespiration is to be measured. The "sample gas" and "reference gas" streams are mixed prior to entry into the reference and leaf chambers. The "$^{13}CO_2$ gas" is provided to the leaf chamber premixed from a tank with flow set by a mass flow controller. A water bath is used to control the water mole fraction to the same level across all three gas streams. The air to be analyzed is directed over the membrane of the MIMS used to measure $^{12}CO_2$ concentrations in the $^{13}CO_2$ background gas mix. *FC* mass flow controller, *FM* flow meter, *V1–V3* valve 1–3. *Arrows* indicate gas streams open to the atmosphere

The valve can be switched to supply CO_2-free air for a zero calibration of the MIMS.

3. Both the "reference" and "sample" gas streams coming from the IRGA are directed through an Erlenmeyer flask filled with water and used as a humidifier. Both humidifiers are submerged in the same water bath set to the temperature corresponding to the desired dew point of the humidity in the gas stream.
4. Connect the tank containing 10,000 μmol/mol $^{13}CO_2$ in air to a mass flow controller and an Erlenmeyer flask used as humidifier (hereafter "$^{13}CO_2$ gas"). The humidifier is placed in the same water bath as the other humidifiers (*see* **Note 6**).
5. "Sample gas" and "$^{13}CO_2$ gas" streams are connected to a four-way valve (Valve 2) right before the inlet of the leaf chamber in a manner that allows fast switching from one gas mix to another (*see* **Note 7**). A flow meter can be placed between Valve 2 and the inlet of the leaf chamber (*see* **Notes 8** and **9**).
6. The "reference gas" stream is connected to the reference cell of the IRGA. The reference cell is used to monitor and log CO_2 and H_2O concentrations supplied by the IRGA and the humidifiers. The gas stream exiting the reference cell can be diverted to the membrane inlet via a four-way valve (Valve 3) to supply a standard of known CO_2 concentration (as measured by the reference chamber) to the MIMS during each measurement.
7. The gas stream exiting the leaf chamber is directed over the membrane of the MIMS (*see* **Notes 9** and **10**).

3.2 Measurement Procedure

The described system is set up in a way so that $^{12}CO_2$ concentrations are measured by the MIMS, as the IRGA is sensitive to both $^{13}CO_2$ and $^{12}CO_2$. All other parameters, such as H_2O concentration, temperature, and light intensity, are recorded by the IRGA. Since we can only measure the CO_2 concentration of one gas stream at a time, this prevents us from concomitantly measuring the CO_2 concentrations in the reference cell and the leaf chamber. We therefore have to compare the CO_2 concentration obtained with a leaf in the chamber with a measurement of an empty chamber. This necessitates that the CO_2 and H_2O concentrations as well as the flow rates do not change during the entire duration of the measurement lasting several hours, and the sensitivity of the sensors (IRGA and MIMS) does not change over time or can be accounted for. We therefore recommend a first empty chamber measurement before measuring a leaf, followed by a second empty chamber measurement. When both empty chamber measurements correspond perfectly, it can be assumed that no drifts in the signal or leaks affected the values of a leaf measured in between the two empty chamber measurements.

We set the MIMS to continuously collect data at a 10 Hz interval. The data output used consisted of 1-s averages of the collected raw data (*see* **Note 11**).

1. Start the MIMS and let it run for sufficient time to minimize subsequent drift in the baseline signal. Put dry ice/ethanol trap in place to dry the gas diffusing through the membrane.
2. Set the mass flow controllers for the N_2 and O_2 tanks to the desired flow rates. Set the CO_2 mixer of the IRGA to the desired CO_2 concentration (*see* **Note 12**).
3. Set the water bath to the desired dew point temperature (*see* **Note 13**).
4. Check for possible diffusion leaks throughout the system.
5. MIMS calibration 1: Before the first empty chamber measurement and with Valve 1 set to "zero gas," go through a sequence of Valve 3 set to "zero gas" for 10 min; Valve 3 set to "reference gas" for five set points of the CO_2 mixer, ranging from a concentration of 0 μmol/mol $^{12}CO_2$ to above the highest expected CO_2 concentration; and Valve 3 set to "zero" for another 10 min. This sequence of IRGA CO_2 concentrations can be used to calibrate the MIMS output for CO_2 concentration (*see* **Note 14**).
6. Empty chamber measurement 1: Once the CO_2 mixer of the IRGA is set to the desired concentration and the signal is stable, perform the first empty chamber measurement using the following sequence (*see* **Note 15**):

0–100 s:	Valve 1 set to "zero gas" (*see* **Note 16**)	
100–200 s:	Valve 1 set to "leaf chamber";	Valve 3 set to "reference gas"
200–700 s:	Valve 1 set to "zero gas";	Valve 3 set to "zero gas" (set for the remaining time)
700–1200 s:	Valve 1 set to "leaf chamber";	Valve 2 set to "sample gas"
1200–1400 s:	Valve 1 set to "leaf chamber";	Valve 2 set to "$^{13}CO_2$ gas" (*see* **Note 17**)
1400–1800 s:	Valve 1 set to "leaf chamber";	Valve 2 set to "sample gas"
1800–2200 s:	Valve 1 set to "zero gas"	

This sequence contains a known CO_2 standard, a steady-state condition with the chosen $^{12}CO_2$ concentration, and a 200-s-period, under which photorespiratory $^{12}CO_2$ flux into $^{13}CO_2$

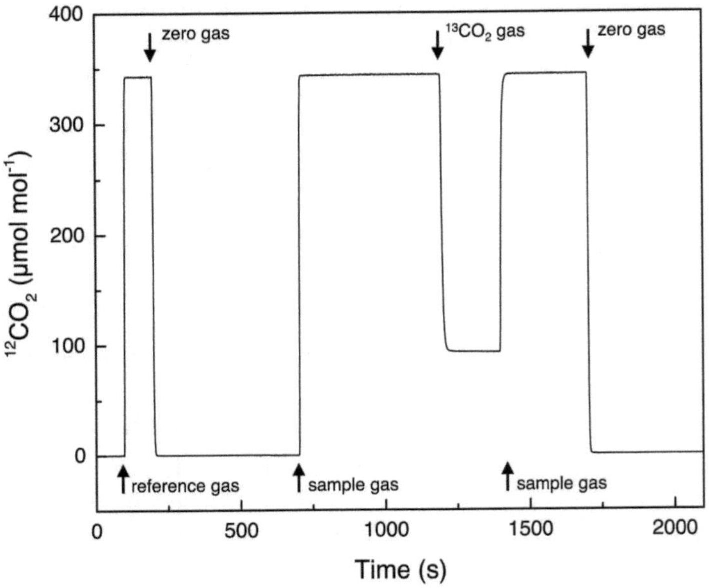

Fig. 3 Sequence of events during the measurement. The measurement is started with two periods of "zero gas," interspersed with a brief period of "reference gas." The leaf has time to come to steady state during the first exposure to the "sample gas." For 200 s the air supply to the leaf chamber is switched to the "$^{13}CO_2$ gas" mix, during which the $^{12}CO_2$ evolution from photorespiration and respiration is observed. The "$^{13}CO_2$ gas" used in this case contains about 85 µmol/mol $^{12}CO_2$. A final period of "zero gas" allows the system to be purged of CO_2 and is used to account for drift in the MIMS baseline

background is measured. An example of the data obtained with the MIMS during this sequence of events is given in Fig. 3.

7. Place leaf in leaf chamber and let it acclimate (*see* **Note 18**).
8. Leaf measurement: Repeat the previous measurement sequence with the leaf in the chamber.
9. Empty chamber measurement 2: Take the leaf out of the chamber and repeat the measurement sequence a second time with an empty chamber (*see* **Note 19**).
10. MIMS calibration 2: After all measurements are taken, repeat the CO_2 calibration procedure for the MIMS (**step 4**).

3.3 Calculation of Photorespiratory Fluxes

Two sets of data have now been obtained: 1-s averages of mV signals corresponding to $^{12}CO_2$ concentrations provided by the MIMS and IRGA data recorded at 4-s intervals.

1. Check and correct for drift in the MIMS baseline. The CO_2 calibration and each individual measurement include several minutes of data recorded with CO_2-free "zero gas." These values can be used to subtract the baseline, which may shift slightly throughout the day (*see* **Note 20**).

2. After subtracting the baseline, convert the corrected mV signal obtained by the MIMS to $^{12}CO_2$ concentrations using a linear regression between mV signal and known CO_2 concentrations (as provided by the IRGA during the calibration measurements; Subheading 3.2, **step 5**). As the sensitivity of the MIMS may change slightly throughout the day, this needs to be done independently for MIMS calibration 1 and 2. The CO_2 concentrations for the measurements sandwiched between the two MIMS calibrations can be calculated accounting for any possible drift in sensitivity. Empty chamber and leaf measurements contain 100 s of a known CO_2 concentration ("reference gas"), which can be used as additional markers to check and account for drift.
3. Now the two empty chamber measurements can be overlaid. $^{12}CO_2$ concentrations before and during the exposure to the $^{13}CO_2$ gas mix should not differ by more than the noise in the signal (*see* **Note 21**). Do the same with the leaf measurement, which should be offset from the empty chamber measurements (Fig. 4).

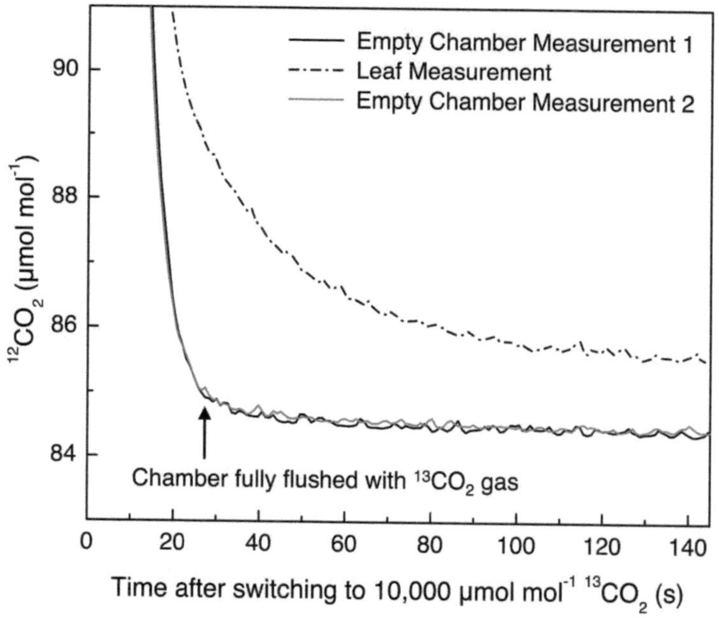

Fig. 4 Comparison of the $^{12}CO_2$ transients obtained from the two empty chamber measurements and the leaf measurement during the supply of "$^{13}CO_2$ gas" mix, after the MIMS mV signal was corrected for drift and converted to CO_2 concentrations. $t = 0$ s denotes the time at which the chamber air supply is switched from "sample gas" to "$^{13}CO_2$ gas", as detected by the MIMS. It takes between 20 and 30 s until the air inside the chamber is completely turned over to the new gas mix. A good correspondence between the two empty chamber measurements indicates reliable measurements and provides confidence in the accuracy of the leaf measurement taken between them

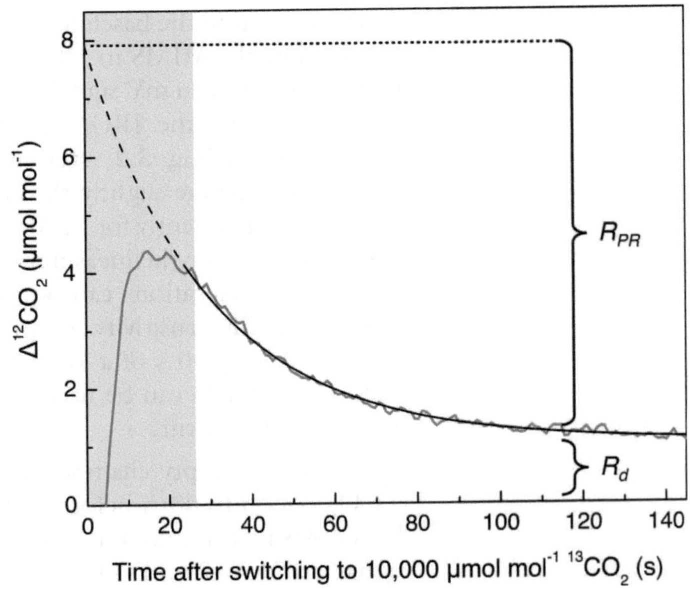

Fig. 5 Measured transient of the change in $^{12}CO_2$ concentration ($\Delta^{12}CO_2$) due to $^{12}CO_2$ release from photorespiration (R_{PR}) and mitochondrial respiration (R_d) during the supply of "$^{13}CO_2$ gas" (*thick grey line*). For the initial period of 20–30 s the leaf chamber has not been fully flushed with "$^{13}CO_2$ gas" yet; therefore these data have to be discarded (*grey shaded area*). A first-order exponential decay function is fitted to the data after that period (*thin black line*) and extrapolated back to the time of the switch (*dashed line*). The value of $\Delta^{12}CO_2$ at $t = 0$ corresponds to the sum of CO_2 emitted from R_{PR} and R_d. We assume that the high CO_2 concentration does not affect the CO_2 release from R_d, which is therefore represented by the offset of the fitted curve. In contrast, CO_2 released by glycine decarboxylation decreases with time as photorespiration is fully suppressed. After about 100 s, the photorespiratory metabolite pools are turned over and $^{12}CO_2$ release from photorespiration is negligible

4. The difference between the leaf measurement and the average of the two empty chamber measurements represents the CO_2 being taken up or released by the leaf. Exposure to 10,000 µmol/mol $^{13}CO_2$ instantaneously reduces Rubisco oxygenation events to negligible amounts, but the CO_2 released from glycine decarboxylation continues for some time with decreasing magnitude (*see* **Note 22**, Fig. 5).
5. For the first 20–30 s (depending on leaf chamber volume and gas flow rate) after switching to the "$^{13}CO_2$ gas mix" the MIMS will detect a mixture of "sample gas" and "$^{13}CO_2$ gas." These data cannot be used for the analysis. Data that need to be discarded can be evaluated from the empty chamber measurements (*see* **Note 23**).

Since we are interested in photorespiratory CO_2 release under the steady-state conditions chosen for the measurement and not under the nonphotorespiratory conditions provided during the exposure to the "$^{13}CO_2$ gas mix," steady-state values have to be extrapolated from the data recorded after the gas in the chamber has fully turned over (Fig. 5). We achieve this by fitting a first-order exponential decay function between the time the chamber was fully flushed with "$^{13}CO_2$ gas" (approx. 30 s) and the end of the exposure to "$^{13}CO_2$ gas" (approx. 200 s), and estimate values at $t = 0$ s with this function (*see* **Note 20**).

6. Photorespiratory flux rates can now be calculated using the difference in steady-state CO_2 concentrations obtained from the MIMS as described above, together with the water mole fractions of the leaf chamber and reference chamber recorded by the IRGA, and the flow rates measured by the mass flow meter (*see* **Note 24**).

4 Notes

1. This is used to condense any water that makes it past the membrane, before the gas enters the inlet of the mass spectrometer. Make sure that during the experiment this dewar always contains enough dry ice.
2. The background gas composition depends on the condition under which photorespiration is going to be measured, i.e., the same N_2 and O_2 concentration as used during the measurement. The high concentration of $^{13}CO_2$ in the absence of $^{12}CO_2$ serves two purposes: It allows only $^{13}CO_2$ to be assimilated, preventing new $^{12}CO_2$ from entering the photorespiratory path, and it prevents any $^{12}CO_2$ released from photorespiration to be reassimilated [13]. The final $^{13}CO_2$ concentration can deviate from 10,000 μmol/mol, but needs to be sufficiently high to suppress photorespiration to a negligible amount. We found this concentration to be a good compromise between the degree of suppression of photorespiration and the consumed amount of $^{13}CO_2$. As the $^{13}CO_2$ concentration is likely too high to be accurately measured in the MIMS, a rough idea of the actual concentration of $^{13}CO_2$ in the premixed gas is sufficient. However, since we are measuring the $^{12}CO_2$ efflux from the leaf, it is important that the concentration of $^{12}CO_2$ in the mix is as low as possible. We recommend

that the proportion of $^{12}CO_2$ should be no more than 1% of the total CO_2 concentration in the mix.

3. This experimental setup provides the most flexibility for mixing the desired gas from tanks of N_2 and O_2 using two mass flow controllers. Alternatively, if photorespiration is to be measured only at 21% oxygen, a tank of CO_2-free air controlled by a single mass flow controller will be sufficient. We recommend not using the built-in soda lime scrub of the IRGA to get rid of the CO_2 in ambient air, as soda lime can exhaust over the time of the experiment, resulting in unstable CO_2 concentrations in the measurement gas stream.
4. It is critical that the humidity in the air stream is kept constant. The water bath therefore has to be capable of keeping a constant temperature and not cycle between a high and a low temperature around the set point.
5. The gas flow coming from the tanks has to be set higher than the flow rate being taken in by the IRGA. The T-connector has to be open to the atmosphere via the four-way valve, so that the pressure at the gas inlet of the IRGA is at ambient pressure.
6. This ensures that all air streams, independent of their flow rate and how they are generated, contain the same water mole fraction. This should be tested before the first measurements. If necessary, the water bath can be stirred to ensure uniform temperature distribution.
7. The quality of the photorespiratory flux measurements depends on the turnover time of the air in the chamber: a small chamber volume combined with a high flow rate and a large leaf area is ideal. We have successfully used the 2 × 3 cm 6400-02B LED light source chamber supplied with the LI-6400XT, which allowed us to automatically control leaf temperature and light intensity while at the same time logging data of all required environmental parameters.
8. The flow rates of the "sample gas" and the "$^{13}CO_2$ gas" stream have to be carefully adjusted to be the same. A flow meter at this location can facilitate this adjustment before the actual measurements.
9. To minimize the turnover time from one gas mix to another, tubing distances should be kept as short as possible between the four-way valve and the mass spectrometer.
10. Ensure that enough tubing is attached after the membrane to prevent any diffusion of room CO_2 back to the exposed surface of the membrane.
11. With our experimental setup, the leaf chamber is fully flushed with the $^{13}CO_2$ gas mix after approximately 30 s. Accurate measurements of the $^{12}CO_2$ concentration during the

remaining 170 s of the exposure to the $^{13}CO_2$ gas mix at 1-s intervals are necessary for reliable data analysis.

12. We suggest a total flow rate of 1 L/min. Mass flow controllers require some time to reach a stable flow rate. Give them sufficient time and make sure that flow rates are constant. We prefer to control the CO_2 mixer of the LI-6400XT based on a mV signal rather than a fixed CO_2 concentration; this ensures a constant CO_2 supply.
13. The water bath has to reach a constant temperature, including the humidifiers immersed in it. A drift in the water bath temperature will cause a drift in the recorded signal. Make sure that with the set dew point the humidity at any point in the system is below the condensation point, particularly with a leaf inside the leaf chamber. Consider heating the experimental setup or the whole room if necessary to avoid condensation.
14. We find that a duration of approximately 5 min at each CO_2 concentration is sufficient time to reach a stable MIMS reading and averaging time.
15. Do not forget to match the IRGAs before each measurement. We record IRGA data in regular intervals of 4 s throughout the measurement, which can be used to verify the stability of the measurement gases as well as photosynthetic parameters of the leaf.
16. It can take several minutes for all CO_2 to be flushed out with "zero gas" to obtain an accurate background reading.
17. The "$^{13}CO_2$ gas" does not have to be flowing for the entire experiment. However, the tank should be opened enough time in advance for the mass flow controller to reach a stable flow rate and the humidifier and tubing to be fully purged with the gas mix. Two to five minutes is a good estimate to go by, depending on the volume of the humidifier, etc.
18. Photosynthetic parameters need to be stable before switching to "$^{13}CO_2$ gas," which includes in particular CO_2 assimilation rate and stomatal conductance. Make sure that no diffusion leaks are present by blowing around the chamber. Use vacuum grease to patch any leaks that may be found.
19. After this point, another leaf may be measured, using the second empty chamber measurement of the first leaf as the first empty chamber measurement of the second leaf. This dramatically reduces the time it takes to obtain more than one full set of measurements.
20. We have used the Peak Analyzer function of the OriginPro software package (OriginLab Corporation, Northampton, MA, USA) for this task.

21. The timing might need to be adjusted between the different measurements. We set $t = 0$ independently for each measurement at the time the first change in CO_2 concentration can be observed by the MIMS after switching to the $^{13}CO_2$ gas mix.
22. We find that the decrease in $^{12}CO_2$ released from glycine decarboxylation can be well described by a first-order exponential decay function.
23. After the switch to "$^{13}CO_2$ gas" the $^{12}CO_2$ concentration recorded by the MIMS will move towards a new steady-state value in the empty chamber measurements. The data prior to stability in the $^{12}CO_2$ values need to be discarded. We assume that the presence of a leaf in the chamber does not alter the time it takes for the gas mix to turn over.
24. Be aware that H_2O transpired by the leaf will affect the absolute CO_2 concentration passing over the membrane. This needs to be taken into account for the final photorespiratory flux calculation. An excellent overview on how to calculate photosynthetic flux rates from measured mole fractions of CO_2 and H_2O can be found in Chapter 1 of the LI-6400XT manual (Li-Cor, Lincoln, NE, USA).

Acknowledgements

F.A.B. is supported by funding from the ARC Centre of Excellence for Translational Photosynthesis awarded to Graham Farquhar.

References

1. Busch FA (2013) Current methods for estimating the rate of photorespiration in leaves. Plant Biol 15(4):648–655. doi:10.1111/j.1438-8677.2012.00694.x
2. Sharkey TD (1988) Estimating the rate of photorespiration in leaves. Physiol Plant 73(1):147–152
3. Farquhar GD, von Caemmerer S, Berry JA (1980) A biochemical model of photosynthetic CO_2 assimilation in leaves of C_3 species. Planta 149(1):78–90
4. Cousins AB, Walker BJ, Pracharoenwattana I, Smith SM, Badger MR (2011) Peroxisomal hydroxypyruvate reductase is not essential for photorespiration in Arabidopsis but its absence causes an increase in the stoichiometry of photorespiratory CO_2 release. Photosynth Res 108(2–3):91–100. doi:10.1007/s11120-011-9651-3
5. Abadie C, Boex-Fontvieille ERA, Carroll AJ, Tcherkez G (2016) *In vivo* stoichiometry of photorespiratory metabolism. Nat Plants 2:15220. doi:10.1038/nplants.2015.220
6. Flexas J, Barbour MM, Brendel O, Cabrera HM, Carriquí M, Díaz-Espejo A, Douthe C, Dreyer E, Ferrio JP, Gago J, Gallé A, Galmés J, Kodama N, Medrano H, Niinemets Ü, Peguero-Pina JJ, Pou A, Ribas-Carbó M, Tomás M, Tosens T, Warren CR (2012) Mesophyll diffusion conductance to CO_2: an unappreciated central player in photosynthesis. Plant Sci 193–194:70–84. doi:10.1016/j.plantsci.2012.05.009
7. von Caemmerer S, Evans JR (2015) Temperature responses of mesophyll conductance differ greatly between species. Plant Cell Environ 38 (4):629–637. doi:10.1111/pce.12449
8. Xiong D, Liu XI, Liu L, Douthe C, Li Y, Peng S, Huang J (2015) Rapid responses of mesophyll conductance to changes of CO_2 concentration, temperature and irradiance are affected by N supplements in rice. Plant Cell Environ 38:2541–2550. doi:10.1111/pce.12558

9. Warren C (2006) Estimating the internal conductance to CO_2 movement. Funct Plant Biol 33(5):431–442. doi:10.1071/fp05298
10. Pons TL, Flexas J, von Caemmerer S, Evans JR, Genty B, Ribas-Carbo M, Brugnoli E (2009) Estimating mesophyll conductance to CO_2: methodology, potential errors, and recommendations. J Exp Bot 60(8):2217–2234. doi:10.1093/jxb/erp081
11. Busch FA, Sage RF (2016) The sensitivity of photosynthesis to O_2 and CO_2 concentration identifies strong Rubisco control above the thermal optimum. New Phytol. doi:10.1111/nph.14258
12. Whitney SM, von Caemmerer S, Hudson GS, Andrews TJ (1999) Directed mutation of the Rubisco large subunit of tobacco influences photorespiration and growth. Plant Physiol 121(2):579–588. doi:10.1104/pp.121.2.579
13. Busch FA, Sage TL, Cousins AB, Sage RF (2013) C_3 plants enhance rates of photosynthesis by reassimilating photorespired and respired CO_2. Plant Cell Environ 36(1):200–212. doi:10.1111/j.1365-3040.2012.02567.x

Chapter 2

Measurement of Transcripts Associated with Photorespiration and Related Redox Signaling

Amna Mhamdi, Pavel I. Kerchev, Patrick Willems, Graham Noctor, and Frank Van Breusegem

Abstract

To study photorespiration and to characterize related components, gene expression analysis is a central approach. An overview of the experimental setup, protocols, and methods we use to investigate photorespiration-associated gene expression is presented. Within this chapter, we describe simple procedures to experimentally alter the photorespiratory flux and provide protocols for transcriptomic analysis with a focus on genes encoding photorespiratory proteins as well as those induced by photorespiratory hydrogen peroxide (H_2O_2). Examples of typical results are presented and their significance to understanding redox signaling is discussed.

Key words Photorespiration, CO_2, Catalase, H_2O_2, Redox signaling, Transcriptomics

1 Introduction

Photorespiratory carbon and nitrogen recycling is a complex metabolic route that involves several intracellular compartments, impacting metabolite pools and redox/energy status at these locations [1, 2]. Classical mutagenesis screens starting in the 1970s established that the photorespiratory recycling pathway and related processes involve specific genes [3, 4]. These genes are characterized by (1) high expression in photosynthetic tissues and (2) a "photosynthetic"-type diurnal rhythm, in which transcripts peak at the end of the night period.

As the isolation of gene-specific mutants has shown, loss of function of many of the proteins involved in photorespiration leads to a "photorespiratory phenotype" characterized by a conditional inability to grow well under conditions favoring substantial rates of photorespiration. Examples of such gene-specific isoforms are GLUTAMINE SYNTHETASE2 (GS2), SERINE HYDROXY-METHYLTRANSFERASE1 (SHM1), and CATALASE2 (CAT2)

Alisdair R. Fernie et al. (eds.), *Photorespiration: Methods and Protocols*, Methods in Molecular Biology, vol. 1653,
DOI 10.1007/978-1-4939-7225-8_2, © Springer Science+Business Media LLC 2017

[5–7]. Genetic redundancy probably explains why no photorespiratory mutants for certain steps were identified using classical screens. For example, it now seems that the *GLYCOLATE OXIDASE1* (*GOX1*) and *GOX2* both have roles in converting glycolate to glyoxylate with the concomitant production of hydrogen peroxide (H_2O_2) [8].

Genes involved in essential photorespiratory reactions in *Arabidopsis thaliana* are summarized in Table 1. Most of them are highly and relatively constitutively expressed [1, 2], especially compared to the marked fluctuations that one can observe in photorespiratory metabolites both in the short and long term. This is probably because (1) photorespiratory flux is mainly controlled by the initiating reaction catalyzed by ribulose-1,5-bisphosphate carboxylase/oxygenase (RuBisCO) and (2) this reaction can occur at very high rates compared to most other plant metabolic pathways, and thus proteins must be in place to transform and transport the metabolites that are produced. Despite this view of photorespiration as a "housekeeping" pathway, evidence continues to emerge suggesting that at least some elements associated with photorespiration play a role in tuning plant responses to the environment [1, 10]. One notable aspect here is the fast H_2O_2 production occurring in photorespiration [11] which can potentially contribute to processes such as plant responses to pathogens [12, 13].

Table 1
List of photorespiration-related genes in *Arabidopsis* and their expression levels in air and high CO_2

Localization	Enzyme name	ID	Transcript level		
			hCO_2	2d air	4d air
Chloroplast	Phosphoglycolate phosphatase				
	(PGLP1)	At5g36700	4393	4356	4657
Peroxisome	Glycolate oxidase				
	GOX1/GOX2	At3g14420/ At3g14415	6321	6943	7150*
Peroxisome	Catalase				
	CAT2	At4g35090	2721	2762	2898
Peroxisome	Glutamate:glyoxylate aminotransferase				
	GGT1	At1g23310	5989	4617*	5453
Peroxisome	Serine:glyoxylate aminotransferase				
	SGAT1	At2g13360	7703	11105*	9972*

(continued)

Table 1
(continued)

Localization	Enzyme name	ID	Transcript level		
			hCO_2	2d air	4d air
Mitochondria	Glycine decarboxylase				
	GDCH1	At2g35370	6090	7933*	7273
	GDCH2	At2g35120	308	279	266
	GLDP1	At4g33010	8242	7610	8983
	GLDP2	At2g26080	1859	1532*	1776
	GDCT	At1g11860	5944	6051	6414
Mitochondria	Serine hydroxymethyl transferase				
	SHM1	At4g37930	9552	7150*	9386
Peroxisome	Hydroxypyruvate reductase				
	HPR1	At1g68010	6002	5689	6566
	HPR2	At1g79870	953	956	966
Chloroplast	Glycerate kinase				
	GLYK	At1g80380	1582	1632	1828
Chloroplast	Glutamine synthetase 2				
	GS2	At5g35630	9351	8773	9800
Chloroplast	Glutamate synthase (GOGAT)				
	GLU1	At5g04140	3204	2022*	3007
Chloroplast	Chloroplast envelope transporters				
	DIT1	At5g12860	5679	5653	5842
	DiT2.1	At5g64290	1659	1736	1773
	DiT2.2	At5g64280	475	611*	573*

Transcript levels in high CO_2 and air are taken from [9]. Data are means of three biological replicates obtained by ATH1 microarray analysis of plants growing at 3000 μL/L CO_2 in short days (high CO_2) for 5 weeks before transfer to 400 μL/L CO_2 at the same day length for 2 or 4 days (2d air, 4d air).

*Indicates significant difference at $P < 0.05$ (Student's t test). Note that *GOX1* and *GOX2* are detected by the same probe set

In this chapter, we describe basic procedures to quantify transcripts in plant tissues by using transcriptomic and targeted methods, first with reference to the photorespiratory pathway in general and, second, with a specific focus on transcripts that can be used to assess stress responses induced by photorespiratory H_2O_2. We also describe protocols that can be applied to alter the photorespiration, including a simple and convenient protocol to promote photorespiratory flux in order to assess the response of plants to this condition.

2 Materials

2.1 Plant Material

1. The *Arabidopsis thaliana* (L.) Heynh. (accession Columbia-0 [Col-0]) seeds.
2. *Catalase2* (*cat2*) mutant seeds.

2.2 Growth Media and Solutions

1. Murashige and Skoog (MS) plant growth medium: For 1 L, weigh 4.3 g MS basal salt mixture, 0.5 g 2-(*N*-morpholino) ethanesulfonic acid (MES), 0.1 g *myo*-inositol, and 10 g sucrose. Add approximately 900 mL deionized water and adjust pH to 5.8 with 1 M KOH. Top up with water to 1 L and add 8 g tissue culture-grade agar. Autoclave at 120 °C, for 15 min. After cooling to 60 °C, pour 100 mL medium in each Petri dish.
2. Carbon dioxide gas: CO_2 (UN1013).
3. TRIzol reagent.
4. Chloroform.
5. Isopropanol.
6. 75% (v/v) ethanol.
7. Diethylpyrocarbonate (DEPC) water.
8. DNase.

2.3 Equipment

1. Plant growth chamber with CO_2 control (e.g., Vötsch or Snijders).
2. Surgical tape (3 M™ Micropore™).
3. Parafilm M®.
4. RNase-free tubes and filter tips.
5. cDNA synthesis kit (such as SuperScript II reverse transcriptase from ThermoFisher Scientific).
6. Petri dishes (diameter 15 cm).
7. Potting mixture (such as Jiffy).
8. Spectrophotometer (Nanodrop or equivalent).

3 Methods

Here we describe approaches that we have used in Arabidopsis to manipulate photorespiration and to quantify transcripts in these conditions. The methods can be adapted to other plants, although the small stature of *Arabidopsis* makes it particularly well suited to some of the procedures.

3.1 Experimental Settings to Conditionally Increase the Metabolic Flux Through the Photorespiratory Pathway

The rate of photorespiratory flux is mainly controlled by the rate of RuBisCO oxygenation. Two factors that can be easily manipulated to vary this rate are (1) the CO_2 concentration surrounding leaves and (2) irradiance. Many studies of photorespiratory mutants have compared responses in air and high CO_2 at typical irradiance found in laboratory growth rooms (100–200 μmol/m^2/s). In such studies, CO_2 levels of 3000 μL/L (7.5-fold current air levels) are sufficient to decrease photorespiration to negligible levels [7]. Other studies seek to supra-stimulate photorespiration. This can be achieved by increasing irradiance or by decreasing CO_2 availability to below typical air levels [14, 15].

3.1.1 In Vitro Photorespiration-Promoting Conditions

The procedure below can be used to promote the rate of photorespiration by placing plants in conditions that promote exhaustion of CO_2 and generation of O_2, processes that can be reinforced by the use of continuous illumination. This protocol has been used to assess the importance of mutations or chemical treatments in wild-type and catalase-deficient Arabidopsis [8, 15].

1. Sow surface-sterilized seeds in Petri dishes containing 100 mL solid MS medium.
2. Evenly space 16 Col-0 seeds in one part of the dish and 16 *cat2* seeds in the other part.
3. Seal the plates with one-layer surgical tape.
4. Place the plates for 3–4 days at 4 °C in the dark to facilitate even germination.
5. Transfer the plates to a controlled growth environment (16-h/8-h light/dark, light intensity, approximately 50 μmol/m^2/s, 21 °C).
6. After 3 weeks, remove the surgical tape and seal the plates with several layers of Parafilm M® to prevent the influx of fresh air.
7. Transfer the plates to a controlled growth environment with continuous light (light intensity approximately 50 μmol/m^2/s, 21 °C) (*see* **Note 1**).

3.1.2 Transfer from CO_2-Enriched Atmosphere to Ambient Air at Moderate or High Irradiance

1. Germinate cold-treated seeds on a potting mixture in a growth chamber with CO_2 control.
2. Maintain CO_2-enriched atmosphere and controlled environmental conditions (*see* **Note 2**).
3. After 3–5 weeks, stop the CO_2 supply and allow the CO_2 levels to equalize between the growth chamber and the ambient air.
4. To compare the effect of CO_2 only, maintain all other conditions identical to those at high CO_2.

5. To drive photorespiration at particularly high rates, increase the irradiance simultaneously with the transfer of the plants to air, if necessary by transfer to another growth chamber with a higher light intensity (*see* **Note 3**).

3.2 Analyses of Photorespiration-Related Transcripts

Besides genes encoding RuBisCO subunits, there are about 20 genes involved in the core photorespiratory pathway (Table 1). Expression levels of these genes can be monitored during transcriptomic analyses with tools, such as the Affymetrix ATH1 microarray or, more frequently now, RNA sequencing. For more targeted specific analyses, quantitative polymerase chain reaction (qPCR) is more convenient. If desired, both can be performed on the same RNA preparation.

In addition to transcripts directly involved in photorespiration, analysis of H_2O_2-sensitive marker genes has been widely used to probe oxidative stress and related signaling processes in model plants deficient in catalase. The photorespiratory *Arabidopsis cat2* mutant allows cellular redox state and gene expression to be manipulated in an easy and predictable manner [16]. Its use has significantly furthered our understanding of interactions in oxidative stress and allowed H_2O_2-sensitive genes to be defined [17–19].

3.2.1 RNA Extraction

For transcriptomic or more targeted studies, use at least three biological replicates.

1. Harvest 50–100 mg leaf material in RNase-free tube and grind tissue under liquid nitrogen to a fine powder.
2. Add 1 mL TRIzol reagent and vortex to dissolve and homogenize the sample.
3. Incubate for 5 min at room temperature.
4. Add 200 μL chloroform, shake by hand for 15 s, and incubate at room temperature for 15 min.
5. Centrifuge for 15 min at 11,000 × *g* at 4 °C.
6. Precipitate RNA by adding 500 μL of isopropanol to the aqueous phase in a fresh RNase-free tube.
7. Incubate at room temperature for 10 min.
8. Centrifuge for 10 min at 11,000 × *g* at 4 °C.
9. Wash the RNA pellet with 1 mL of 75% (v/v) ethanol.
10. Centrifuge for 5 min at room temperature.
11. Dry the pellet for 10 min at room temperature.
12. Redissolve the RNA in 40 μL DEPC water and incubate at 65 °C for 10 min.
13. Leave the sample on ice overnight (*see* **Note 4**).
14. Centrifuge the sample for 10 min at 11,000 × *g* and 4 °C and transfer the supernatant to a new RNase-free tube.

3.2.2 Transcriptomic Analysis

Transcriptomic analysis is typically performed by specialized platforms. Until recently, hybridization techniques that cover a large part of the transcriptome, such as microarray chips, have been used most widely. Data processing of Affymetrix microarrays is relatively standardized, with the preferred method being Robust Multi-chip Average for normalization [20, 21] followed by differential expression analysis with R-packages such as limma [22].

Presently, high-throughput RNA sequencing (RNA-seq) is the method of choice. Compared to microarrays, RNA-Seq is more efficient for detecting low-abundance transcripts, separating different splice forms, identifying genetic variants, and offering a broad dynamic range. However, a good experimental design is a crucial prerequisite for an informative RNA-seq analysis and dependent on the aim of the study. In previous RNA-seq studies of photorespiration in *Arabidopsis*, we sequenced 20–30 million 75-bp single-end reads per sample with three replicates per condition [8]. Compared to microarray data processing, RNA-seq analysis requires more bioinformatics expertise and resources. A simplified workflow to determine differential gene expression by RNA-seq is described below.

1. Obtain raw reads and perform quality controls, using tools such as FastQC (http://www.bioinformatics.babraham.ac.uk/projects/fastqc/) and NGSQC [23].
2. Trim adaptor sequences and remove low-quality reads or bases using tools as Trimmomatic [24].
3. Align preprocessed reads to a genome using spliced aligners such as TopHat [25] or STAR [26].
4. Quantify on gene level using tools such as htseq-count [27] or featureCounts [28].
5. Perform differential expression analysis R-packages such as EdgeR [29] or DESeq2 [30].
6. Define differentially expressed genes (DEGs) by a fold change and significance threshold.
7. Inspect lists of DEGs manually or by bioinformatics analyses, such as gene set enrichment analysis and motif discovery.

Analyses using the ATH1 microarray chip revealed that most of the photorespiratory transcripts are not very sensitive to transfer from high CO_2 to air, at least within the 4-day timescale of our experiment (Table 1). Nevertheless, enzyme and protein activities may respond to altered photorespiratory rates at other levels. For example, extractable activities of catalase increase about twofold within 4 days after transfer to air, an effect that is dependent on expression of the photorespiratory CAT2 isoform [7].

Figure 1 shows an example of microarray data of selected marker genes for which the expression is driven by H_2O_2 in the

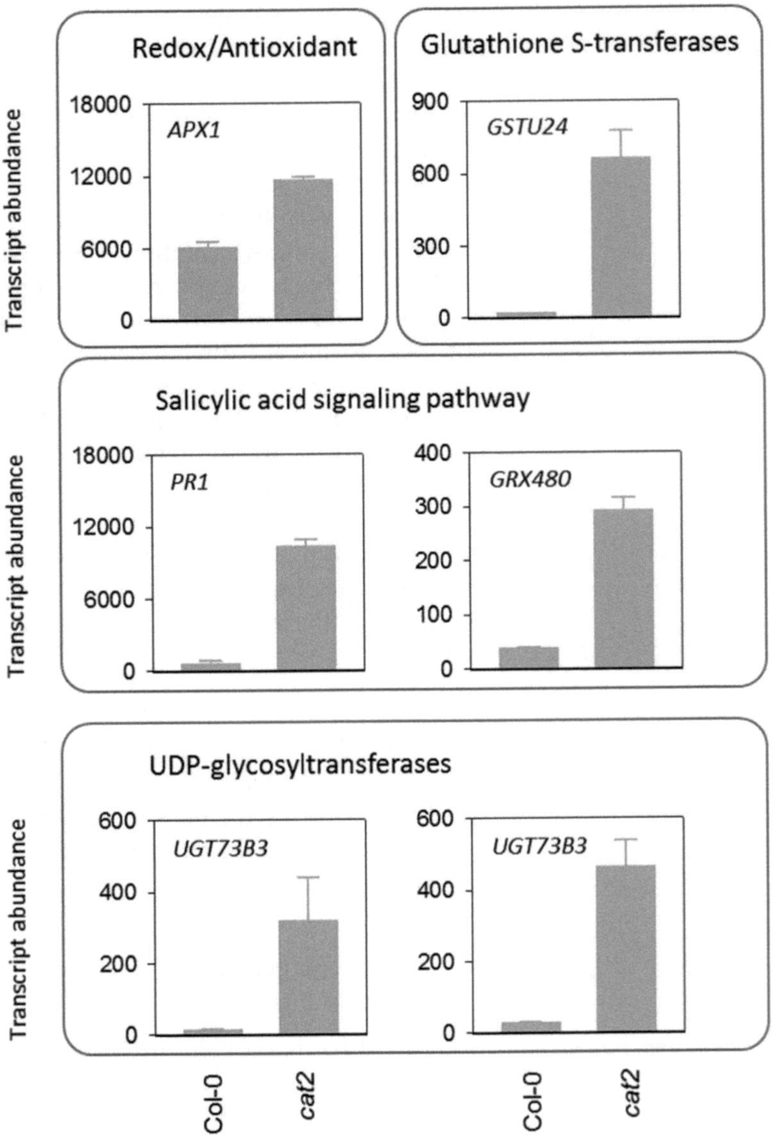

Fig. 1 Example of H_2O_2-inducible genes. Transcript abundance in Col-0 and *cat2* plants grown from seeds for 3 weeks under long-day air conditions. Data are means ± SE of three biological replicates extracted from microarray data. Gene-specific primers for qPCR analyses are given in Table 2

cat2 mutant. These include genes that respond rapidly (within hours) to transfer from nonphotorespiratory conditions to air (specific glutathione *S*-transferases (GSTs) and UDP-glucuronosyltransferases (UGTs)) as well as ASCORBATE PEROXIDASE1 APX1, encoding a cytosolic ascorbate peroxidase that is important in oxidative stress responses [19, 31]. Under some circumstances, catalase-deficient plants growing in photorespiratory conditions strongly activate pathogenesis-related (PR) responses, including marked accumulation of salicylic acid and induction of related genes [12]. *PR1* expression is an excellent marker of this response; although this gene is induced much more slowly by photorespiratory H_2O_2 than the others shown in Fig. 1, typically several days are required in *cat2* growing at moderate irradiance. While catalase-deficient plants are useful systems to dissect the role of oxidative stress in plant responses to the environment, the precise significance of photorespiration in plant responses to pathogens in catalase-depleted plants is yet to be established. Interestingly, many of the genes shown in Fig. 1 can also be significantly induced by growing wild-type plants at high CO_2, where photorespiration is repressed [32].

3.2.3 qPCR Analysis of Transcripts That Are Strongly Induced by Photorespiratory H_2O_2

The *cat2* photorespiratory mutant is one of several excellent model systems that have been used to identify reliable markers for oxidative stress [14, 33, 34]. This is crucial given the difficulties surrounding direct measurement of reactive oxygen species (ROS; [35]). Quantification of ROS-responsive genes is an essential part of the armory available to evaluate oxidative stress responses. With this objective, at least two of the genes shown in Fig. 1 can be analyzed by targeted qPCR using the primer sequences given in Table 2, according to the following protocol.

1. Grow *Arabidopsis* Col-0, *cat2*, and related lines in parallel in conditions that drive photorespiration. In our hands, growth of plants in air from seeds in short-day or long-day conditions at a light intensity of 200 µmol/m^2/s is sufficient to activate H_2O_2 signaling in *cat2* without stressing the wild type (*see* **Note 5**).
2. Harvest 50–100 mg of leaf material rapidly by plunging it into liquid nitrogen and store at −80 °C for a maximum of 6 months. Use three to four biological replicates (three or four independent extractions using different samples).
3. Extract RNA as above (*see* Subheading 3.2.1).
4. Measure the RNA concentration with a spectrophotometer (*see* **Note 6**). Reads at optical density (OD) 260 and 280 are required.
5. Remove DNA contamination by treating 1 µg of the RNA sample with DNase.
6. Prepare cDNA from DNase-treated samples using one the cDNA synthesis kits commercially available and according to the manufacturer's instructions.
7. Perform the qPCR to quantify H_2O_2-inducible transcripts using two or more of the gene-specific primer pairs (Table 2) (*see* **Note 7**).

Table 2
Primer sequences for qPCR analyses of transcripts for enzymes involved in photorespiratory H_2O_2 metabolism and genes induced by increased photorespiratory H_2O_2

Transcript	Transcript ID	Oligonucleotide name[a]	Sequence
Photorespiratory enzymes			
CATALASE			
CAT2	At4g35090	qCAT2for	5′-TGCTGGAAACTACCCTGAATGG-3′
		qCAT2rev	5′-TCAACACCATACGTCCAACAGG-3′
GLYCOLATE OXIDASE			
GOX1	At3g14420	qGO1for	5′-TCTCATTGGCAGCTGAAGGA-3′
		qGO1rev	5′-GAGTGTCCCATTCGGTGGTA-3′
GOX2	At3g14415	qGO2for	5′-TTTGCACTAGCTGCTGAAGGA-3′
		qGO2rev	5′-ATAACCTGGGCAAATGGCGT-3′
Photorespiratory H_2O_2-inducible genes			
ASCORBATE PEROXIDASE 1			
APX1	At1g07890	qAPX1for	5′-GCACTATTGGACGACCCTGT-3′
		qAPX1rev	5′-GCAAACCCAAGCTCAGAAAG-3′
PATHOGENESIS-RELATED 1			
PR1	At2g14610	qPR1for	5′-AGGCTAACTACAACTACGCTGCG-3′
		qPR1rev	5′-GCTTCTCGTTCACATAATTCCCAC-3′
GLUTAREDOXIN CC-TYPE			
GRX480	At1g28480	qGRX480for	5′-GCGGTCCTTGAGATTGATG-3′
		qGRX480rev	5′-AAACCGCCGGTAACTTCAC-3′
GLUTATHIONE *S*-TRANSFERASE U 24			
GSTU24	At1g17170	qGSTU24for	5′-GGCGAGTATGTTTGGGATG-3′
		qGSTU24rev	5′-TTCATCTCGAGGAGCAAGG-3′
UDP-GLYCOSYL TRANSFERASE 74E2			
UGT74E2	At1g05680	qUGT74E2for	5′-GAATCGTCCTCATACCCGAAT-3′
		qUGT74E2rev	5′-GCTTTGGACCCATTTCAACA-3′
UGT73B3	At4g34131	qUGT73B3for	5′-CCTCACCACACCTCTCAACTC-3′
		qUGT73B3rev	5′-TCTGGTAACCCGAGATCCAC-3′

[a]*for* forward, *rev* reverse

8. Run the amplification in duplicate or triplicate technical repeats.
9. To analyze the data, produce a single mean value for each extract by averaging the technical repeats, if they show acceptable levels of reproducibility. From the single values generated for each extract, obtain the final means ($n = 3$ or 4) for each genotype or condition and compare the statistical significance using $P < 0.05$ (*see* **Note 8**).

4 Notes

1. To ensure the efficiency of the RGCL assay, inspect the integrity of the parafilm every day and, if cracked, add new layers.
2. For *Arabidopsis* plants, 3000 μL/L CO_2 is sufficient to largely abolish photorespiration. Growth conditions could be adjusted to day length of choice, light intensity of approximately 120–200 $\mu mol/m^2/s$, and 65% relative humidity.
3. Note that photorespiration saturates with similar kinetics to photosynthesis, so that an irradiance of 600 $\mu mol/m^2/s$ may be sufficient to achieve maximum rates of photorespiration in *Arabidopsis*.
4. Samples should be incubated for at least for 2 h.
5. As detailed in Subheading 3.1, other strategies can be used to promote photorespiration.
6. RNA concentrations could be measured with 1–2 μL of the sample, in case a Nanodrop is used, but for other spectrophotometers, larger amounts might be needed.
7. We use *ACTIN2* as a reference gene. If necessary, a second reference gene can be used (such as *PYROPHOSPHATASE2* or *UBIQUITIN*), but in our conditions, *ACTIN2* transcript signals do not show significant variability and similar results are obtained with different reference genes.
8. qPCR is often better able to detect differences but effects should be qualitatively similar to those observed in microarray analyses.

Acknowledgments

We thank Martine De Cock for help in preparing the revised version of the manuscript. This work was supported by the Agence Nationale de la Recherche project CYNTHIOL (no. ANR12–BSV6–0011) to G.N. and by grants from Ghent University Multidisciplinary Research Partnership “Sustainable BioEconomy” (project 01MRB510W), the Interuniversity Attraction Poles Program (IUAP P7/29), and the Research Foundation-Flanders (grant no. G0D7914N) to F.V.B., and also by the Scientific Exchange program Flanders-France (grant Tournesol T.2008.21) to G.N. and F.V.B.P.K. is a recipient of Omics@VIB Marie Curie COFUND fellowship.

References

1. Foyer CH, Bloom AJ, Queval G et al (2009) Photorespiratory metabolism: genes, mutants, energetics, and redox signaling. Annu Rev Plant Biol 60:455–484
2. Obata T, Florian A, Timm S, Bauwe H, Fernie AR (2016) On the metabolic interactions of (photo)respiration. J Exp Bot 67:3003–3014
3. Somerville CR (1986) Analysis of photosynthesis with mutants of higher plants and algae. Annu Rev Plant Physiol 37:467–507
4. Leegood RC, Lea PJ, Adcock MD et al (1995) The regulation and control of photorespiration. J Exp Bot 46:1397–1414
5. Blackwell RD, Murray AJS, Lea PJ (1987) Inhibition of photosynthesis in barley with decreased levels of chloroplastic glutamine synthetase. J Exp Bot 38:1799–1809
6. Engel N, Ewald R, Gupta KJ, Zrenner R, Hagemann M, Bauwe H (2011) The presequence of Arabidopsis serine hydroxymethyltransferase SHM2 selectively prevents import into mesophyll mitochondria. Plant Physiol 157:1711–1720
7. Queval G, Issakidis-Bourguet E, Hoeberichts FA et al (2007) Conditional oxidative stress responses in the Arabidopsis photorespiratory mutant *cat2* demonstrate that redox state is a key modulator of daylength-dependent gene expression and define photoperiod as a crucial factor in the regulation of H_2O_2-induced cell death. Plant J 52:640–657
8. Kerchev P, Waszczak C, Lewandowska A et al (2016) Lack of GLYCOLATE OXIDASE1, but not GLYCOLATE OXIDASE2, attenuates the photorespiratory phenotype of CATALASE2-deficient Arabidopsis. Plant Physiol 171:1704–1719
9. Queval G, Neukermans J, Vanderauwera S et al (2012) Day length is a key regulator of transcriptomic responses to both CO_2 and H_2O_2 in *Arabidopsis*. Plant Cell Environ 35:374–387
10. Kangasjärvi S, Neukermans J, Li S et al (2012) Photosynthesis, photorespiration, and light signalling in defence responses. J Exp Bot 63:1619–1636
11. Noctor G, Veljovic-Jovanovic S, Driscoll S et al (2002) Drought and oxidative load in the leaves of C_3 plants: a predominant role for photorespiration? Ann Bot 89:841–850
12. Chaouch S, Queval G, Vanderauwera S et al (2010) Peroxisomal hydrogen peroxide is coupled to biotic defense responses by ISOCHORISMATE SYNTHASE1 in a daylength-related manner. Plant Physiol 153:1692–1705
13. Rojas CM, Senthil-Kumar M, Wang K et al (2012) Glycolate oxidase modulates reactive oxygen species-mediated signal transduction during nonhost resistance in *Nicotiana benthamiana* and *Arabidopsis*. Plant Cell 24:336–352
14. Vanderauwera S, Zimmermann P, Rombauts S et al (2005) Genome-wide analysis of hydrogen peroxide-regulated gene expression in Arabidopsis reveals a high light-induced transcriptional cluster involved in anthocyanin biosynthesis. Plant Physiol 139:806–821
15. Kerchev P, Mühlenbock P, Denecker J et al (2015) Activation of auxin signalling counteracts photorespiratory H_2O_2-dependent cell death. Plant Cell Environ 38:253–265
16. Mhamdi A, Queval G, Chaouch S et al (2010) Catalase function in plants: a focus on *Arabidopsis* mutants as stress-mimic models. J Exp Bot 61:4197–4220
17. Mhamdi A, Hager J, Chaouch S et al (2010) Arabidopsis GLUTATHIONE REDUCTASE1 plays a crucial role in leaf responses to intracellular hydrogen peroxide and in ensuring appropriate gene expression through both salicylic acid and jasmonic acid signaling pathways. Plant Physiol 153:1144–1160
18. Tognetti VB, Van Aken O, Morreel K et al (2010) Perturbation of indole-3-butyric acid homeostasis by the UDP-glucosyltransferase *UGT74E2* modulates *Arabidopsis* architecture and water stress tolerance. Plant Cell 22:2660–2679
19. Vanderauwera S, Suzuki N, Miller G et al (2011) Extranuclear protection of

chromosomal DNA from oxidative stress. Proc Natl Acad Sci U S A 108:1711–1716
20. Bolstad BM, Irizarry RA, Åstrand M et al (2003) A comparison of normalization methods for high density oligonucleotide array data based on variance and bias. Bioinformatics 19 185–193
21. Irizarry RA, Hobbs B, Collin F et al (2003) Exploration, normalization, and summaries of high density oligonucleotide array probe level data. Biostatistics 4:249–264
22. Smyth GK (2004) Linear models and empirical Bayes methods for assessing differential expression in microarray experiments. Stat Appl Genet Mol Biol 3:Article 3
23. Dai M, Thompson RC, Maher C et al (2010) NGSQC: cross-platform quality analysis pipeline for deep sequencing data. BMC Genomics 11(Suppl 4):S7
24. Bolger AM, Lohse M, Usadel B (2014) Trimmomatic: a flexible trimmer for Illumina sequence data. Bioinformatics 30:2114–2120
25. Trapnell C, Pachter L, Salzberg SL (2009) TopHat: discovering splice junctions with RNA-Seq. Bioinformatics 25:1105–1111
26. Dobin A, Davis CA, Schlesinger F et al (2013) STAR: ultrafast universal RNA-seq aligner. Bioinformatics 29:15–21
27. Anders S, Pyl PT, Huber W (2015) HTSeq—a Python framework to work with high-throughput sequencing data. Bioinformatics 31:166–169
28. Liao Y, Smyth GK, Shi W (2014) featureCounts: an efficient general purpose program for assigning sequence reads to genomic features. Bioinformatics 30:923–930
29. Robinson MD, McCarthy DJ, Smyth GK (2010) edgeR: a Bioconductor package for differential expression analysis of digital gene expression data. Bioinformatics 26:139–140
30. Love MI, Huber W, Anders S (2014) Moderated estimation of fold change and dispersion for RNA-seq data with DESeq2. Genome Biol 15:550
31. Davletova S, Rizhsky L, Liang H et al (2005) Cytosolic ascorbate peroxidase 1 is a central component of the reactive oxygen gene network of Arabidopsis. Plant Cell 17:268–281
32. Mhamdi A, Noctor G (2016) High CO_2 primes plant biotic stress defences through redox-linked pathways. Plant Physiol 172:929–942
33. Gadjev I, Vanderauwera S, Gechev TS et al (2006) Transcriptomic footprints disclose specificity of reactive oxygen species signaling in Arabidopsis. Plant Physiol 141:436–445
34. Willems P, Mhamdi A, Stael S et al (2016) The ROS wheel: refining ROS transcriptional footprints. Plant Physiol 171:1720–1733
35. Noctor G, Mhamdi A, Foyer CH (2016) Oxidative stress and antioxidative systems: recipes for successful data collection and interpretation. Plant Cell Environ 39:1140–1160

Chapter 3

Measurement of Enzyme Activities

Hermann Bauwe

Abstract

The determination of enzyme activities in organ or organellar extracts is an important means of investigating metabolic networks and allows testing the success of enzyme-targeted genetic engineering. It also delivers information on intrinsic enzyme parameters such as kinetic properties or impact of effector molecules. This chapter provides protocols on how to assess activities of the enzymes of the core photorespiratory pathway, from 2-phosphoglycolate phosphatase to glycerate 3-kinase.

Key words Enzyme assays, Glycerate 3-kinase, Glycine decarboxylase, Glycolate oxidase, Glyoxylate aminotransferase, 2-Phosphoglycolate phosphatase, Photorespiration, Serine hydroxymethyltransferase

1 Introduction

The photorespiratory recycling of 2-phosphoglycolate (2PG) to 3-phosphoglycerate (3PGA) involves many enzymes and transporters that are distributed over four compartments of photosynthesizing plant cells [1]. In the chloroplast, an oxygen-dependent side reactivity of ribulose 1,5-bisphosphate carboxylase generates 2-phosphoglycolate (2PG) that is hydrolyzed to glycolate by 2PG phosphatase (PGLP). Next, in the peroxisome, glycolate oxidase (GOX) uses O_2 to convert glycolate into glyoxylate (the produced H_2O_2 is detoxified by catalase), which becomes transaminated to glycine by the parallel action of glutamate:glyoxylate aminotransferase (GGAT) and serine:glyoxylate aminotransferase (SGAT). Glycine then moves into the mitochondrion where the glycine cleavage multienzyme system (GCS for glycine cleavage system or GDC for glycine decarboxylase complex) and serine hydroxymethyltransferase (SHMT) convert two molecules of glycine to one molecule of serine, which goes back to the peroxisome to return its amino group in the SGAT reaction. The produced 3-hydroxypyruvate (3HP) is reduced to glycerate by 3HP reductase (HPR). The peroxisomal HPR (HPR1) can be efficiently bypassed

Alisdair R. Fernie et al. (eds.), *Photorespiration: Methods and Protocols*, Methods in Molecular Biology, vol. 1653,
DOI 10.1007/978-1-4939-7225-8_3, © Springer Science+Business Media LLC 2017

via a cytosolic HPR (HPR2). Eventually, the glycerate becomes phosphorylated by the chloroplastidal enzyme glycerate 3-kinase (GLYK) to yield 3PGA, which can be used to regenerate ribulose 1,5-bisphosphate. In addition to this core of photorespiratory carbon metabolism, several more enzymes are required for example to re-assimilate the photorespiratory NH_3 in the photorespiratory glutamate synthase cycle or to remove inhibitory 5-formyl tetrahydrofolate produced in a side reaction of SHMT.

The procedures below are based on protocols for measuring individual enzyme activities in extracts of *Arabidopsis thaliana* (Arabidopsis) rosette leaves; modifications of the extraction procedure by the addition of protection reagents such as insoluble PVP or desalting the extract may be necessary for other plants. Usage of robot-based platforms to measure multiple enzyme activities [2] is not a topic of this chapter. Furthermore, the distinction between isoforms of a specific enzyme that are or are not involved in photorespiration or different enzymes that catalyze the same reaction may require specific protocols.

2 Materials

2.1 General Materials

1. Use analytical grade chemicals. Prepare all aqueous solutions in ultrapure water (*see* **Note 1**) and store in a fridge unless indicated otherwise.
2. Bradford reagent for protein quantification.
3. Scintillation cocktail (e.g., Ultima Gold, Perkin Elmer).
4. Plastic scintillation vials (6 ml volume, with screw caps).
5. Liquid scintillation counter (e.g., Perkin Elmer TriCarb 2810 TR).
6. Spectrophotometer with selectable wavelength and with a temperature-controlled cuvette holder (preferably multiple cuvettes).
7. Dry-block thermostat or temperature-controlled water bath.
8. Heating blocks for microtubes, Falcon tubes, and scintillation vials.
9. Stopwatch.

2.2 Materials for Protein Extraction from Leaves and Mitochondria

1. Liquid nitrogen.
2. Refrigerated centrifuge.
3. Extraction buffer: 50 mM Hepes-KOH (pH 7.5), 5 mM $MgCl_2$, 0.2 mM 5′-pyridoxalphosphate (PLP) for PLP-dependent enzymes, 1 mM EGTA, 1 mM dithiothreitol, 0.25% (w/v) BSA, 0.1% (v/v) Triton X-100, 10% (v/v) glycerol, 1 mM phenylmethylsulfonyl fluoride (PMSF; *see* **Note 2**).

4. Lysis buffer: 5 mM Mops, 5 mM Tris, 1 mM β-mercaptoethanol, 1 mM EGTA, 20 μM PLP, 1 mM serine, 1 mM octyl-β-D-glucopyranoside (pH 7.0; *see* **Note 3**).

2.3 Materials for 2-Phosphoglycolate Phosphatase

1. Spectrophotometer set to 820 nm.
2. PGLP assay buffer: 50 mM Hepes-KOH (pH 7.5), 2 mM $MgCl_2$.
3. Ascorbate: 10% (w/v) ascorbic acid in water (prepare fresh).
4. Acid molybdate: 1 N Sulfuric acid, 0.42% (w/v) ammonium molybdate tetrahydrate $(NH_4)_6Mo_7O_{24}·4H_2O)$, store at room temperature.
5. Phosphate calibration solution: 5 mM NaH_2PO_4 in water, store at −20 °C.
6. 2PG: 20 mM 2-phosphoglycolate in water, store at −20 °C.

2.4 Materials for Glycolate Oxidase

1. Liquid-phase oxygen electrode system such as the Oxygraph Plus (Hansatech Instruments) linked to a temperature-controlled water bath set to 25 °C.
2. GOX assay buffer: 75 mM Triethanolamine–HCl (pH 7.8), 0.75 mM oxidized glutathione, 0.1 mM riboflavin-5′-phosphate (FMN), store in the dark (*see* **Notes 4** and **5**).
3. Na-glycolate: 100 mM in water, store at −20 °C.

2.5 Materials for Photorespiratory Aminotransferases

1. Spectrophotometer with a temperature-controlled holder at 25 °C, 340 nm.
2. Semi-microabsorption cuvettes (*see* **Note 6**).
3. Enzyme assay buffer: 100 mM Hepes-KOH (pH 7.5).
4. NADH: 3 mM NADH in water, store at −20 °C.
5. PLP: 20 mM 5′-Pyridoxal phosphate in water, store at −20 °C.
6. Ammonium chloride: 2 M NH_4Cl in water, store at room temperature.
7. Na-glyoxylate: 20 mM Na-glyoxylate in water, store at −20 °C.
8. Na-glutamate: 200 mM Na-glutamate in water, store at −20 °C.
9. Serine: 200 mM L-serine in water, store at −20 °C.
10. GLDH: Bovine liver glutamate dehydrogenase (e.g., Sigma Aldrich, 35 U mg/μl).
11. HPR: Hydroxypyruvate reductase (*see* **Note 7**).

2.6 Materials for Hydroxypyruvate Reductase

1. Spectrophotometer with a temperature-controlled holder at 25 °C, 340 nm.
2. Semi-microabsorption cuvettes (*see* **Note 6**).

3. Enzyme assay buffer: 100 mM Hepes-KOH (pH 7.5).
4. DTT: 100 mM Dithiothreitol in water, aliquot and store at −20 °C.
5. NADH: 3 mM NADH in water, store at −20 °C.
6. Na-hydroxypyruvate: 20 mM in water, store at −20 °C.

2.7 Materials for Glycerate 3-Kinase

1. Spectrophotometer with a temperature-controlled holder at 25 °C, 340 nm.
2. Semi-microabsorption cuvettes (*see* **Note 6**).
3. Enzyme assay buffer: 100 mM Hepes-KOH (pH 7.5).
4. NADH: 3 mM Na-NADH in water, store at −20 °C.
5. ATP: 100 mM Na-ATP in water, store at −20 °C.
6. $MgCl_2$: 100 mM $MgCl_2$ in water.
7. PEP: 50 mM Na-phospho*enol*pyruvate in water, store at −20 °C.
8. Glycerate: 100 mM Na-glycerate in water, store at −20 °C.
9. PK: Pyruvate kinase (e.g., Sigma Aldrich, 500 U/mg).
10. LDH: Lactate dehydrogenase (e.g., Sigma Aldrich, 550 U/mg).

2.8 Materials for the Glycine Cleavage System

2.8.1 Glycine-Dependent NAD^+ Reduction (for the Whole GCS Reaction)

1. Spectrophotometer with a temperature-controlled holder at 25 °C, 340 nm.
2. Quartz or UV-transparent disposable micro plastic cuvettes (e.g., Brand UV microcuvettes, 70–550 μl).
3. Turnover flange stoppers (e.g., 9.8 mm bottom and 10.7 mm diameter to fit onto microtubes and cuvettes).
4. Hamilton syringes.
5. Hypodermic needles.
6. Supply of nitrogen gas.
7. GCS assay buffer: 30 mM K-phosphate (pH 7.2), 1.5 mM EGTA, 1.5 mM $MgCl_2$, 15 μM pyridoxal 5′-phosphate, 1.5 mM dithiothreitol. Degas at a vacuum line and flush with nitrogen before use.
8. NAD^+: 20 mM NAD in water, store at −20 °C.
9. THF: 20 mM Tetrahydrofolate·3HCl in oxygen-free 10 mM Tris–HCl (pH 7), 0.1 M β-mercaptoethanol. Prepare fresh and maintain under nitrogen on ice (*see* **Note 8**).
10. Glycine: 400 mM Glycine in water, aliquot and store at −20 °C.

2.8.2 P-Protein Glycine-^{14}C-Bicarbonate Exchange

1. Dry-block thermostat or temperature-controlled water bath.
2. Heating block at 80 °C (for scintillation vials).
3. Recombinant lipoylated H-protein at about 20 μM (*see* **Note 9**).

4. P-protein assay buffer: 0.5 M Na-phosphate (pH 6.0).
5. ^{14}C-Bicarbonate: 30 mM $NaH^{14}CO_3$ (500 Bq/nmol; *see* **Note 10**).
6. TCA: 50% (v/v) trichloroacetic acid in water.
7. PLP: 1 mM Pyridoxal 5′-phosphate in water, store at −20 °C.
8. DTT: 10 mM Dithiothreitol in water, aliquot and store at −20 °C.
9. Glycine: 200 mM glycine in water, aliquot and store at −20 °C.

2.8.3 Dihydrolipoamide Dehydrogenase

1. Spectrophotometer with a temperature-controlled holder at 25 °C, 340 nm.
2. UV-transparent (>320 nm or better) optical glass or semi-microdisposable plastic cuvettes.
3. DHLD assay buffer: 0.2 M K-phosphate buffer (pH 6.3), 3 mM EDTA, 1.2 mg bovine serum albumin/ml (add fresh prior to use).
4. NADH: 3 mM Na-NADH in water, store at −20 °C.
5. Lipoamide: 20 mM α-Lipoamide in ethanol, aliquot and store at −20 °C.
6. EDTA: 30 mM Na-EDTA in water (pH 7.0).

2.9 Materials for Serine Hydroxymethyltransferase

1. Dry-block thermostat or temperature-controlled water bath.
2. Heating block for microtubes.
3. Heating block at 95 °C (for Falcon tubes).
4. Turnover flange stoppers (e.g., 9.8 mm bottom and 10.7 mm diameter to fit onto microtubes).
5. Hamilton syringes.
6. Hypodermic needles.
7. Supply of nitrogen gas.
8. SHMT assay buffer: 100 mM K-phosphate (pH 7.4), 1.5 mM EDTA, 1.5 mM dithiothreitol, store at 4 °C. Degas at a vacuum line and flush with nitrogen before use.
9. PLP: 20 mM PLP in water, store at −20 °C.
10. KCl: 50 mM KCl in water, store at 4 °C.
11. THF: 20 mM Tetrahydrofolate·3HCl in oxygen-free 10 mM Tris–HCl (pH 7), 0.1 M β-mercaptoethanol. Prepare fresh and maintain under nitrogen on ice (*see* **Note 8**).
12. 200 mM L-Serine in water, store at −20 °C.
13. 3-^{14}C-labeled L-serine (about 2 GBq/mmol; *see* **Note 11**).
14. Na-acetate: 1 M Sodium acetate in water, acetic acid to pH 4.5, store at 4 °C.

15. Formaldehyde: 0.1 M Formaldehyde in water. Prepare fresh by dilution from commercial concentrated solution and store on ice during measurements (*see* **Note 12**).
16. Dimedone: 0.4 M Dimedone in 50% (v/v) ethanol. Prepare fresh by dissolving in ethanol and then adding the water as the solution is not very stable.
17. Toluene.

3 Enzyme Assays

3.1 Leaf Total Soluble Proteins

1. Harvest into a microtube and immediately freeze a minimum of about 100 mg leaf material in liquid nitrogen, preferably at the middle of the light period (*see* **Note 13**). Store at −80 °C if necessary.
2. Add one steel ball (4 mm diameter, precooled in liquid nitrogen) per microtube and grind the leaf material down to a fine powder using a Retsch mill.
3. Add 300 μl of extraction buffer to the powder and mix thoroughly. Centrifuge for 10 min at 15,000 × *g* and 4 °C. Transfer supernatant to a new microtube and store on ice until all samples were processed.
4. Determine protein concentration following the Bradford [3] method.

3.2 Mitochondrial Matrix Proteins

1. Isolate mitochondria from about 50–100 g leaf material according to published protocols, e.g., [4]. The procedure suggested by Keech et al. [5] works very well for Arabidopsis. Determine protein concentration [3]. Store mitochondria at −80 °C until extraction (*see* **Notes 14** and **15**).
2. Dilute mitochondria in lysis buffer to about 1 mg/ml protein and release matrix proteins by lysis in three successive freeze-thaw cycles (liquid nitrogen and water bath at 25–30 °C).
3. Centrifuge at 4 °C, 15,000 × *g*, for 30 min.
4. Concentrate supernatant by using Vivaspin or similar columns with a 10 kDa exclusion limit (e.g., Sartorius) and adjust to a protein concentration of about 1 mg/ml [3].
5. Store on ice or for long-term storage at −80 °C (*see* **Note 15**).

3.3 2-Phosphoglycolate Phosphatase

PGLP (EC 3.1.3.18) belongs to the large haloalkanoic acid dehalogenase protein superfamily with members acting on a variety of substrates including phosphorylated proteins (for example [6]). Historically, PGLP was first purified by Richardson and Tolbert [7]. Catalysis requires Mg^{2+} and comprises two partial reactions: phosphoryl transfer from the substrate to a conserved aspartate followed by hydrolysis of the phospho-aspartyl intermediate to regenerate the enzyme. The metal cofactor Mg^{2+} mediates close approximation of the aspartate carboxyl group to the substrate phosphoryl group [8]. PGLP is highly specific for 2PG with a K_m of about 25 μM and a broad pH optimum [9]. The low-activity cytosolic PGLP2 isoform is not involved in photorespiration [10].

3.3.1 Assay

The assay below follows phosphate release via formation of phosphomolybdenum blue. The procedure is adapted from [9, 11]. It is important that all tubes and bottles are carefully cleaned to avoid phosphate contamination.

1. Bring dry-block thermostat or temperature-controlled water bath to 25 °C. Set heating block to 45 °C.
2. Set spectrophotometer to 820 nm.
3. Prepare ascorbic-molybdate reagent by mixing 100 μl ascorbate and 600 μl acid molybdate each in six 1.5-ml microtubes, one for every time point (0, 2, 4, 6, 8, 10 min).
4. Mix about 100 μg total soluble protein with PGLP assay buffer to a total volume of 1800 μl in a 2 ml microtube. Equilibrate to 25 °C.
5. Add 200 μl of 20 mM 2PG to start the reaction.
6. At every time point (e.g., 0, 2, 4, 6, 8, and 10 min), transfer 300 μl to a microtube containing the ascorbic-molybdate reagent and mix to stop the reaction.
7. Incubate at 45 °C for 20 min.
8. Transfer solution to a semi-microdisposable cuvette and measure absorbance at 820 nm.
9. Measure calibration curves by diluting 0, 1, 2, 5, 10, 15, and 20 μl (0–100 nmol) of the 5 mM phosphate calibration solution with water to a total volume of 300 μl, add to a mixture prepared as above from 100 μl ascorbic acid solution and 600 μl acid molybdate solution, and treat as above. Plot absorbance at 820 nm against nmol phosphate.

3.4 Glycolate Oxidase

GOX (EC 1.1.3.15; also known as short-chain L-2-hydroxy acid oxidase [12]) is a flavin mononucleotide (FMN)-dependent enzyme, which was first described by Clagett et al. [13]. The enzyme uses molecular O_2 as the terminal oxidant. In the first part of a two-step reaction, glycolate reduces the enzyme-bound flavin to flavin hydroquinone and becomes oxidized to glyoxylate, which leaves the active site. Next, in the oxidative half-reaction, a single electron is transferred from the reduced flavin to molecular oxygen to yield the radical pair superoxide anion and flavin semiquinone, followed by formation of H_2O_2 and the fully reoxidized flavin quinone [14, 15].

3.4.1 Assay

The procedure below follows oxygen consumption and is adapted from Hackenberg et al. [16]. It is useful for purified GOX. Alternative assays that follow glyoxylate (*see* Chapter 10) [17, 18] or H_2O_2 production (for example [13, 18]) should be used for leaf protein extracts.

1. Calibrate oxygen electrode to zero- and air-oxygen at 25 °C (*see* **Note 17**).
2. Pipette 900 μl of GOX assay buffer into the chamber and keep the electrode chamber in the dark or dimmed light (*see* **Note 5**).
3. Add 50 μg total soluble protein and close the chamber with the plunger. Switch on the stirrer and monitor the oxygen concentration for about 3 min to determine background oxygen consumption (*see* **Note 18**).
4. Add 50 μl of 100 mM Na-glycolate through the plunger (avoid introducing any air bubbles) and monitor oxygen consumption for another 5–10 min.
5. Subtract background from glycolate-dependent oxygen consumption and calculate rate (full scale corresponds to 250 nmol/ml oxygen).

3.5 Photorespiratory Aminotransferases

Two aminotransferases, glutamate:glyoxylate aminotransferase (GGAT, EC 2.6.1.4) and serine:glyoxylate aminotransferase (SGAT, EC 2.6.1.45), convert glyoxylate to glycine for the subsequent conversion of the two-carbon skeleton of glycolate into the three-carbon skeleton of glycerate [19, 20]. As nearly all aminotransferases, GGAT and SGAT require pyridoxal 5′-phosphate (PLP) and have a relaxed substrate specificity (discussed in [21]). Catalysis includes transaldimination of the PLP cofactor. First, the substrate amino acid binds and replaces the lysine-bound PLP to form an external aldimine that becomes hydrolyzed to pyridoxamine phosphate and the product 2-oxoacid, which leaves the active site. Next, the substrate 2-oxoacid binds and receives the amino group of pyridoxamine to yield the product amino acid and regenerate PLP, which rebinds to the enzyme as an internal aldimine [22].

Established assays couple oxoacid production to reductive amination of 2-oxoglutarate to glutamate by glutamate dehydrogenase (GGAT) or reduction of 3HP to glycerate by NADH-dependent hydroxypyruvate reductase (SGAT). The procedures below are adapted from Igarashi et al. [23] for GGAT and from Liepman and Olsen [24] for SGAT.

3.5.1 Assay of Glutamate: Glyoxylate Aminotransferase

1. Set spectrophotometer to 340 nm at 25 °C.
2. Pipette into cuvette (*see* **Notes 6, 16, and 19**):

(a) 500 μl Enzyme assay buffer (100 mM Hepes-KOH).
(b) 50 μl NADH (3 mM).
(c) 50 μl PLP (20 mM).
(d) 50 μl Na-glyoxylate (20 mM).
(e) 50 μl Ammonium chloride (2 M).
(f) 0.5 U GLDH (15 μg, depending on supplier).

3. Add water and 10–50 μl protein sample (less for the purified enzyme) to a total volume of 900 μl. Mix and measure baseline for about 2 min until stable.
4. Start reaction with 100 μl Na-glutamate (200 mM) and follow change in absorbance to determine the stable initial rate (2–8 min).
5. Calculate enzyme activity with the molar extinction coefficient $\varepsilon = 6220\ M^{-1}\ cm^{-1}$ of NADH at 340 nm (*see* **Note 20**).

3.5.2 Assay of Serine: Glyoxylate Aminotransferase

1. Set spectrophotometer to 340 nm at 25 °C.
2. Pipette into cuvette (*see* **Notes 6, 16, and 19**):
   (a) 500 μl Enzyme assay buffer (100 mM Hepes-KOH).
   (b) 50 μl NADH (3 mM).
   (c) 50 μl PLP (20 mM).
   (d) 50 μl Na-glyoxylate (20 mM).
   (e) 0.5 U HPR.
3. Add water and 10–50 μl protein sample (less for the purified enzyme) to a total volume of 900 μl. Mix and measure baseline for about 2 min until stable.
4. Start reaction with 100 μl serine (200 mM) and follow change in absorbance to determine the stable initial rate (2–8 min).
5. Calculate enzyme activity with the molar extinction coefficient $\varepsilon = 6220\ M^{-1}\ cm^{-1}$ of NADH at 340 nm (*see* **Note 20**).

3.6 Hydroxypyruvate Reductase

HPR (EC 1.1.1.29, D-glycerate dehydrogenase, HPR1 for the peroxisomal HPR) was first identified by Stafford et al. [25] and its location in the peroxisome discovered by Tolbert et al. [26]. The enzyme shows relaxed substrate specificity, also accepting NADPH and glyoxylate as substrates although with much lower activity. For that reason, the enzyme was also purified and crystallized as NADH-dependent glyoxylate reductase [27, 28]. On the other hand, several other enzymes including the cytosolic HPR2 [29] and glyoxylate reductases catalyze the EC 1.1.1.29 reaction [30] though with much lower activity at least in leaves.

Established assays follow NADH oxidation at 340 nm similar to the aminotransferase assays above. The procedure below is adapted from Tolbert et al. [26] and Timm et al. [29].

3.6.1 Assay

1. Set spectrophotometer to 340 nm at 25 °C.
2. Pipette into cuvette (*see* **Notes 6, 16, and 19**):
   (a) 500 μl Enzyme assay buffer (100 mM Hepes-KOH).
   (b) 10 μl DTT (100 mM).
   (c) 50 μl NADH (3 mM).
3. Add water and 10–50 μl protein sample to a total volume of 950 μl. Mix and measure baseline for about 2 min until stable.
4. Start reaction with 50 μl Na-hydroxypyruvate (20 mM) and follow change in absorbance to determine the stable initial rate (2–8 min).
5. Calculate enzyme activity with the molar extinction coefficient $\varepsilon = 6220\ M^{-1}\ cm^{-1}$ of NADH at 340 nm (*see* **Note 20**).

Glyoxylate and NADPH at the same concentrations can be substituted for hydroxypyruvate and NADH, respectively, to measure glyoxylate reductase. Due to generally lower leaf activities, these assays would typically require more protein for reliable measurements.

3.7 D-Glycerate 3-Kinase

GLYK (EC 2.7.1.31) must not be confused with the phylogenetically distinct D-glycerate 2-kinase (EC 2.7.1.165) of bacteria, archaea, and animals [31, 32]. GLYK catalysis follows a random-order sequential mechanism with very high specificity for D-glycerate as phosphate acceptor [33].

An often-used assay couples ADP generation to NADH oxidation via pyruvate kinase and lactate dehydrogenase. This assay however does not distinguish between the 3-kinase and the 2-kinase activity which may be present in some organisms. If distinction is essential, 3PGA formation should be followed as described for example in Kleczkowski et al. [34] or Bartsch et al. [31]. For proper assay, it must also be noted that GLYK of some C_4 grasses, but not of C_3 plants, is auto-inhibited in the absence of thiols [34, 35].

3.7.1 Assay

ADP generation is measured by coupling it to NADH formation via pyruvate kinase (PK) and lactate dehydrogenase (LDH) modified from Kleczkowski et al. [34].

1. Set spectrophotometer to 340 nm at 25 °C.
2. Pipette into cuvette (*see* **Notes 6, 16, and 19**):
   (a) 500 μl Enzyme assay buffer (100 mM Hepes-KOH).
   (b) 50 μl NADH (3 mM).
   (c) 50 μl ATP (100 mM).
   (d) 50 μl $MgCl_2$ (100 mM).
   (e) 50 μl PEP (50 mM).
   (f) 2.5 U Each of PK and LDH.

3. Add water and 10–50 μl protein sample (less for the purified enzyme) to a total volume of 950 μl. Mix and measure baseline for about 2 min until stable.
4. Start reaction with 50 μl Na-glycerate (100 mM) and follow change in absorbance to determine the stable initial rate (2–8 min).
5. Calculate enzyme activity with the molar extinction coefficient $\varepsilon = 6220\ M^{-1}\ cm^{-1}$ of NADH at 340 nm (*see* **Note 20**).

3.8 Glycine Cleavage System

Glycine serves both as a one-carbon donor and as a one-carbon acceptor to produce serine in the mitochondria. In course of this process, the GCS [36] provides methylene tetrahydrofolate to SHMT. The multistep glycine cleavage reaction was first discovered in bacteria [37]. The whole reaction cycle comprises three reactions, which are catalyzed by the PLP-dependent enzyme P-protein (glycine decarboxylase; EC 1.4.4.2), the tetrahydrofolate (THF)-dependent enzyme T-protein (aminomethyltransferase; EC 2.1.2.10), and the NAD^+-dependent enzyme L-protein (dihydrolipoamide dehydrogenase; EC 1.8.1.4). The reactions are coupled by a shared lipoylated substrate protein, H-protein (H for hydrogen carrier protein). The lipoyl arm in the dithiolan form serves as an oxidant during the oxidative decarboxylation of glycine by the P-protein and conveys the produced aminomethylene entity to the T-protein, which forms methylene THF (CH_2-THF) as the eventual product of glycine decarboxylation. Thereafter, the lipoyl arm is reduced and must be reoxidized by L-protein, reducing NAD^+ to NADH. All reactions of the GCS reaction cycle are basically reversible (then referred to as glycine synthase).

The four proteins occur in very large amounts (about 130 mg/ml matrix space corresponding to about 32% of the total matrix protein) in green-leaf mitochondria, where they are organized in a multienzyme complex, which greatly facilitates large-scale enzyme preparation and the assay of the combined and individual reactions.

3.8.1 Glycine-Dependent NADH Formation (Entire GCS Reaction Cycle)

This assay is an adaptation of the methods in Bourguignon et al. [4] and Besson et al. [38].

1. Prepare mitochondrial matrix extracts as described in Subheading 3.2.
2. Set spectrophotometer to 340 nm at 25 °C.
3. The GCS buffer and all the stock solutions should be maintained under (a slow stream of) nitrogen to exclude oxygen.
4. Flush the cuvette and pipette tip with nitrogen.
5. Pipette into the cuvette (*see* **Notes 8** and **16**):
   (a) 266 μl of GCS assay buffer plus 34 μl of water.

(b) 50 μl Matrix protein extract (can be varied, calculate the amount of water above).

(c) 20 μl NAD^+ (20 mM).

6. Apply turnover flange stopper and flush with nitrogen through a hypodermic needle (insert second hypodermic needle as outlet).
7. Add 10 μl THF (20 mM) with a Hamilton syringe. Pipette up and down to mix and measure baseline at 340 nm for about 2 min until stable.
8. Add 20 μl glycine (400 mM) with a Hamilton syringe to start reaction and follow change in absorbance to determine the stable initial rate (2–8 min).
9. Calculate enzyme activity with the molar extinction coefficient $\varepsilon = 6220\ M^{-1}\ cm^{-1}$ of NADH (*see* **Note 20**).

3.8.2 Glycine-^{14}C-Bicarbonate Exchange (P-Protein)

This assay follows the artificial reverse reaction catalyzed by P-protein in the presence of H-protein and bicarbonate [39] and is an adaptation of the methods in Bourguignon et al. [4] and Hasse et al. [40] (*see* **Note 21**).

1. Prepare mitochondrial matrix extracts as described in Subheading 3.2 or use the purified enzyme.
2. Bring dry-block thermostat or temperature-controlled water bath to 25 °C.
3. Prepare four liquid scintillation vials per reaction by pipetting 80 μl of 50% TCA onto the bottom of each vial.
4. Place one 1.5 ml microtube per reaction into block thermostat.
5. Have a stopwatch ready and a pipetting scheme that reminds you what to do when in case you are handling multiple samples at the same time.
6. Pipette into each microtube (this mix is sufficient for four time points; *see* **Note 22**):

(a) 50 μl P-protein assay buffer.

(b) 50 μl DTT (10 mM).

(c) 50 μl PLP (1 mM).

(d) 150 μg Lipoylated H-protein (*see* **Note 23**).

(e) 75 μl Glycine (200 mM, 75 μl water for blank reactions).

(f) Water to a total volume (including protein from next step) of 450 μl.

7. Add 10–50 μg mitochondrial matrix protein.
8. Start reaction with 50 μl ^{14}C-bicarbonate (30 mM) and immediately transfer 100 μl into the TCA drop in a scintillation vial and mix to stop the reaction.

9. At each further time point (for example 10, 20, 30 min), remove another 100 µl of the reaction mixture and stop the reaction as before.
10. Dry content of all vials overnight (keep vials open) under a fume hood at 80 °C.
11. Add 500 µl water to each vial, cap, and vortex to completely dissolve the residue.
12. Prepare a positive (1 µl 30 mM $NaH^{14}CO_3$ into 500 µl 0.01 M NaOH) and a negative control (500 µl water). Also include control reactions without glycine.
13. Add 5 ml liquid scintillation cocktail per vial, cap, and vortex to emulsify.
14. Count disintegrations per time and calculate rates from the specific radioactivity of the bicarbonate substrate, considering dilution factor(s) and the controls from **step 12**.

The method above may be modified by measuring two time points only (including the zero-time blank control) once linearity has been established.

3.8.3 Dihydrolipoamide Dehydrogenase (L-Protein)

This assay measures lipoate reduction, which is the reverse direction. Details on how to measure the forward reaction can be found for example in Yan et al. [41], from where this method was adapted. Note that dihydrolipoamide dehydrogenase is also present in plastids.

Assay

1. Prepare mitochondrial matrix extracts as described in Subheading 3.2.
2. Set spectrophotometer to 340 nm at 25 °C.
3. Pipette into the cuvette (*see* **Notes 16** and **19**):
   (a) 500 µl DHLD assay buffer (200 mM K-phosphate).
   (b) 50 µl Lipoamide (20 mM).
   (c) 50 µl NADH (3 mM).
   (d) 50 µl EDTA (30 mM).
   (e) Add water to a total volume of 990 µl. Mix and measure baseline for about 2 min until stable.
4. Add 10 µl matrix protein extract (less for the purified enzyme) to the cuvette and follow the change in absorption for another 2–8 min.
5. Calculate enzyme activity with the molar extinction coefficient $\varepsilon = 6220\ M^{-1}\ cm^{-1}$ of NADH (*see* **Note 20**).

3.9 Serine Hydroxymethyltransferase

Serine hydroxymethyltransferase (EC 2.1.2.1) catalyzes a reversible PLP-dependent reaction. Similar to the aminotransferases and the P-protein, the PLP initially bound as an "internal" aldimine to the ε-amino group of a conserved lysine residue reacts with the α-amino group of the respective substrate amino acid in a transaldimination reaction to form an "external aldimine." Binding of glycine and CH_2-THF basically follows a random-order sequential mechanism. In the ternary complex, the methylene group of CH_2-THF becomes released from the THF's N^{10}-nitrogen atom followed by the addition of H_2O and transfer of the resulting hydroxymethyl moiety from the THF's N^5-nitrogen atom to the α-carbon of glycine [42]. In the photorespiratory pathway, the more than twofold higher activity of the GCS reactions relative to SHMT ensures a high CH_2-THF/THF ratio in the mitochondrial matrix, pushing the SHMT reaction in the thermodynamically non-favored direction towards the synthesis of serine [43].

Two out of five SHMTs, SHMT1 and SHMT2, are located in the mitochondria of Arabidopsis [44] and function as redundant enzyme components of the glycine-serine cycle of one-carbon metabolism in most cells. Quantitatively, SHMT1 dominates in the mitochondria of photosynthesizing cells, and only this isoform is essential for photorespiration at least in Arabidopsis [45, 46].

3.9.1 Assay

Several methods have been used to determine SHMT activity including enzyme-coupled assays [47] and an HPLC-based fluorometric folate assay [48]. The method below is an adaptation of the protocol proposed by Taylor and Weissbach [49] as modified by Eisenhut et al. [50]. The ^{14}C-labeled hydroxymethylene group of the serine is transferred to THF to form CH_2-THF. After termination of the reaction, the radiolabeled moiety is released as formaldehyde and trapped in a dimedone-formaldehyde adduct (*see* **Note 21**).

1. Prepare mitochondrial matrix extracts as described in Subheading 3.2 or use the purified enzyme.
2. Bring dry-block thermostat or temperature-controlled water bath to 25 °C. Set heating block for 15 ml Falcon tubes to 95 °C.
3. Have a stopwatch ready and a pipetting scheme that reminds you what to do when in case you are handling multiple samples at the same time.
4. The SHMT buffer and all the stock solutions should be maintained under (a slow stream of) nitrogen to exclude oxygen.
5. Pipette into a Falcon tube (for four assays, *see* **Notes 8, 16, and 19**):
   (a) 3.75 ml SHMT assay buffer (100 mM K-phosphate, 1.5 mM EGTA, 1.5 mM DTT).

(b) 250 μl KCl (50 mM).

(c) 60 μl PLP (20 mM).

(d) 100 μl Serine (200 mM).

(e) 165 μl Water.

(f) Flush with and then maintain under a slow but continuous stream of nitrogen.

6. Add 150 μl Na-acetate each into 15 ml Falcon tubes, one for each time point (e.g., 0, 5, 10, 15 min).
7. Flush 2 ml microtube (one per assay), pipette tip with nitrogen, and add 865 μl of the above mixture plus 100 μl matrix protein (variable but adjusted to a total volume of 965 μl). Seal microtube with turnover flange stopper and incubate for 5 min at 25 °C under a slow stream of nitrogen applied through a hypodermic needle (insert second hypodermic needle as outlet).
8. Add 10 μl ^{14}C-L-serine with a Hamilton syringe and mix. Remove a 100 μl zero-time aliquot and mix with the Na-acetate in one of the Falcon tubes.
9. Add 50 μl THF (20 mM) with a Hamilton syringe and mix to start the reaction.
10. At each time point (e.g., 5, 10, 15 min), transfer a 100 μl aliquot into the Na-acetate drop in one of the Falcon tubes to stop the reaction.
11. Successively add 100 μl 0.1 M formaldehyde and 150 μl 0.4 M dimedone to each Falcon tube. Vortex shortly, close tube, and incubate at 95 °C for 10 min to accelerate formation of the formaldehyde-dimedone complex. Afterwards store on ice for 5 min. Treat zero-time sample exactly the same way.
12. Add 3 ml toluene per tube, vortex, and centrifuge at 5000 × *g* for 5 min.
13. Mix 2 ml supernatant with 2 ml liquid scintillation cocktail. Count disintegrations per time and calculate rates with the specific radioactivity of the serine in the assay. Note that the specific radioactivity must be calculated since it is not equal but much less than the specific radioactivity given by your supplier due to dilution with unlabeled serine.

    The method above may be modified by measuring two time points only (including the zero-time blank control once linearity of the assay has been established).

4 Notes

1. Ultrapure water has an electrical conductivity of 5.5×10^{-8} S/cm corresponding to 18 MΩ cm at 25 °C.

2. PMSF is unstable in aqueous solution and must be added just before extraction from a 0.1 M stock in isopropanol.
3. Protease inhibitors such as leupeptin, Pefabloc, or inhibitor cocktails may also be added.
4. The Good buffers based on the piperazine ring including HEPES form radicals in the presence of H_2O_2 and hence are not suitable for measuring GOX activity. The oxidized glutathione stabilizes the reaction by protecting it against reducing agents in the extract [17].
5. FMN is light sensitive. Prepare a 10 mM stock solution in water and store aliquots at −20 °C. Add to buffer prior to the measurements and cover the tube with alumina foil.
6. Standard semi-micro (micro for the GCS assay) UV-transparent (>320 nm or better) optical glass or disposable plastic cuvettes.
7. Hydroxypyruvate reductase (glycerate dehydrogenase) is presently available only from US Biological as human recombinant glyoxylate reductase/hydroxypyruvate reductase. You may choose to produce recombinant plant HPRl as described in Timm et al. [29].
8. THF is very oxygen sensitive and very unstable in solution particularly at neutral or basic pH! Always handle it under anaerobic conditions and flush GCS or SHMT buffer with oxygen-free gas before dissolving or diluting the THF. Using an anaerobic bench is helpful for weighing and aliquoting THF. Folate poly-glutamylation lowers the K_m but does not strongly affect V_{max} [38]; hence poly-glutamylated THF is not necessary for this assay. A commercial supplier of good-quality THF and an informative product information sheet is Schircks Laboratory, Bauma, Switzerland. Please notice that commercial THF usually is an about equimolar mixture of two diastereomers and that only the (6S)-THF acts as a substrate of GCS and SHMT. The molar extinction coefficient of THF is about 31,900 M^{-1} cm^{-1} at 298 nm [51].
9. H-protein is not commercially available. You may have to produce recombinant H-protein, which also needs lipoylation by lipoate-protein ligase, as described in Hasse et al. [40].
10. Specific radioactivity is given by your supplier, e.g., Perkin Elmer, 40-60 mCi (1.48–2.22 GBq) per mmol, and should be diluted to your needs in a bicarbonate solution of appropriate concentration. Ensure that dilution solvent is basic. Aliquoting the solution into an acidic solvent will result in significant radioactivity loss of gaseous carbon dioxide. Since the Na-bicarbonate solutions are thermally sensitive, follow the procedure recommended by your supplier to avoid incomplete recovery.

11. C-3-labeled serine gives more reproducible results in comparison with U-labeled serine.
12. Formaldehyde is toxic and must be handled under a fume hood whenever possible.
13. For reproducible results, plants should be grown in controlled environment to a defined developmental stage; for example fully developed rosette leaves of Arabidopsis correspond to growth stage 5.1 according to Boyes et al. [52]. Diverse mixtures and media can be used for reproducibly growing Arabidopsis (https://abrc.osu.edu/seed-handling). Because Arabidopsis is a facultative long-day plant, flowering is delayed and vegetative growth favored under short days (<12 h). Typical short-day conditions are a 10-h photoperiod, 20/18 °C day/night temperature, 120–150 μmol photons/m^2/s, and 60–70% relative humidity. Growth of most photorespiratory mutants requires that the air CO_2 concentration is elevated to 0.5% [53] or even 1% (PGLP mutants) in CO_2-controlled growth cabinets. Knockout mutants for photorespiratory enzymes that are essential in other metabolism, for example GCS component proteins, cannot be grown even under elevated CO_2.
14. Reliable assays of GCS and SHMT require preparation of mitochondria. Here, or generally when organelles must be prepared from leaves and there is no specific reason to use light-adapted material, leaves should be harvested (not frozen) at the end of the night to minimize chloroplast disruption by starch granules because broken chloroplasts interfere with mitochondria isolation.
15. Freezing disrupts the mitochondria—they are no longer "intact." Freezing of the lysate may damage proteins: add 30% (v/v) glycerol to maintain enzymatic activity at −80 °C.
16. Store sensitive substrates on ice during the experiment. If they were stored frozen, thaw on ice, invert one or two times, and collect the solution by short centrifugation.
17. Oxygen concentration in a stirred sample of air-saturated deionized water held at 25 °C at a standard atmospheric pressure of 100 kPa is 250 nmol/ml [54]. Correction for different pressures is usually not necessary.
18. It is very important that the stirring flea spins smoothly and that there are no air bubbles present. At least 3–4 mm of the solution must be visible in the plunger to isolate the chamber from the environmental air.
19. This mix can be prepared for several assays in advance and stored on ice but must be brought to 25 °C before adding the protein sample.

20. An absorbance change of −0.622 corresponds to the oxidation of 100 (40) nmol NADH in 1 (0.4) ml. An absorbance change between 0.05 and 1 per minute is about perfect for data evaluation.
21. It is important to observe all legal and institutional regulations for work with radioisotopes.
22. Always include a control reaction without glycine.
23. Addition of extra lipoylated H-protein is necessary due to dilution of the H-protein present in the matrix extract (150 μg per 0.5 ml corresponds to about 20 μM H-protein).

Acknowledgments

This work was supported by grants of the Deutsche Forschungsgemeinschaft particularly in the DFG Forschergruppe 1186 Photorespiration: Origin and Metabolic Integration in Interacting Compartments (Promics). Stefan Timm (PGLP and optical tests), Maria Wittmiß (GCS), and Ralph Ewald (SHMT), all at Rostock University, kindly double-checked the provided protocols.

References

1. Bauwe H, Hagemann M, Fernie AR (2010) Photorespiration: players, partners and origin. Trends Plant Sci 15:330–336
2. Gibon Y, Blaesing OE, Hannemann J, Carillo P, Höhne M, Hendriks JHM, Palacios N, Cross J, Selbig J, Stitt M (2004) A robot-based platform to measure multiple enzyme activities in arabidopsis using a set of cycling assays: comparison of changes of enzyme activities and transcript levels during diurnal cycles and in prolonged darkness. Plant Cell 16:3304–3325
3. Bradford MM (1976) A rapid and sensitive method for the quantitation of microgram quantities of protein utilizing the principle of protein-dye binding. Anal Biochem 72:248–254
4. Bourguignon J, Neuburger M, Douce R (1988) Resolution and characterization of the glycine cleavage reaction in pea leaf mitochondria. Properties of the forward reaction catalysed by glycine decarboxylase and serine hydroxymethyltransferase. Biochem J 255:169–178
5. Keech O, Dizengremel P, Gardeström P (2005) Preparation of leaf mitochondria from *Arabidopsis thaliana*. Physiol Plant 124:403–409
6. Seifried A, Knobloch G, Duraphe PS, Segerer G, Manhard J, Schindelin H, Schultz J, Gohla A (2014) Evolutionary and structural analyses of mammalian haloacid dehalogenase-type phosphatases AUM and chronophin provide insight into the basis of their different substrate specificities. J Biol Chem 289:3416–3431
7. Richardson KE, Tolbert NE (1961) Phosphoglycolic acid phosphatase. J Biol Chem 236:1285–1290
8. Allen KN, Dunaway-Mariano D (2009) Markers of fitness in a successful enzyme superfamily. Curr Opin Struct Biol 19:658–665
9. Hardy P, Baldy P (1986) Corn phosphoglycolate phosphatase—purification and properties. Planta 168:245–252
10. Schwarte S, Bauwe H (2007) Identification of the photorespiratory 2-phosphoglycolate phosphatase, PGLP1, in Arabidopsis. Plant Physiol 144:1580–1586
11. Somerville CR, Ogren WL (1979) A phosphoglycolate phosphatase-deficient mutant of *Arabidopsis*. Nature 280:833–836
12. Clagett CO, Tolbert NE, Burris RH (1949) Oxidation of α-hydroxy acids by enzymes from plants. J Biol Chem 178:977–987
13. Dellero Y, Mauve C, Boex-Fontvieille E, Flesch V, Jossier M, Tcherkez G, Hodges M (2015) Experimental evidence for a hydride transfer

mechanism in plant glycolate oxidase catalysis. J Biol Chem 290:1689–1698
14. Pennati A, Gadda G (2011) Stabilization of an intermediate in the oxidative half-reaction of human liver glycolate oxidase. Biochemistry 50:1–3
15. Hackenberg C, Kern R, Hüge J, Stal LJ, Tsuji Y, Kopka J, Shiraiwa Y, Bauwe H, Hagemann M (2011) Cyanobacterial lactate oxidases serve as essential partners in N_2 fixation and evolved into photorespiratory glycolate oxidases in plants. Plant Cell 23:2978–2990
16. Yamaguchi K, Nishimura M (2000) Reduction to below threshold levels of glycolate oxidase activities in transgenic tobacco enhances photoinhibition during irradiation. Plant Cell Physiol 41:1397–1406
17. Feierabend J, Beevers H (1972) Developmental studies on microbodies in wheat leaves: I. Conditions influencing enzyme development. Plant Physiol 49:28–32
18. Macheroux P, Massey V, Thiele DJ, Volokita M (1991) Expression of spinach glycolate oxidase in *Saccharomyces cerevisiae*: purification and characterization. Biochemistry 30:4612–4619
19. Kisaki T, Tolbert NE (1969) Glycolate and glyoxylate metabolism by isolated peroxisomes or chloroplasts. Plant Physiol 44:242–250
20. Kisaki T, Tolbert NE (1970) Glycine as substrate for photorespiration. Plant Cell Physiol 11:247–258
21. Hirotsu K, Goto M, Okamoto A, Miyahara I (2005) Dual substrate recognition of aminotransferases. Chem Rec 5:160–172
22. Dunathan HC (1966) Conformation and reaction specificity in pyridoxal phosphate enzymes. Proc Natl Acad Sci U S A 55:712–716
23. Igarashi D, Tsuchida H, Miyao M, Ohsumi C (2006) Glutamate:glyoxylate aminotransferase modulates amino acid content during photorespiration. Plant Physiol 142:901–910
24. Liepman AH, Olsen LJ (2001) Peroxisomal alanine:glyoxylate aminotransferase (AGT1) is a photorespiratory enzyme with multiple substrates in *Arabidopsis thaliana*. Plant J 25:487–498
25. Stafford HA, Magaldi A, Vennesland B (1954) The enzymatic reduction of hydroxypyruvic acid to D-glyceric acid in higher plants. J Biol Chem 207:621–629
26. Tolbert NE, Yamazaki RK, Oeser A (1970) Localization and properties of hydroxypyruvate and glyoxylate reductases in spinach leaf particles. J Biol Chem 245:5129–5136
27. Zelitch I (1953) Oxidation and reduction of glycolic and glyoxylic acids in plants: II. Glyoxylic acid reductase. J Biol Chem 201:719–726
28. Zelitch I (1955) The isolation and action of crystalline glyoxylic acid reductase from tobacco leaves. J Biol Chem 216:553–575
29. Timm S, Nunes-Nesi A, Pärnik T, Morgenthal K, Wienkoop S, Keerberg O, Weckwerth W, Kleczkowski LA, Fernie AR, Bauwe H (2008) A cytosolic pathway for the conversion of hydroxypyruvate to glycerate during photorespiration in *Arabidopsis*. Plant Cell 20:2848–2859
30. Allan WL, Clark SM, Hoover GJ, Shelp BJ (2009) Role of plant glyoxylate reductases during stress: a hypothesis. Biochem J 423:15–22
31. Bartsch O, Hagemann M, Bauwe H (2008) Only plant-type (GLYK) glycerate kinases produce D-glycerate 3-phosphate. FEBS Lett 582:3025–3028
32. Kern R, Bauwe H, Hagemann M (2011) Evolution of enzymes involved in the photorespiratory 2-phosphoglycolate cycle from cyanobacteria via algae toward plants. Photosynth Res 109:103–114
33. Kleczkowski LA, Randall DD, Zahler WL (1985) The substrate specificity, kinetics, and mechanism of glycerate-3-kinase from spinach leaves. Arch Biochem Biophys 236:185–194
34. Kleczkowski LA, Randall DD (1985) Light and thiol activation of maize leaf glycerate kinase—the stimulating effect of reduced thioredoxins and ATP. Plant Physiol 79:274–277
35. Bartsch O, Mikkat S, Hagemann M, Bauwe H (2010) An autoinhibitory domain confers redox regulation to maize glycerate kinase. Plant Physiol 153:832–840
36. Kikuchi G, Motokawa Y, Yoshida T, Hiraga K (2008) Glycine cleavage system: reaction mechanism, physiological significance, and hyperglycinemia. Proc Jpn Acad Ser B Phys Biol Sci 84:246–263
37. Sagers RD, Gunsalus IC (1961) Intermediary metabolism of *Diplococcus glycinophilus*: I. Glycine cleavage and one-carbon interconversion. J Bacteriol 81:541–549
38. Besson V, Rebeille F, Neuburger M, Douce R, Cossins EA (1993) Effects of tetrahydrofolate polyglutamates on the kinetic parameters of serine hydroxymethyltransferase and glycine decarboxylase from pea leaf mitochondria. Biochem J 292:425–430
39. Klein SM, Sagers RD (1966) Glycine metabolism: I. Properties of the system catalyzing the exchange of bicarbonate with the carboxyl-group of glycine in *Peptococcus glycinophilus*. J Biol Chem 241:197–205
40. Hasse D, Mikkat S, Hagemann M, Bauwe H (2009) Alternative splicing produces an H-

protein with better substrate properties for the P-protein of glycine decarboxylase. FEBS J 276:6985–6991

41. Yan L-J, Thangthaeng N, Forster MJ (2008) Changes in dihydrolipoamide dehydrogenase expression and activity during postnatal development and aging in the rat brain. Mech Ageing Dev 129:282–290
42. Schirch V, Szebenyi DM (2005) Serine hydroxymethyltransferase revisited. Curr Opin Chem Biol 9:482–487
43. Rebeille F, Neuburger M, Douce R (1994) Interaction between glycine decarboxylase, serine hydroxymethyltransferase and tetrahydrofolate polyglutamates in pea leaf mitochondria. Biochem J 302:223–228
44. McClung CR, Hsu M, Painter JE, Gagne JM, Karlsberg SD, Salome PA (2000) Integrated temporal regulation of the photorespiratory pathway. Circadian regulation of two *Arabidopsis* genes encoding serine hydroxymethyltransferase. Plant Physiol 123:381–392
45. Voll LM, Jamai A, Renné P, Voll H, McClung CR, Weber APM (2006) The photorespiratory *Arabidopsis shm1* mutant is deficient in SHM1. Plant Physiol 140:59–66
46. Engel N, Ewald R, Gupta KJ, Zrenner R, Hagemann M, Bauwe H (2011) The presequence of Arabidopsis serine hydroxymethyltransferase SHM2 selectively prevents import into mesophyll mitochondria. Plant Physiol 157:1711–1720
47. Sopitthummakhun K, Maenpuen S, Yuthavong Y, Leartsakulpanich U, Chaiyen P (2009) Serine hydroxymethyltransferase from *Plasmodium vivax* is different in substrate specificity from its homologues. FEBS J 276:4023–4036
48. Zhang Y, Sun K, Roje S (2008) An HPLC-based fluorometric assay for serine hydroxymethyltransferase. Anal Biochem 375:367–369
49. Taylor RT, Weissbach H (1965) Radioactive assay for serine transhydroxymethylase. Anal Biochem 13:80–84
50. Eisenhut M, Kahlon S, Hasse D, Ewald R, Lieman-Hurwitz J, Ogawa T, Ruth W, Bauwe H, Kaplan A, Hagemann M (2006) The plant-like C_2 glycolate pathway and the bacterial-like glycerate cycle cooperate in phosphoglycolate metabolism in cyanobacteria. Plant Physiol 142:333–342
51. Uyeda K, Rabinowitz JC (1963) Fluorescence properties of tetrahydrofolate and related compounds. Anal Biochem 6:100–108
52. Boyes DC, Zayed AM, Ascenzi R, McCaskill AJ, Hoffman NE, Davis KR, Gorlach J (2001) Growth stage-based phenotypic analysis of Arabidopsis: a model for high throughput functional genomics in plants. Plant Cell 13:1499–1510
53. Timm S, Bauwe H (2013) The variety of photorespiratory phenotypes – employing the current status for future research directions on photorespiration. Plant Biol (Stuttg) 15:737–747
54. Delieu T, Walker DA (1972) An improved cathode for the measurement of photosynthetic oxygen evolution by isolated chloroplasts. New Phytol 71:201–225

Chapter 4

In Vitro Alkylation Methods for Assessing the Protein Redox State

Flavien Zannini, Jérémy Couturier, Olivier Keech, and Nicolas Rouhier

Abstract

Cysteines are important residues for protein structure, function, and regulation. Owing to their modified reactivity, some cysteines can undergo very diverse redox posttranslational modifications, including the reversible formation of disulfide bonds, a widespread protein regulatory process as well exemplified in plant chloroplasts for Calvin-Benson cycle enzymes. Both core- and peripheral-photorespiratory enzymes possess conserved cysteines, some of which have been identified as being subject to oxidative modifications. This is not surprising considering their presence in subcellular compartments where the production of reactive species can be important. However, in most cases, the types of modifications and their biochemical effect on protein activity have not been validated, meaning that the possible impact of these modifications in a complex physiological context, such as photorespiration, remains obscure.

We here describe a detailed set of protocols for alkylation methods that have been used so far to (1) study the protein cysteine redox state either in vitro by submitting purified recombinant proteins to reducing/oxidation treatments or in vivo by western blots on protein extracts from plants subject to environmental constraints, and its dependency on the two major reducing systems in the cell, i.e., the thioredoxin and glutathione/glutaredoxin systems, and (2) determine two key redox parameters, i.e., the cysteine pK_a and the redox midpoint potential.

Key words Alkylation, Cysteine, Oxidative modification, pK_a, Redox potential

1 Introduction

Cysteine (Cys) is one of the least abundant amino acids in proteins; however, it is very important for protein structure and function being for instance found in protein active sites and/or subject to co- or posttranslational modifications including lipidation (prenylation or palmitoylation) but also disulfide bond formation, glutathionylation or nitrosylation to only cite the major ones. In specific protein environments, e.g., in active sites, the pK_a of some cysteine residues is decreased towards acidic values compared to the pK_a ~ 8.3 of free cysteines [1]. Thus, under certain physiological conditions, e.g., around neutral pH, the more reactive thiolate (S^-) form will predominate as compared to the thiol

Alisdair R. Fernie et al. (eds.), *Photorespiration: Methods and Protocols*, Methods in Molecular Biology, vol. 1653,
DOI 10.1007/978-1-4939-7225-8_4, © Springer Science+Business Media LLC 2017

(SH) form. On the other hand, such a chemical property implies that these residues are usually more sensitive to oxidative modifications. Hence, in subcellular compartments or under specific conditions, the cysteine residues will be oxidized by various types of molecules/agents leading notably to the formation of sulfenic acids (SOH), disulfide bonds (S–S), and glutathionylated (S–SG) or nitrosylated (S–NO) forms [2]. All these aforementioned post-translational modifications are reversible, mostly through the action of the thioredoxin (TRX) and/or glutathione (GSH)/glutaredoxin (GRX) reducing systems. If not condensed with another cysteine, e.g., in the absence of a vicinal thiol group or a reductant, sulfenic acids may eventually be irreversibly oxidized to sulfinic (SO_2H) and sulfonic (SO_3H) acids. The only documented case of sulfinic acid reduction concerns the ATP- and Mg^{2+}-dependent reduction of sulfinylated peroxiredoxins catalyzed by sulfiredoxins [3]. Depending on the physiological context, the oxidative modification of cysteine residues may have diverse impacts on protein function, for which many examples of protein activation or inactivation have been reported. A well-documented example of redox regulation concerns the Calvin-Benson cycle enzymes that are activated by light and incidentally by the ferredoxin/ferredoxin-thioredoxin reductase/TRX system, through dithiol-disulfide exchanges [4]. Besides these regulatory mechanisms, a transient and reversible inactivation by oxidation can also be seen as a protective mechanism preventing irreversible protein over-oxidation.

A targeted in vitro approach to study cysteine sensitivity to oxidation within a given protein is to perform cysteine alkylation using maleimide-containing reagents, which specifically bind thiol groups. The cysteine derivatization will result in a mobility shift on nonreducing SDS-PAGE that is dependent on the number of cysteines oxidized. One of the first compounds used was the 4-acetamido-4′-maleimidylstilbene-2,2′-disulfonic acid (AMS), which induced an electrophoretic migration shift of ~0.5 kDa per free thiol group [5–7]. However, as this was sometimes insufficient for separating a mixture of different species, especially for proteins with a high molecular weight, it has sometimes been replaced by maleimides linked to polymers, such as polyethylene glycol (PEG) of different sizes ranging generally from 1 to 30 kDa [8, 9]. The drawback of these PEG-maleimide compounds is that PEG polymerization likely generates a heterogeneous mixture that often results in broad bands on SDS-PAGE. Moreover, the fact that these polymers are not charged prevents linear mobility shifts on SDS-PAGE. Therefore, researchers in both academic and industrial settings have developed chemically synthesized maleimide probes coupled to peptides [10] or nucleic acids [11].

Such an alkylation method can be used with both recombinant proteins and protein extracts, under the condition that a specific antibody is available for the western blot detection. Besides evaluating the sensitivity of a given protein to a specific oxidative condition, this method also has the ability to determine which of the TRX or GSH/GRX systems can reduce a given oxidized form [9, 12]. Using protein extracts made from plants grown under specific controlled conditions, this method enables the assessment of changes in the levels of oxidation over time-course experiments, and is partially quantitative. That being said, without using chemicals specific to a given modification in tandem with mass spectrometry analyses, this method will neither indicate the type(s) of modification, nor the identity of which cysteine(s) is/are modified. The method described here has been successfully and elegantly tested for detecting the activation of Calvin-Benson cycle enzymes in response to various light intensities and/or in specific genetic backgrounds as *trx* knock-out plants [13].

By analyzing the lists of potential redox-regulated proteins identified by large-scale redox proteomic analyses, we recently pointed to the fact that nearly all photorespiratory enzymes from the canonical pathway, as well as most of the known photorespiration-associated enzymes, may be subject to redox regulation, even though the type of modifications and the effect on activity have in fact almost never been investigated [14]. Hence, we describe here three complementary methods that aim at assessing the cysteine redox properties and redox states into a protein of interest. The first method describes the analysis of a protein redox state, which can be applied both in vitro and in vivo by anyone interested in the regulation of the activity of photorespiratory enzymes and of course of any other protein susceptible to redox regulation. The two other in vitro methods also rely on cysteine alkylation and allow determining cysteine pK_a and the redox midpoint potential of protein disulfides.

2 Materials

2.1 Required Instrumentation

Bench size shaker, benchtop centrifuge, dialysis membrane tubing and/or desalting columns, fluorometer, gel electrophoresis system, power supplier, UV-visible spectrophotometer.

2.2 Buffers and Solutions

2.2.1 Compound Solubility for Stock Solutions

Stock solutions of hydrogen peroxide (H_2O_2), iodoacetamide (IAM), reduced glutathione (GSH), oxidized glutathione (GSSG), nitrosoglutathione (GSNO), 1,4-dithiothreitol (DTT_{red}), trans-4,5-dihydroxy-1,2-dithiane (DTT_{ox}), tris(2-carboxyethyl) phosphine hydrochloride (TCEP), and trichloroacetic acid (TCA) are prepared in distilled or Milli-Q water. For GSH, GSSG, and GSNO solutions, the pH has to be readjusted to pH 7.0 or 8.0 using sodium hydroxide. Stock solutions of

5-iodoacetamidofluorescein (fluo-IAM), of monobromobimane (mBBr) and of 5,5'-dithiobis-(2-nitrobenzoic Acid) (DTNB) are prepared, respectively, at 9.7 mM in dimethylsulfoxide (DMSO), at 73 mM in acetonitrile and at 10 to 20 mM in ethanol.

2.2.2 Preparation of pH 2.0–10.0 Buffers for the Determination of Cysteine pK_a

1. 100 mM Citrate buffers between pH 2.0 and 6.0 are prepared by mixing 100 mM citric acid and sodium citrate solutions.
2. 100 mM Phosphate buffers between pH 6.5 and 8.0 are prepared by mixing 100 mM potassium phosphate monobasic and dibasic solutions.
3. 100 mM Borate buffers between pH 8.5 and 10.0 are prepared by mixing 100 mM boric acid and potassium chloride solutions and by adjusting pH with sodium or potassium hydroxide 100 mM.

2.2.3 Other Solutions

1. Alkylation solution: 100 mM Tris–HCl pH 8.0, 1% SDS, 2 mM alkylating agent (maleimide polyethylene glycol of 2 kDa (2 kDa mPEG-maleimide) or 4-acetamido-4′-maleimidylstilbene-2,2′-disulfonic acid (AMS)).
2. 2× Non-reducing Laemmli buffer: 120 mM Tris–HCl pH 6.8, 4% sodium dodecyl sulfate (SDS), 20% glycerol, and 0.02% (w/v) bromophenol blue.

3 Methods

All procedures can be carried out at room temperature unless otherwise stated.

3.1 Examination of the Cysteine Redox State and Susceptibility to Oxidative Treatments

The Ellman's reagent (DTNB) is often used to examine the number of free thiol groups of a purified recombinant protein subject to oxidative treatments. However, this titration is often hampered or imprecise when there is a high number of cysteines. Moreover, the interpretation can be made difficult when disulfide-linked oligomers are formed because the results are expressed as moles of SH per moles of monomer. The alkylation method described here associated to SDS-PAGE separation constitutes a good complement, giving access in particular to the oligomerization state of the protein. Moreover such an approach can also be employed to evaluate the capacity of known oxidoreductases to reduce or oxidize the protein of interest. Before studying the sensitivity of a protein towards different oxidative compounds, a prerequisite is to obtain a completely reduced protein. Procedures for reduction, oxidation, and alkylation of the target proteins are described below together with a case example.

3.1.1 Reducing Treatment

1. Add a tenfold molar excess of reductant (DTT_{red} or TCEP) to the purified recombinant protein (*see* **Note 1**).

2. Incubate for 30 min (*see* **Note 2**).
3. Dialyze the protein against the desired buffer using membrane tubing or small desalting columns, preferentially at 4 °C (*see* **Note 3**).

3.1.2 Oxidizing Treatment

1. Add a fivefold molar excess of oxidizing reagents (GSSG, GSNO, H_2O_2, DTT_{ox}) to the pre-reduced sample (*see* **Note 4**).
2. Incubate for 2 h (*see* **Note 5**).
3. Dialyze the protein against the desired buffer using membrane tubing or small desalting columns, preferentially at 4 °C (optional, *see* **Note 6**).

3.1.3 Protein Alkylation for SDS-PAGE Analysis

1. Add 50 μL of 20% TCA to 50 μL of sample containing 5–10 μg of protein (either untreated, oxidized, or reduced) (*see* **Notes 7** and **8**).
2. Gently mix by inverting the tube several times and place on ice for 15 min.
3. Centrifuge for 10 min at 12,100 × *g* in a benchtop centrifuge.
4. Remove carefully and discard the supernatant (*see* **Note 9**).
5. Wash the pellet with 100% acetone kept at −20 °C without resuspending it and centrifuge for 10 min at 12,100 × *g*.
6. Remove carefully and discard the supernatant (*see* **Note 9**).
7. Dry the pellet at ambient air for 5 min (*see* **Note 10**).
8. Add 10 μL of alkylation solution (*see* **Note 11**).
9. Gently shake for 15 min at room temperature (*see* **Note 12**).
10. Add 10 μL of 2× non-reducing Laemmli buffer (*see* **Note 13**).
11. At this step, samples can be stored at −25 °C or subject to a non-reducing SDS-PAGE (*see* **Notes 14–16**).

3.1.4 Possible Result Obtained Upon Oxidative Treatment of a Protein of 12 kDa with a Single Cysteine

Firstly, it is important to verify that the oxidation/reduction of a protein does not impact per se its mobility on non-reducing SDS-PAGE, which might be the case when it induces conformational rearrangements as observed for glutathione peroxidase-like proteins [15]. Secondly, it is strongly advised to include controls for fully reduced and fully oxidized (if possible) proteins, which will give both extremities of the scale, as long as there is no disulfide-linked oligomer. In addition to delimiting the extremities of the scale, having alkylated and non-alkylated reduced proteins on the same gel allows in principle to control that there is no atypical shift relatively to the expected number of cysteines. For instance, we observed previously that treating a recombinant poplar TRX z possessing two cysteines with the 2 kDa mPEG maleimide generated a shift of 15 kDa instead of 4 kDa [8].

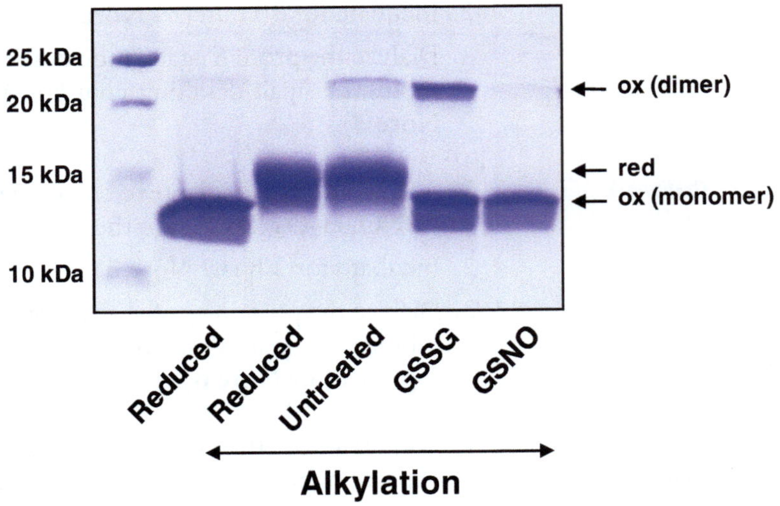

Fig. 1 Protein sensitivity to oxidative treatments. Around 5 μg of pre-reduced, untreated, or oxidized proteins obtained after 2-h oxidative treatments with GSSG or GSNO were alkylated with 2 kDa mPEG maleimide prior to separation on non-reducing SDS–PAGE [9]

Afterwards, one can indeed compare reduced, untreated, and oxidized proteins. In the example shown in Fig. 1, the absence of shift at the level of the monomer upon GSSG or GSNO treatments is interpreted here as GSSG or GSNO oxidizing the cysteine, with GSSG also promoting the formation of a dimer linked by an intermolecular disulfide. The untreated protein is mostly present as a reduced monomer with a small proportion of disulfide-linked dimer. The fact that this band is also seen in the non-alkylated reduced protein but not in the alkylated reduced protein suggests that some oxidation of the untreated protein occurred during sample preparation/migration, as explained in **Note 14**.

3.2 Estimation of Cysteine pK_a

The pK_a of cysteines into proteins is influenced by several factors, in particular their amino acid environment. Considering that thiolates are more reactive than thiols, measuring this parameter will give clue about their capacity to act or not as a nucleophile in a thiol-disulfide exchange reaction or to be susceptible to oxidation. Determining the pK_a value of cysteine residues using this method can be done only with proteins possessing a single cysteine or eventually two cysteines with sufficiently different pK_a.

3.2.1 Protein Alkylation for SDS-PAGE Analysis

1. Incubate around 10 μM of pre-reduced protein (*see* Subheading 3.1.1) with a 100-fold excess IAM for 15 min under gentle stirring in buffers ranging from pH 2.0 to 10.0 (every 0.5 unit for precision) in a final volume of 50 μL (*see* **Notes 17** and **18**).

2. Add 50 μL of 20% TCA to 50 μL of reaction mixture.
3. The following steps are similar to **steps 2–11** of Subheading 3.1.3.

3.2.2 Protein Alkylation for Fluorescence Detection

1. Incubate around 10 μM of pre-reduced protein (*see* Subheading 3.1.1) with a 20-fold excess of fluo-IAM for 15 min in the dark under gentle stirring in buffers of different pHs (every 0.5 unit for precision) in a final volume of 200 μL (*see* **Notes 17** and **18**).
2. Add 200 μL of 20% TCA to 200 μL of reaction mixture.
3. Place on ice for 15 min in the dark (*see* **Note 19**).
4. Centrifuge for 10 min at 12,100 × *g* in a benchtop centrifuge.
5. Remove carefully and discard the supernatant (*see* **Note 9**).
6. Wash the pellet with 100% acetone kept at −20 °C and centrifuge for 10 min at 12,100 × *g*.
7. Remove carefully and discard the supernatant (*see* **Note 9**).
8. Dry the pellet at ambient air for 5 min in the dark (*see* **Note 19**).
9. Add 400 μL of 100 mM Tris–HCl pH 8.0 and 1% SDS.
10. Gently shake for 15 min at room temperature (*see* **Note 12**).
11. Measure the fluorescence emission of the resulting solution at 515 nm after excitation at 492 nm (*see* **Note 20**).

3.2.3 Possible Result Obtained for a Protein of 12 kDa with a Single Cysteine and Interpretation

The IAM specifically binds to deprotonated (i.e., thiolate forms) but not to protonated (i.e., thiol forms) cysteines. Hence, subsequent alkylation will only happen on thiol forms. This can be visualized on non-reducing SDS-PAGE by a retarded migration when the cysteine is in a thiol form (Fig. 2a). By performing alkylation at different pHs, the cysteine pK_a can be approximately determined (in this example around 7.5) as the pH value where

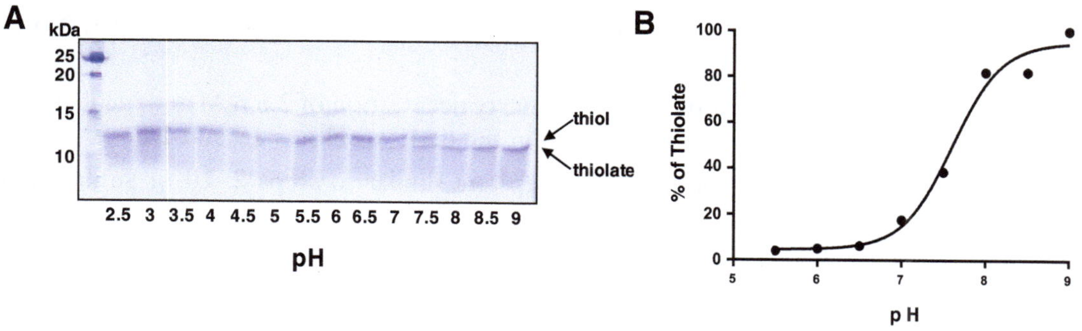

Fig. 2 pK_a determination of cysteine sulfhydryl group. This was done either by coupling reaction with a nonfluorescent alkylating reagent and protein separation on SDS-PAGE (**a**) or by measuring fluorescence after reaction with a fluorescent alkylating reagent (**b**)

50% of each form is encountered. Performing similar experiment by using fluo-IAM enables a more precise determination of the pK_a value (Fig. 2b). The fluorescence intensity is proportional to the quantity of cysteine under thiolate form. In this example, fitting the data to the following nonlinear regression Y= Bottom + (Top − Bottom)/(1 + 10^((LogIC50 − X × HillSlope)) gives a pK_a value of 7.6 which is comparable to the value deduced with nonfluorescent IAM.

3.3 Determination of the Redox Midpoint Potential of a Reversible Oxidative Modification

Measuring the redox midpoint potential will give clue about the capacity of a species to accept or donate electrons which in the case of a dithiol-disulfide couple translates into the capacity to be oxidized or reduced. Since a single protein can undergo several redox modifications involving the same or different cysteines and unless the redox potentials of two modifications found in a given protein are enough different to be distinguished, it is important to start with a homogenous oxidized protein sample, i.e., with a large majority having the same oxidative modification. In the absence of indication about the midpoint potential of the redox modification, it is worth trying either glutathione or DTT to generate reference redox potentials because they cover a different range from −100 mV to −250 mV and from −220 mV to −400 mV, respectively.

3.3.1 Protein Alkylation for SDS-PAGE Analysis

1. Incubate around 10 μM of oxidized protein in 100 mM HEPES pH 7.0 for 2 h in 50 μL reaction mixtures adjusted to different redox potentials by varying ratios between oxidized and reduced DTT or glutathione at a final concentration of 2 mM (*see* Tables 1 and 2) (*see* **Notes 21** and **22**).
2. The following steps are similar to **steps 2–11** of Subheading 3.1.3.

3.3.2 Protein Alkylation for Fluorescence Detection

1. Incubate around 10 μM of oxidized protein in 100 mM HEPES pH 7.0 for 2 h in 200 μL reaction mixtures adjusted to different redox potentials by varying ratios between oxidized and reduced DTT or glutathione (*see* Tables 1 and 2) (*see* **Notes 22** and **23**).
2. Add mBBr to a final concentration of 2.5 mM and incubate for a further 20 min in the dark (*see* **Notes 19** and **24**).
3. The following steps are similar to **steps 2–10** of Subheading 3.2.2.
4. Measure the fluorescence emission of the resulting solution at 472 nm after an excitation at 380 nm (*see* **Note 25**).

Table 1
Redox potentials obtained by mixing DTT_{ox}/DTT_{red} solutions

DTT_{ox} 2 mM (μL)	DTT_{red} 2 mM (μL)	Final volume (μL)	Final $[DTT_{ox}]$ (mM)	Final $[DTT_{red}]$ (mM)	Ratio $[DTT_{ox}]/[DTT_{red}]$	Redox potential E_h (mV)
499.88	0.12	500	1.99952	0.00048	4166	−220
499.74	0.26	500	1.99896	0.00104	1922	−230
499.43	0.57	500	1.99772	0.00228	876	−240
498.76	1.24	500	1.99504	0.00496	402	−250
497.3	2.7	500	1.98920	0.00108	184	−260
494.16	5.84	500	1.97664	0.02336	85	−270
487.45	12.55	500	1.94980	0.05020	39	−280
473.45	26.55	500	1.89380	0.10620	18	−290
445.57	54.43	500	1.78228	0.21772	8	−300
394.9	105.1	500	1.57960	0.42040	4	−310
316.5	183.5	500	1.26600	0.73400	2	−320
220.93	279.07	500	0.88372	1.11628	0.8	−330
113.27	386.73	500	0.45308	1.54692	0.29	−343
71.48	428.52	500	0.28592	1.71408	0.16	−350
35.57	464.43	500	0.14228	1.85772	0.077	−360
16.97	483.03	500	0.06788	1.93212	0.035	−370
7.94	492.06	500	0.03176	1.96824	0.016	−380
3.67	496.33	500	0.01468	1.98532	0.007	−390
1.69	498.31	500	0.00676	1.99324	0.003	−400

The 2 mM oxidized and reduced DTT solutions are prepared in 100 mM HEPES pH 7.0
Calculations were made accordingly to the Nernst equation by taking an E_m value for the DTT_{ox}/DTT_{red} couple of −327 mV at pH 7.0

3.3.3 Possible Result Obtained for a Protein of ~20 kDa with Two Cysteines Forming an Intramolecular Disulfide Bridge and Interpretation

Using the SDS-PAGE analysis, the redox midpoint potential of the intramolecular disulfide, i.e., the value where there is 50% of reduced forms and 50% of oxidized forms, can be roughly estimated to be between −290 mV and −300 mV (Fig. 3A). For the same protein, the use of mBBr allows to calculate a redox midpoint potential of −298 ± 1 mV at pH 7.0 (Fig. 3B) after fitting fluorescence variations as a function of E_h values using the following nonlinear regression: Y = Bottom + (Top − Bottom)/(1 + 10^((LogIC50 – X × HillSlope)).

Table 2
Redox potentials obtained by mixing GSH and GSSG solutions

GSSG 2 mM (μL)	GSH 2 mM (μL)	Final volume (μL)	Final [GSSG] (mM)	Final [GSH] (mM)	Ratio [GSSG]/[GSH]^2	Redox potential E_h (mV)
454.8	45.2	500	1.8192	0.1808	55,652	−100
434.7	65.3	500	1.7388	0.2612	25,486	−110
406.7	93.3	500	1.6268	0.3732	11,680	−120
368.8	131.2	500	1.4752	0.5248	5356	−130
319.6	180.4	500	1.2784	0.7216	2455	−140
259.7	240.3	500	1.0388	0.9612	1124	−150
193.5	306.5	500	0.7740	1.2260	515	−160
129.5	370.5	500	0.5180	1.4820	236	−170
77.3	422.7	500	0.3092	1.6908	108	−180
41.6	458.4	500	0.1664	1.8336	49.5	−190
20.8	479.2	500	0.0832	1.9168	22.6	−200
9.98	490.02	500	0.0399	1.9601	10.4	−210
4.7	495.3	500	0.0188	1.9812	4.8	−220
2.2	497.8	500	0.0088	1.9912	2.2	−230
0.996	499.004	500	0.0040	1.9960	1	−240
0.457	499.543	500	0.0018	1.9982	0.46	−250

The 2 mM GSH and GSSG solutions are prepared in 100 mM HEPES pH 7.0
Calculations were made accordingly to the Nernst equation by taking an E_m value for the GSSG/GSH couple of −240 mV at pH 7.0

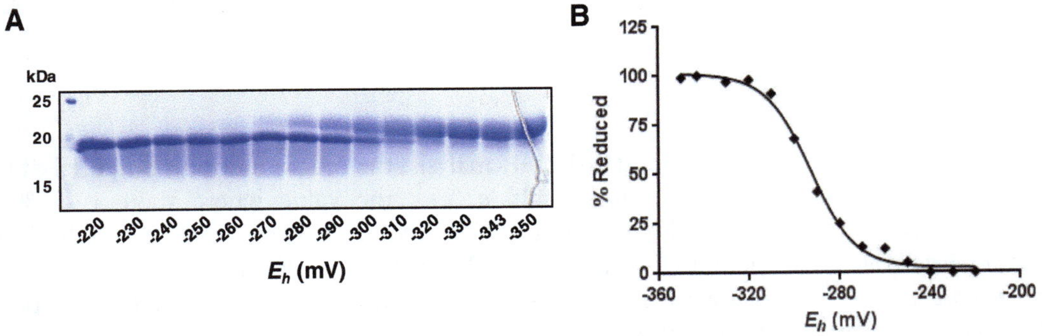

Fig. 3 Determination of the redox midpoint potential of a disulfide bond. This was done by alkylation with 2 kDa mPEG-maleimide and separation on SDS-PAGE (**A**) or by alkylation with mBBr (**B**). The 0% and 100% reduced indicate the proportion of reduced proteins and correspond to the minimum and maximum values of fluorescence, respectively

4 Notes

1. The procedures for producing and purifying recombinant proteins are out of the scope and will not be detailed here.
2. This time can be increased for proteins, the disulfides of which are more recalcitrant to reduction by DTT or TCEP.
3. While desalting columns are all more or less equivalent, the membrane tubing exists in different diameters and more importantly pore sizes that you have to adapt depending on the molecular weight of the protein of interest. Desalting columns are preferable because it is more rapid which thus could avoid reoxidation that might occur upon removal of the reductant.
4. Performing experiments by varying the ratio between concentrations of protein and oxidizing agents allows evaluating and comparing protein sensitivity to these agents.
5. Performing time-course experiments by varying the incubation time allows evaluating and comparing protein sensitivity to oxidizing agents.
6. The chemical treatment of the proteins (oxidation or reduction) before alkylation can be performed without subsequent dialysis as TCA specifically precipitates proteins but not chemical reagents that remain in the supernatant after centrifugation. In the case where proteins are initially reduced for the subsequent examination of their sensitivity to oxidizing reagents, they have to be dialyzed.
7. Any sample/TCA combination which leads to a final 10% TCA solution should be fine.
8. It is worth noting that a similar alkylation procedure can be applied to a cellular protein extract by starting at this step. The preparation of this extract will not be detailed here but it has to be done in the presence of the alkylating agent to avoid modifications of the redox state of the proteins occurring during sample preparation. If the samples do not have to be concentrated, TCA precipitation can eventually be replaced by a denaturation step with buffers containing 8 M urea or 6 M guanidine-HCl. By this approach, you can then directly proceed to **step 9**.
9. When removing the supernatant, do not touch the pellet.
10. It is important to remove all acetone traces before resuspending pellets of precipitated proteins to avoid subsequent perturbation of protein migration.
11. The molecular weight and the number of cysteine residues present in the protein of interest can influence the choice of

the alkylating reagent. Companies usually offer a large panel of mPEG maleimide compounds with variable molecular weight (1–30 kDa) which might be very convenient. That said, the fact that it generates broad bands on SDS-PAGE may prevent the separation between reduced and oxidized forms particularly for proteins with high molecular weights. In this case, DNA-maleimide forms are certainly more adapted since it gives more focused bands on SDS-PAGE and one currently commercialized version generates shift around 15 kDa per cysteine. Finally, despite generating a smaller migration shift (0.5 kDa), AMS may still be preferred for small-molecular-weight proteins because bands remain focused and in the same gel area.

12. Alternatively, after adding the alkylation solution, vortex for several seconds and wait for 15 min.
13. Any sample/Laemmli buffer combination which leads to a 1× final concentration is fine.
14. Oxidation of non-alkylated samples can take place during migration since it is performed in non-reducing conditions.
15. Prior to efficient protein separation, you have to adapt the type and the percentage of acrylamide of your gel to maximize separation in the molecular weight range of interest. Hence, for proteins smaller than 25 kDa, a Tricine gel may be required, whereas for bigger proteins, standard 15% acrylamide glycine SDS-PAGE may be sufficient to separate protein bands.
16. Depending on whether the experiment has been done with recombinant protein or protein extract, protein bands can be detected by Coomassie blue staining or by Western blotting with specific antibody, respectively.
17. For recombinant proteins, the concentration used has to be adapted according to its size to avoid using more than 10 μg of protein. Indeed, using a too important protein amount may prevent an efficient separation on SDS-PAGE.
18. For protein with multiple cysteines, it is usually necessary to perform site-directed mutagenesis to generate protein variants with a single cysteine.
19. As fluorescent alkylating reagents such as fluo-IAM and mBBr are photolabile, it is important to protect the reaction mixture from light.
20. Fluorescence emission of alkylating reagents depends on pH. In order to make accurate pK_a determinations, it is absolutely necessary to precipitate and resuspend protein samples in the same buffer to make sure that differences in fluorescence values are related to the thiol/thiolate state of cysteine residue and not to the pH value of initial solutions used for incubation.

21. A 2-h incubation time is often sufficient to reach equilibration but you can increase this time up to 4 h. In some cases, owing to the fact that the formation of a disulfide bond can sometimes induce a large conformational rearrangement, the two oxidation states can be visualized on a non-reducing SDS-PAGE without alkylation step. In contrast with the determination of pK_a value of cysteine residues, the presence of additional cysteines should not interfere if their redox state does not change along the incubation period.
22. Higher concentrations (for example 5 mM) of both oxidizing and reducing reagents can be used. It is necessary to recalculate the corresponding redox potential of each reaction mixture accordingly.
23. Redox potentials depend on pH value. Similar experiments can be performed at different pH but it will be necessary to employ different buffers and to recalculate the corresponding redox potentials of each reaction mixture accordingly. By doing measurements at different pHs, you can determine from the slope (either ~30 or ~60 mV pH/unit) whether the two-electron reduction process is accompanied by the uptake of one or of two protons over a given pH range.
24. mBBr reacts with all thiol groups, both from the protein cysteine residues and from the reducing agent. Hence, the final mBBr concentration has to be adapted to be slightly higher than to the total concentration of thiol groups present in the reaction mixture.
25. Measure first the fluorescence emission of the most reduced sample (i.e., incubated in the solution with the lowest redox potential) to determine whether the excitation intensity used is adequate and does not lead to signal saturation.

Acknowledgments

The UMR1136 is supported by a grant overseen by the French National Research Agency (ANR) as part of the "Investissements d'Avenir" program (ANR-11-LABX-0002-01, Lab of Excellence ARBRE). The Swedish Research Council "VetenskapsRådet" (grant: 621-2014-4688) as well as the Kempe Foundations are acknowledged for their financial support to O.K.

References

1. Poole LB (2015) The basics of thiols and cysteines in redox biology and chemistry. Free Radic Biol Med 80:148–157
2. Couturier J, Chibani K, Jacquot JP, Rouhier N (2013) Cysteine-based redox regulation and signaling in plants. Front Plant Sci 4:105

3. Biteau B, Labarre J, Toledano MB (2003) ATP-dependent reduction of cysteine-sulphinic acid by *S. cerevisiae* sulphiredoxin. Nature 425:980–984
4. Michelet L, Zaffagnini M, Morisse S, Sparla F, Perez-Perez ME, Francia F et al (2013) Redox regulation of the Calvin-Benson cycle: something old, something new. Front Plant Sci 4:470
5. Appenzeller-Herzog C, Ellgaard L (2008) *In vivo* reduction-oxidation state of protein disulfide isomerase: the two active sites independently occur in the reduced and oxidized forms. Antioxid Redox Signal 10:55–64
6. Cho SH, Porat A, Ye J, Beckwith J (2007) Redox-active cysteines of a membrane electron transporter DsbD show dual compartment accessibility. EMBO J 26:3509–3520
7. Lu H, Allen S, Wardleworth L, Savory P, Tokatlidis K (2004) Functional TIM10 chaperone assembly is redox-regulated *in vivo*. J Biol Chem 279:18952–18958
8. Chibani K, Tarrago L, Schürmann P, Jacquot JP, Rouhier N (2011) Biochemical properties of poplar thioredoxin z. FEBS Lett 585:1077–1081
9. Couturier J, HC W, Dhalleine T, Pégeot H, Sudre D, Gualberto JM et al (2014) Monothiol glutaredoxin-BolA interactions: redox control of *Arabidopsis thaliana* BolA2 and SufE1. Mol Plant 7:187–205
10. Parent A, Elduque X, Cornu D, Belot L, Le Caer JP, Grandas A et al (2015) Mammalian frataxin directly enhances sulfur transfer of NFS1 persulfide to both ISCU and free thiols. Nat Commun 6:5686
11. Hara S, Nojima T, Seio K, Yoshida M, Hisabori T (2013) DNA-maleimide: an improved maleimide compound for electrophoresis-based titration of reactive thiols in a specific protein. Biochim Biophys Acta 1830:3077–3081
12. Shaikhali J, Davoine C, Brännström K, Rouhier N, Bygdell J, Björklund S, Wingsle G (2015) Biochemical and redox characterization of the mediator complex and its associated transcription factor GeBPL, a GLABROUS1 enhancer binding protein. Biochem J 468:385–400
13. Yoshida K, Hara S, Hisabori T (2015) Thioredoxin selectivity for thiol-based redox regulation of target proteins in chloroplasts. J Biol Chem 290:14278–14288
14. Keech O, Gardeström P, Kleczkowski LA, Rouhier N (2017) The redox control of photorespiration: from biochemical and physiological aspects to biotechnological considerations. Plant Cell Environ. 40:553–569
15. Navrot N, Collin V, Gualberto J, Gelhaye E, Hirasawa M, Rey P et al (2006) Plant glutathione peroxidases are functional peroxiredoxins distributed in several subcellular compartments and regulated during biotic and abiotic stresses. Plant Physiol 142:1364–1379

Chapter 5

Dimethyl-Labeling-Based Quantification of the Lysine Acetylome and Proteome of Plants

Ines Lassowskat, Markus Hartl, Fabian Hosp, Paul J. Boersema, Matthias Mann, and Iris Finkemeier

Abstract

Photorespiratory enzymes in different cellular compartments have been reported to be posttranslational modified by phosphorylation, disulfide formation, S-nitrosylation, glutathionylation, and lysine acetylation. However, not much is known yet about the function of these modifications to regulate the activities, localizations, or interactions of the proteins in this metabolic pathway. Hence, it will be of great importance to study these modifications and their temporal and conditional occurrence in more detail. Here, we focus on the analysis of lysine acetylation as a relatively newly discovered modification on plant metabolic enzymes. The acetylation of lysine residues within proteins is a highly conserved and reversible posttranslational modification which occurs in all living organisms. First discovered on histones and implied in the regulation of gene expression, lysine acetylation also occurs on a diverse set of cellular proteins in different subcellular compartments and is particularly abundant on metabolic enzymes. Upon lysine acetylation, the function of proteins can be modulated due to the loss of the positive charge of the lysine residue. Lysine acetylation was also discovered on proteins involved in photosynthesis and novel tools are needed to study the regulation of this modification in dependence on the environmental conditions, tissues, or plant genotype. This chapter describes a method for the identification and relative quantification of lysine-acetylated proteins in plant tissues using a dimethyl labeling technique combined with an anti-acetyl lysine antibody enrichment strategy. Here, we describe the protein purification, labeling of trypsinated peptides, as well as immuno-enrichment of lysine-acetylated peptides followed by liquid chromatography tandem mass spectrometry (LC-MS/MS) data acquisition and analysis.

Key words Posttranslational modifications, Lysine acetylation, *Arabidopsis*, Photosynthesis, Photorespiration, Dimethyl-labeling, Quantitative proteomics, LC-MS/MS, MaxQuant

1 Introduction

Depending on the environmental conditions, plant tissues and organs require a fast regulation of cellular metabolic processes. Protein-based signaling networks are activated and tuned within seconds when the cellular homeostasis is perturbed, and they orchestrate cellular physiology, by regulating gene expression, translation, and protein turnover. Posttranslational modifications (PTMs) of proteins

Alisdair R. Fernie et al. (eds.), *Photorespiration: Methods and Protocols*, Methods in Molecular Biology, vol. 1653, DOI 10.1007/978-1-4939-7225-8_5, © Springer Science+Business Media LLC 2017

like phosphorylation, acetylation, ubiquitination, or methylation play a pivotal role in all of these processes [1, 2]. Especially lysine residues show a high diversity with at least 20 different PTMs competing for the ε-amino group.

More than 50 years ago, lysine acetylation was first discovered on histones where it renders an important role in the regulation of gene transcription by modification of core histone tails [3–5]. In comparison to phosphorylation, which has been extensively studied, technical challenges hindered detailed analyses of lysine acetylation for a long time [6, 7]. In the past years, major advances in high-resolution mass spectrometry combined with efficient antibody-based enrichment techniques enabled the progress in acetylation research [8]. The number of acetylation sites identified in various organisms ranging from bacteria to humans and plants rapidly approaches that of phosphorylation [9–14].

Aside from histone acetylation, it becomes apparent that this modification also frequently occurs on proteins and enzymes involved in central metabolic pathways such as glycolysis, gluconeogenesis, TCA cycle, oxidative phosphorylation, and β-oxidation, and additionally in plants on Calvin cycle, photosynthesis, and photorespiration (e.g., [15–18]). This highlights the importance of this highly dynamic and reversible modification, which is modulated by lysine acetyltransferases (KATs) and lysine deacetylases (KDAC) in dependence on the cellular acetyl-CoA availability. In contrast to lysine methylation, acetylation results in a loss of the positive charge of the lysine residue. This can have a strong impact on protein functions, thereby regulating enzyme activities, protein interactions, protein stabilities, and metabolic pathways for example [19]. Particularly, plant mitochondrial and chloroplast proteins show a high abundance of lysine-acetylated proteins [13, 16, 18, 20–24] and several proteins involved in photorespiration have been identified as lysine acetylated in different plant species (Fig. 1). Photorespiratory enzymes have already been shown to be modified by other posttranslational modifications such as phosphorylation, disulfide formation, glutathionylation, and S-nitrosylation [25–30]. However not much is known about the impact of these modifications on the regulation of the photorespiratory cycle, except that nitrosylation was shown to inhibit several enzymes of this pathway (for recent reviews, *see* [31, 32]). In contrast to S-nitrosylation, which occurs nonenzymatically on cysteine residues during oxidative stress conditions, protein phosphorylation is catalyzed by dedicated enzymes, which are called kinases. During protein phosphorylation a phosphate group from ATP is added to specific amino acids, most commonly to serine, threonine, and tyrosine, thereby adding a negative charge to the protein. Phosphorylated peptides were identified on several photorespiratory enzymes in *Arabidopsis* leaves and cell suspension cultures and under different conditions including resupply and deficiency of

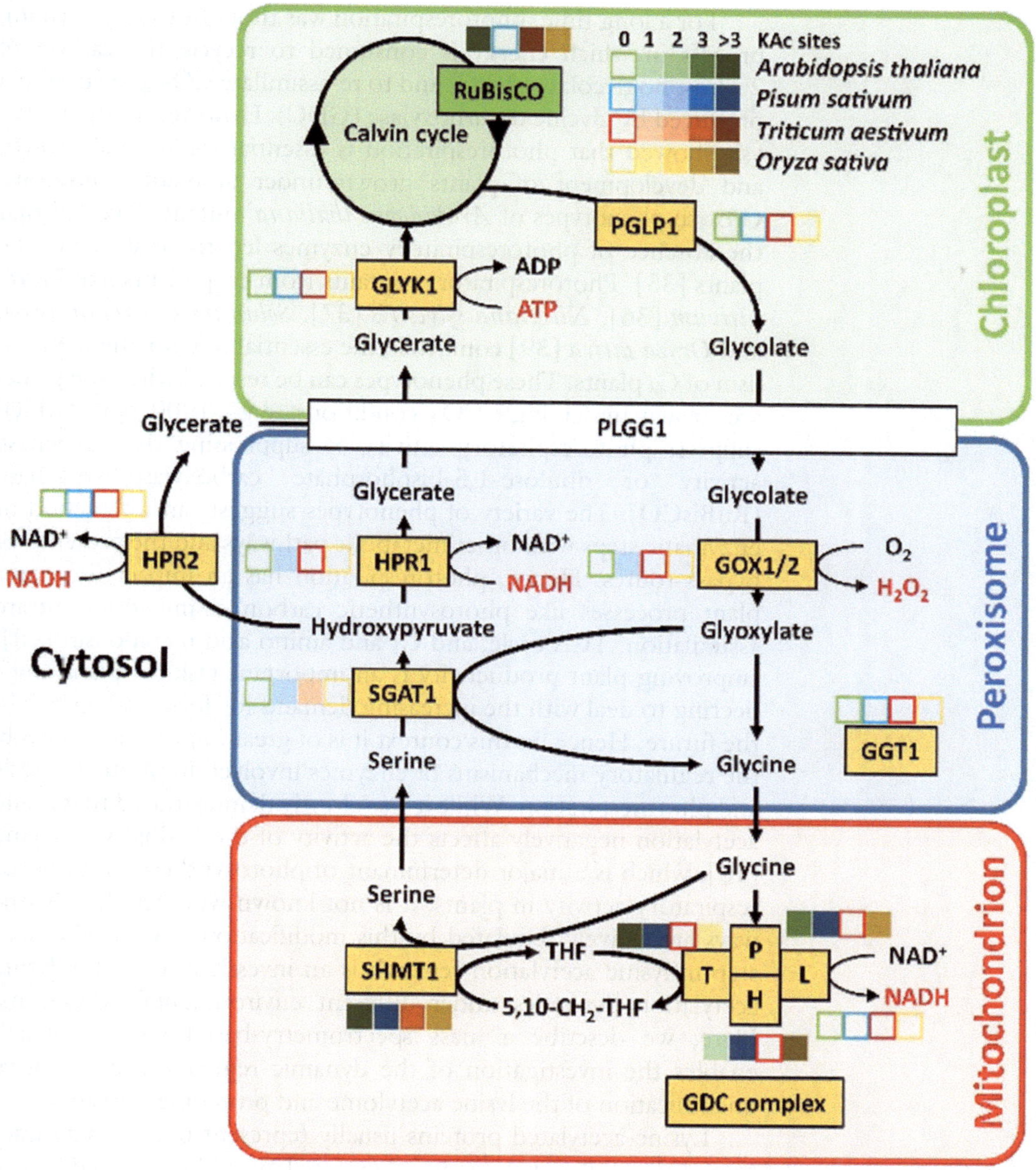

Fig. 1 Lysine-acetylated proteins involved in photorespiration from different plant species. Lysine acetylation sites identified in different studies are plotted according to the number of sites per protein within the photorespiratory cycle. Organisms used are *Arabidopsis thaliana* in *green* [13, 16], *Pisum sativum* in *blue* [22], *Triticum aestivum* in *red* [24], and *Oryza sativa* in *yellow* [20, 21, 23]. Uncharacterized proteins were not considered

nitrate as well as anoxia for example [31, 33, 34]. Although phosphorylation similarly to lysine acetylation has been found on most of the photorespiratory as well as related enzymes, their role in the regulation of this pathway is still unclear and has to be further analyzed in more detail.

For a long time, photorespiration was thought to be a wasteful process in which energy is consumed to recycle the carbon of 2-phosphoglycolate (2-PG) and to re-assimilate CO_2 and ammonia produced by glycine decarboxylase (GDC). However, mutant analysis showed that photorespiration is essential for normal growth and development of plants grown under ambient conditions. Growth phenotypes of *Arabidopsis thaliana* mutants revealed that the absence of photorespiratory enzymes led to small, chlorotic plants [35]. Photorespiratory mutants from crop plants like *Pisum sativum* [36], *Nicotiana sylvestris* [37], *Solanum tuberosum* [38], and *Oryza sativa* [39] confirmed the essential role for the metabolism of C_3 plants. These phenotypes can be rescued when the plants are grown under high CO_2 conditions (e.g., 1000 ppm) which suppress photorespiratory activity by suppressing the oxygenase activity of ribulose-1,5-bisphosphate carboxylase/oxygenase (RuBisCO). The variety of phenotypes suggests an interaction of enzymatic steps with other metabolic pathways and the presence of bypass routes. Hence, photorespiration has an impact on major plant processes like photosynthetic carbon assimilation, nitrate assimilation, TCA cycle, and C_1 and amino acid metabolism [40]. Improving plant productivity is an important task for plant engineering to deal with the increasing demand for food and energy in the future. Hence, in this context it is of great importance to study the regulatory mechanisms of enzymes involved in photosynthesis and photorespiration. While it was already demonstrated that lysine acetylation negatively affects the activity of the RuBisCO enzyme [16], which is a major determinant of photosynthesis and photorespiratory activity in plants, it is not known whether these pathways are actively regulated by this modification. Hence, the next step in lysine acetylation research is an investigation of the lysine acetylation dynamics under different environmental conditions. Here, we describe a mass spectrometry-based method which enables the investigation of the dynamic regulation and relative quantification of the lysine acetylome and proteome in plants.

Lysine-acetylated proteins usually represent only a small fraction of all cellular proteins and therefore have to be enriched before analysis. The enrichment step is most successful at the peptide level, using an anti-acetyllysine antibody, rather than at the protein level. Hence, several purification and enrichment steps are necessary before the sample can be analyzed via high-resolution mass spectrometry (MS). For quantification purposes, it is recommended to differentially label and combine the protein or peptide samples from different extracts that have to be compared, in order to keep the technical variation during the purification steps at a minimum. Several labeling techniques for such MS-based comparisons are available, which include metabolic and chemical labeling. For samples like tissues that cannot be metabolically labeled in comparison to cell cultures for example, chemical labeling at the peptide level is usually applied. The trypsinated peptides are labeled with unique mass tags and peptide samples can then be mixed for subsequent analyses. Multiple isobaric mass tags are available that allow multiplexing of up to ten samples, but co-fragmentation of co-eluting peptides frequently affects quantification accuracy and precision. Here, we present a cheap and efficient dimethyl labeling technique, which allows duplexing or even triplexing for peptides derived from Arabidopsis leaf extracts or any other protein extracts. Different methods and types of MS instruments are available to analyze such acetylated protein samples. Top-down MS is suitable for the analysis of intact proteins and their PTMs, which has been shown for histone proteins [41, 42]. More commonly, bottom-up proteomics is used for the analysis of acetylated proteins, in which trypsinated peptides are separated by nanoflow liquid chromatography (nano-LC) and transferred via electrospray ionization into the MS where the spectra are collected.

Our sample preparation procedure for the identification of lysine-acetylated peptides comprises three major steps before nano-LC-MS/MS and subsequent data analysis (Fig. 2):

1. Extensive solubilization and subsequent digestion of proteins: Given that many membrane proteins are essential for various functions and metabolism, their solubilization is crucial for a deep proteome coverage. In this protocol, we use an adapted filter-assisted sample preparation (FASP) method [43]. In this method, proteins are solubilized using sodium dodecyl sulfate (SDS) which is subsequently removed, without the need of precipitating.
2. Isotopic dimethyl labeling for quantification of peptide ratios from two different samples (duplexing) adapted from the protocol of [44].
3. Enrichment of lysine-acetylated peptides: Since the abundance of acetylated peptides is rather low in the bulk of non-acetylated peptides, an enrichment is absolutely necessary. For the enrichment of lysine-acetylated peptides anti-acetyllysine antibodies from different vendors are available.

For accurate peptide and protein quantification it is necessary that a high labeling efficiency can be achieved. With the isotopic dimethyl labeling technique we achieve label efficiencies of more than 99%, which makes this method suitable for the relative quantification of peptides (Fig. 3a). However, it is important to include a label swap in the experimental design, to avoid any labeling bias between samples, which usually occurs for a few peptides due to a retention time shift during the elution of the deuterated peptides. In the majority of cases, all differentially isotopic labeled peptides of the same sequence co-elute more or less at the same retention time (Fig. 3b).

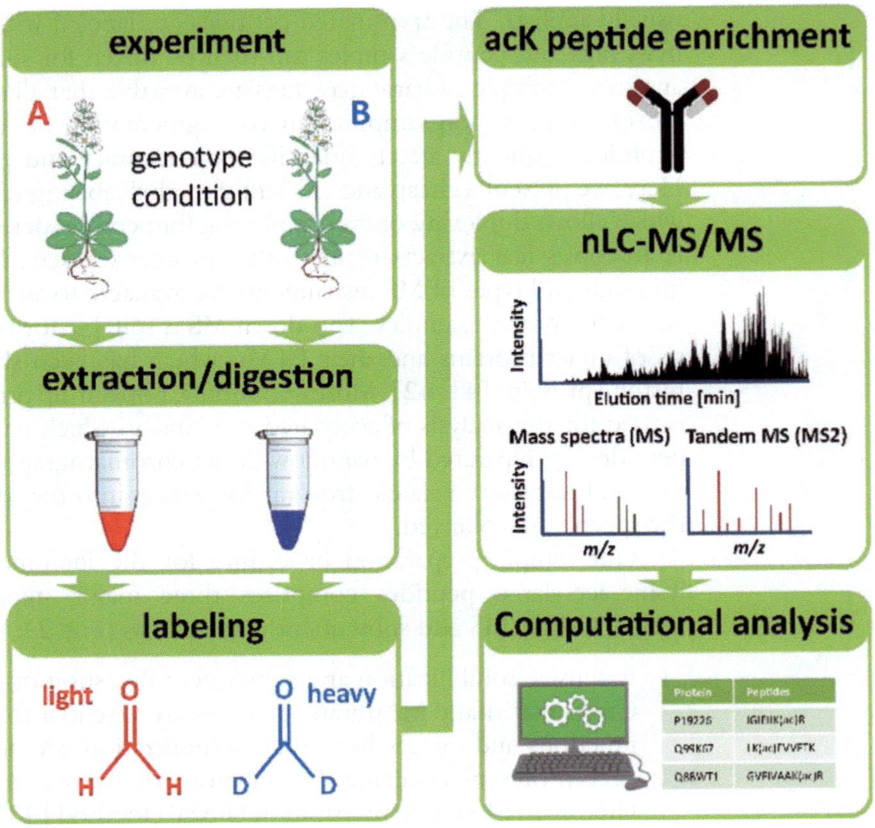

Fig. 2 Workflow for the identification and quantification of lysine-acetylated peptides using LC-MS/MS. After cell lysis, proteins are digested enzymatically, cleaned up on C18 columns, and subsequently isotopically dimethyl-labeled. Peptide samples are eluted from C18 columns and combined. Lysine-acetylated peptides are enriched using an anti-acetyllysine antibody prior to MS analysis. Enriched peptides as well as total peptides are analyzed using nano-LC-MS/MS. The raw data are computationally processed to identify and quantify lysine-acetylated peptides and proteins

2 Materials

2.1 Protein Extraction

1. SDT-lysis buffer: 4% (w/v) SDS, 0.1 M dithiothreitol (DTT) in 100 mM Tris–HCl pH 7.6. Prepare 2.5 volumes per gram of material.
2. Ultrasonic bath.
3. Benchtop centrifuge for microfuge tubes.
4. 660 nm Protein assay with compatibility reagent (Pierce—Thermo Scientific).

2.2 Filter-Assisted Sample Preparation (FASP)

1. Centrifugal filter device (CFD): Amicon Ultra-4 30k MWCO (Millipore) or similar (*see* **Note 1**).
2. Benchtop centrifuge, accommodating 15 mL conical-bottomed tubes, ideally in fixed-angle rotor at a speed of 4000 × *g*.

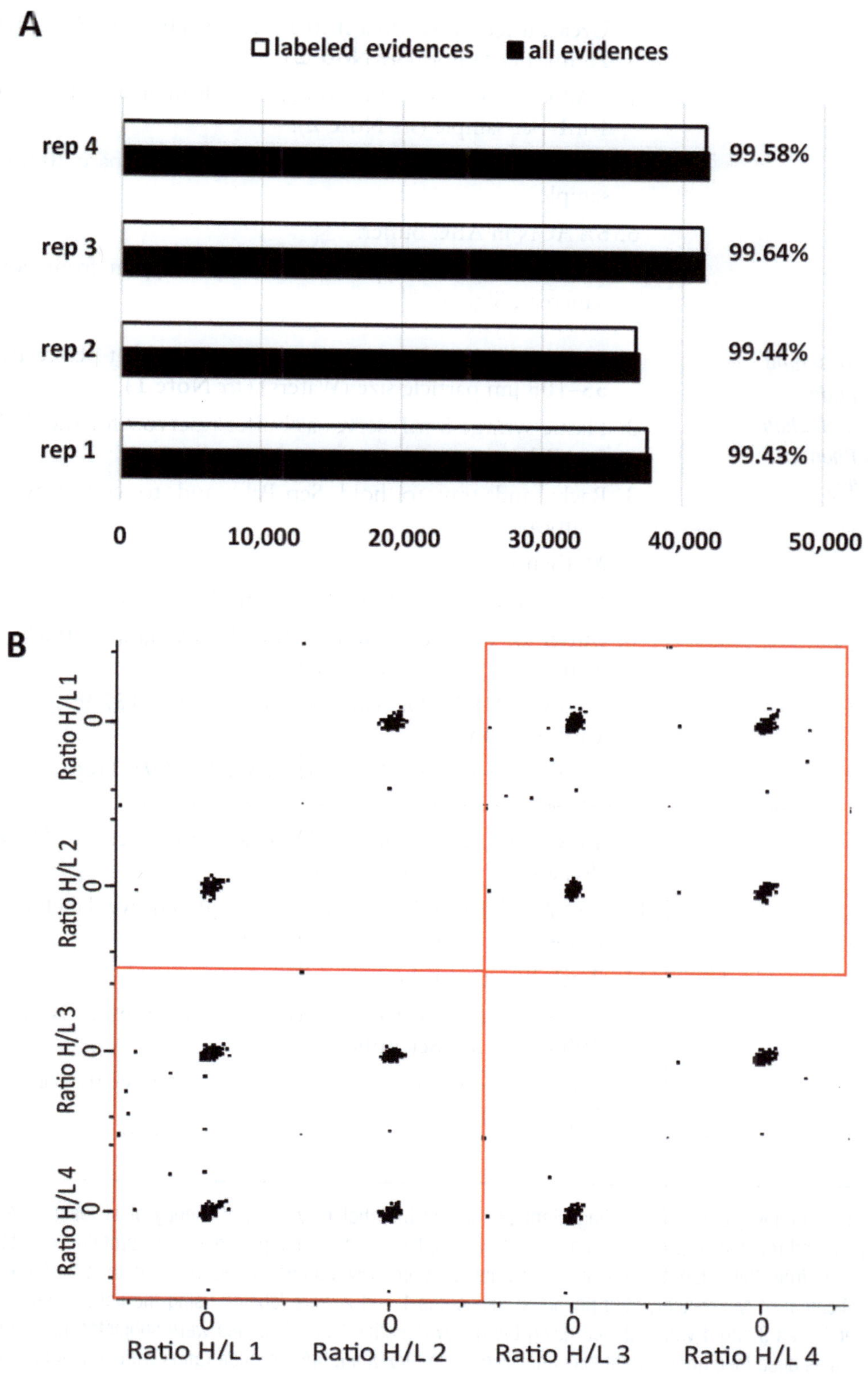

Fig. 3 Quality assessment of the dimethyl labeling efficiency (**a**), and relative quantification of heavy- and light-labeled lysine-acetylated peptides (**b**) from *Arabidopsis thaliana*. (**a**) Efficiency of the stable isotopic

3. Urea buffer: 8 M Urea in 0.1 M Tris–HCl pH 8.5. Prepare 25 mL per sample (*see* **Note 2**).
4. CAA solution: 0.05 M Chloroacetamide in urea buffer. Prepare 1 mL per sample (*see* **Note 2**).
5. ABC buffer: 0.05 M NH_4HCO_3 in water. Prepare 20 mL per sample.
6. 5% ACN in ABC buffer.
7. Proteomics-grade trypsin (for example trypsin from porcine pancreas, Sigma).

2.3 Desalting of Peptides in Combination with Chemical Labeling

1. Sep-Pak C18 classic cartridge, 360 mg sorbent per cartridge, 55–105 μm particle size (Waters) (*see* **Note 1**).
2. Plastic syringe 5 mL serves as buffer reservoir for the 360 mg Sep-Paks.
3. Racks and tray to hold Sep-Paks and to collect washing solutions.
4. Methanol.
5. Buffer A: 0.5% (v/v) formic acid in ultrapure water.
6. Buffer B: 80% (v/v) acetonitrile (HPLC grade), 0.5% (v/v) formic acid in ultrapure water.
7. 50 mM NaH_2PO_4 (monohydrate: MW: 137.99): Prepare 1 mL per sample.
8. 50 mM Na_2HPO_4 (dihydrate: MW: 177.99): Prepare 3.5 mL per sample.
9. 4% CH_2O (light), or CD_2O (heavy) in water (*see* **Note 3**): Prepare 0.25 mL per sample.
10. 0.6 M $NaBH_3CN$ in water (for light/heavy labels); MW: 62.84: Prepare 0.25 mL per sample.
11. 2 mL Reaction tubes.
12. Microvolume UV-VIS photometer (for example NanoDrop 2000, Thermo Scientific).
13. Vacuum concentrator (for example Concentrator 5301, Eppendorf).

Fig. 3 (Continued) dimethyl labeling. Peptides should be labeled with an efficiency of at least 99% in all samples and replicates. (**b**) Quantification accuracy of lysine-acetylated peptides. Four protein extracts were generated from the same Arabidopsis leaf sample and processed as described in the method for the relative quantification of lysine-acetylated peptides. Replicates 1 and 2 were labeled using the light dimethyl label, and replicates 3 and 4 were labeled using the heavy dimethyl label. Samples were subjected to LC-MS/MS analysis as described above. Data was analyzed using MaxQuant [45]. Peptide ratios were log2-transformed. The label-swapped samples are indicated in *red squares*. The majority of quantified peptides have a distribution around 0. Peptide ratios that deviate from this distribution are affected by a retention time shift of the heavy label in one of the samples

2.4 Immuno-Enrichment of Lysine-Acetylated Peptides

1. Ice for cooling samples.
2. Tris-buffered saline buffer (TBS): 50 mM Tris–HCl, 150 mM NaCl, pH 7.6, sterile filtrated, stored at 4 °C.
3. Ultrasonic bath.
4. Universal pH indicator paper, pH 1–11.
5. Microvolume UV-VIS photometer.
6. Acetyllysine antibody, immobilized to agarose beads (Immunechem, ICP0388).
7. Benchtop centrifuge for microfuge tubes.
8. Rolling wheel at 4 °C.
9. Ultrapure water, stored at 4 °C.
10. 1% TFA.
11. GELoader® tips (Eppendorf) or similar gel-loading pipette tips.

2.5 Purification of Enriched Peptides

1. 200 μL Pipette tips.
2. SDB-RPS Empore™ SPE Disks.
3. Self-made tip preparation tool (*see* **Note 4**).
4. Acetonitrile.
5. Buffer C: 30% Methanol, 1% TFA in ultrapure water.
6. Buffer D: 0.2% TFA in ultrapure water.
7. Buffer 1: 100 mM NH_4HCO_2, 40% acetonitrile, 0.5% FA in ultrapure water.
8. Buffer 2: 150 mM NH_4HCO_2, 60% acetonitrile, 0.5% FA in ultrapure water.
9. Buffer 3: 5% Ammonia, 80% acetonitrile.
10. Buffer A*: 2% Acetonitrile (HPLC grade), 0.1% TFA in ultrapure water.
11. Benchtop centrifuge for microfuge tubes.
12. Vacuum concentrator.

3 Methods

3.1 Preparation of Peptides Using FASP

All steps are carried out at room temperature unless indicated else (*see* **Note 5**).

3.1.1 Protein Extraction

1. Leaves are harvested and ground in liquid nitrogen to a fine powder. The desired amount is weighed into a liquid nitrogen-cooled 2 mL (up to 300 mg) or 15 mL reaction tube.
2. Prepare a water bath at 95 °C and preheat the necessary amount of SDT-lysis buffer (2.5 volumes) to 95 °C (*see* **Note 6**).

3. Mix the powder with 2.5 volumes hot SDT-lysis buffer, vortex until the powder is resuspended, place for 5 min in the water bath at 95 °C, and vortex two times for 20 s in between.
4. Sonicate sample in an ultrasonic bath for 10 min.
5. If you started with 15 mL reaction tubes shortly spin down and transfer the suspension with a cut tip to 2 mL reaction tube. Centrifuge for 30 min in a benchtop centrifuge at top speed (15,000–21,000 × *g*) and room temperature.
6. Transfer the supernatant to a new tube without disrupting any pelleted material.
7. Repeat **steps 4** and **5** once.
8. Determine the amount of protein using the 660 nm protein assay with compatibility reagent according to the manufacturer's instructions.

3.1.2 FASP

1. Add 2 mL urea buffer to the CFD and centrifuge for 5 min to condition the membrane (*see* **Note** 7).
2. Dilute the sample with urea buffer to a final SDS concentration lower than 0.5% and transfer to the CFD. Multiple loading might be necessary; 4 mL is the maximum container capacity.
3. Centrifuge at 4000 × *g* for 15 min (or until at least tenfold concentration).
4. Discard flow-through, add another 4 mL urea buffer, and repeat centrifugation.
5. Discard flow-through, add 1 mL CAA solution, mix gently for 1 min with a pipette without touching the membrane, and incubate for 30 min at room temperature in darkness.
6. Centrifuge the CFD at 4000 × *g* for 15 min, and discard flow-through.
7. Add 4 mL of urea buffer to the CFD and centrifuge at 4000 × *g* for 15 min or until at least tenfold concentration. Discard the flow-through and repeat this step twice.
8. Add 4 mL ABC buffer to the CFD and centrifuge at 4000 × *g* for 15 min or until at least tenfold concentration. Discard the flow-through and repeat this step twice.
9. Transfer the CFD to new 15 mL collection tubes.
10. Add 1 mL ABC buffer containing trypsin at an enzyme-to-protein ratio of 1:100 (e.g., 10 μg trypsin for 1 mg protein) and gently mix with a pipette (do not touch the membrane while pipetting) (*see* **Note 8**). If necessary, cover the membrane with addition of ABC buffer.
11. Tighten the lid of the tube well and incubate the CFD at 37 °C overnight.

12. Centrifuge the CFD at 4000 × *g* for 15 min. The eluate now contains the peptides.
13. Add 1 mL ABC buffer to the CFD (rinsing the filters when adding the buffer) and repeat the centrifugation.
14. Repeat **step 13**.
15. Repeat **step 13** with 1 mL of 5% ACN in ABC buffer.
16. Assess the peptide yield on a microvolume UV-VIS photometer at 280 nm, assuming an absorption of 1 for a peptide concentration of 1 mg/mL. Typical yields are approximately 50–60% of the starting amount.
17. Acidify the eluate to a final concentration of 1% TFA.

3.1.3 Desalting of Peptides and Stable Isotopic Dimethyl Labeling

Work in the fume hood, as most of the reagents are toxic! Take care of appropriate waste disposal.

1. Place Sep-Paks in a rack with a tray beneath. In case of 360 mg Sep-Paks assemble Sep-Pak with syringe (without piston) before placing in the rack.
2. In all steps a 5 or 10 mL pipette can be used to increase flow rates moderately.
3. Prepare labeling solutions. Per sample light label: 3.5 mL 50 mM Na_2HPO_4, 1 mL 50 mM NaH_2PO_4, 0.25 mL 4% CH_2O, 0.25 mL 0.6 M $NaBH_3CN$. Per sample heavy label: 3.5 mL 50 mM Na_2HPO_4, 1 mL 50 mM NaH_2PO_4, 0.25 mL 4% CD_2O, 0.25 mL 0.6 M $NaBH_3CN$. Include at least one label swap.
4. Condition a C18 Sep-Pak cartridge by flushing it with 3 mL methanol, 3 mL buffer B, and then 3 mL buffer A.
5. Slowly load peptide sample on the cartridge (*see* **Note 9**).
6. Wash with 1× 1 mL and 1× 3 mL buffer A.
7. Flush with 5 mL of labeling reagent (should take at least 10 min, *see* **Note 9**).
8. Wash with 1× 1 mL and 1× 3 mL buffer A.
9. Elute and collect samples with 2× 700 μL buffer B into 2 mL reaction tubes (*see* **Note 9**).
10. Assess the peptide yield on a microvolume UV-VIS photometer at 280 nm, assuming an absorption of 1 for a peptide concentration of 1 mg/mL (*see* **Note 10**).
11. Combine equal amounts of the differentially labeled samples.
12. Evaporate the sample to complete dryness on a vacuum concentrator. After drying, peptides can be stored for short term at −20 °C or for long term at −80 °C (*see* **Note 11**).

13. Redissolve dried peptides in 50 μL 20% acetonitrile, vortex vigorously, and let rest for 10 min. Use 0.5 μg of peptide to check the labeling efficiency by mass spectrometry analysis before proceeding with the protocol.

3.1.4 Enrichment of Lysine-Acetylated Peptides

All steps involving the anti-acetyllysine antibody should be carried out on ice or at 4 °C.

1. Add 950 μL TBS to the rest of the redissolved peptides and vortex (*see* **Note 12**).
2. Check the pH of the solution by pipetting 0.5 μL of sample on pH indicator paper (*see* **Note 13**).
3. Assess the peptide yield on a microvolume UV-VIS photometer at 280 nm, assuming an absorption of 1 for a peptide concentration of 1 mg/mL.
4. Take an aliquot of 10 μg and store it at −20 °C for measuring the total proteome (input). The input might be further fractionated for a deeper proteome coverage.
5. Transfer 25 μL antibody bead slurry per mg of peptide to a 1.5 mL reaction tube with a cut 200 μL pipette tip (*see* **Note 14**).
6. Wash beads 3 × 5 min with 1 mL TBS on a rolling wheel at 4 °C. Centrifuge at 1000 × *g* in between washes and carefully remove the supernatant (*see* **Note 15**).
7. Add the redissolved peptides to the beads and incubate overnight at 4 °C on a rolling wheel.
8. Wash beads 4 × 5 min with 1 mL TBS on a rolling wheel at 4 °C.
9. Wash beads 2× shortly with ultrapure water to reduce the salt concentration, centrifuging in between at 1000 × *g* for 1 min. Remove supernatant completely and carefully with gel-loading tips.
10. Elute 3 × 5 min with initial bead slurry volume 1% TFA, mixing the beads gently with the pipet and letting them rest for the remaining time period. Centrifuge in between at 2000 × *g* and try to recover as much liquid as possible without transferring beads using gel-loading tips.
11. Assess the peptide concentration of the combined eluates on a microvolume UV-VIS photometer at 280 nm as in **step 3**. Usually the concentration is very low (only a few μg yield in total). Higher concentrations might indicate massive background contamination.

3.1.5 Purification of Enriched Peptides

1. Clean up input and enriched peptides.
2. Assemble the pipette tip (*see* **Note 16**) with tip-to-reaction tube adapter to a 2 mL reaction tube.
3. Condition a tip by flushing it with 100 μL acetonitrile, 100 μL buffer C, and then 100 μL buffer D. Centrifuge each step at 1800 × *g* for 2 min.
4. Slowly load peptide sample on the cartridge by centrifuging at 700 × *g* for 10 min. Check flow-through, and adjust centrifugal force and time if necessary.
5. Wash with 100 μL buffer D. Centrifuge at 1800 × *g* for 2 min.
6. Transfer tip with adapter to new 1.5 mL reaction tube.
7. Elute in three steps with 20 μL buffer 1, 2, and 3 by centrifuging at 700 × *g* for 5 min. Check flow-through, and adjust centrifugal force and time if necessary (*see* **Note 17**).
8. Evaporate the samples to complete dryness on a vacuum concentrator.
9. Dissolve samples in A* for MS measurements (*see* **Note 18**).

3.2 Guidelines for Measuring the Samples on LC-MS/MS

In general, every high-resolution nano-UHPLC-MS/MS setup used for shotgun proteomics can be used for measuring these samples. Typically, we run them on self-packed capillary columns with 75 μm diameter with 15–50 cm C18 reversed-phase material (for example: 1.9 μm ReproSil-Pur C18-AQ, Dr. Maisch GmbH). The column is kept at 50 °C in a column oven throughout the run. We inject max. 0.5 μg peptides and separate them at a flow rate of 300 nL/min with a 5-min linear gradient at 5% buffer B in buffer A (buffer A: 0.1% (v/v) formic acid; buffer B: 80% (v/v) acetonitrile, 0.1% (v/v) formic acid), followed by a 60-min linear gradient to 20% buffer B, 25-min linear gradient to 35% B, 10-min linear gradient to 55% B, 5-min linear gradient to 98% B, and then a 10-min linear gradient at 98% buffer B. We typically detect the peptides on a Q-Exactive HF MS (Thermo Scientific) in positive mode with a scanned mass range of 300–1750 *m/z*, at a resolution of 60,000 and an AGC target value of 3e6. A lock mass of 445.12003 is enabled. The 15 most intense ions are selected for MS2 at a resolution of 15,000 with an isolation window of 1.3 *m/z* and an AGC target of 1e5. Peptides with a charge of +1, 6–8, or >8 or with unassigned charge state are excluded from fragmentation for MS2; dynamic exclusion prevents repeated selection of selected masses for 30 s. We process raw data (input/enriched samples) using MaxQuant software [45] and search against The Arabidopsis Information Resource protein database (http://www.arabidopsis.org), with trypsin specificity and a maximum of two/four missed cleavages (input/enriched samples) at a protein and PSM false discovery rate of 1%. Multiplicity is set to 2 or 3, respectively, for light (Dimethyl0) and heavy (Dimethyl4) labels. Carbamidomethylation of cysteine residues is set as fixed, oxidation of methionine, N-terminal acetylation, and lysine acetylation (enriched samples) as variable modifications. Match between runs and re-quantify is enabled with standard settings.

4 Notes

1. Amicon Ultra-4 can be used for up to 3 mg of protein.
2. Urea and CAA solutions have to be prepared on the day of use. Prepare the urea buffer early in advance. It needs time to dissolve but should not be heated above room temperature to prevent formation of isocyanate which results in carbamylation of proteins. The CAA solution should be stored in the dark until use.
3. Be aware that the stock concentrations of the formaldehyde solutions differ.
4. Use, e.g., Grace replacement needle 16GA SS for TLL type to cut the material and a Hamilton syringe stamp to remove plug into tip.
5. Placing SDS- or urea-containing buffers on ice or at 4 °C can lead to precipitation.
6. Usually a glass beaker placed on a magnetic stirrer with heating works fine. Use magnetic stirring to distribute heat equally.
7. Stop the centrifuge after approximately 1 min and check the buffer retention. In rare cases unusually fast flow-through occurs, which indicates a leak in the membrane. Such damaged units cannot be used and must be replaced.
8. To reach more complete digestion, it is possible to additionally digest the sample with LysC. For example, LysC is added at an enzyme-to-protein ratio of 1:100 and the sample is incubated for 2 h at room temperature, before adding trypsin and incubating overnight as specified.
9. Load sample slowly, just by gravity flow if possible.
10. High amounts of acetonitrile can influence the microvolume measurement through fast evaporation. Take 10 μL of the eluted sample, evaporate to dryness on vacuum concentrator, and dissolve in 2% acetonitrile and 0.1% TFA. Use in microvolume measurement.
11. Complete evaporation removes acidic components from sample which eases the pH adjustment before enrichment of lysine-acetylated peptides.

12. If the peptides do not dissolve well, the sample can also be placed in an ultrasonic bath for 5 min. Higher amounts of 20% acetonitrile can also help to redissolve the pellet; however bear in mind that the antibody tolerates only low amounts of acetonitrile.
13. Formic acid remaining after desalting can lower the pH substantially. After adding the TBS buffer check whether the pH is in the desired range and, if necessary, correct by adding small amounts of 1 M Tris–HCl pH 7.6. Usually, when the correct pH is set the peptides dissolve completely.
14. A ratio of 25 μL antibody bead slurry per 1 mg of peptides results in a reasonable compromise between the number of sites identified and the costs for the antibody. The antibody is quite sensitive to temperature changes and cannot be reused in our experience.
15. Take care not to lose beads while washing. Gel-loading tips are useful for taking off the remaining buffer close to the beads.
16. Assemble pipette tips with filter disks. Use Grace replacement needle to cut a three-layer plug and place needle in tip. Use Hamilton syringe stamp to push plug in tip. Fix plug by pushing slightly with stamp.
17. Transfer eluent directly after elution to new 0.5 mL reaction tube. Reuse 1.5 mL reaction tube in the next step of elution.
18. Peptide concentration of the input has to be assessed and adjusted to desired concentration. The enriched sample usually can be injected completely.

References

1. Uy R, Wold F (1977) Posttranslational covalent modification of proteins. Science 198 890–896
2. Chuh KN, Pratt MR (2015) Chemical methods for the proteome-wide identification of posttranslationally modified proteins. Curr Opin Chem Biol 24:27–37
3. Phillips D (1963) The presence of acetyl groups in histones. Biochem J 87:258–263
4. Allfrey VG, Faulkner R, Mirsky AE (1964) Acetylation and methylation of histones and their possible role in the regulation of RNA synthesis. Proc Natl Acad Sci U S A 51:786–794
5. Berger SL (2007) The complex language of chromatin regulation during transcription. Nature 447:407–412
6. L'Hernault SW, Rosenbaum JL (1985) Chlamydomonas alpha-tubulin is posttranslationally modified by acetylation on the epsilon-amino group of a lysine. Biochemistry 24:473–478
7. Gu W, Roeder RG (1997) Activation of p53 sequence-specific DNA binding by acetylation of the p53 C-terminal domain. Cell 90:595–606
8. Choudhary C, Mann M (2010) Decoding signalling networks by mass spectrometry-based proteomics. Nat Rev Mol Cell Biol 11:427–439
9. Zhang J, Sprung R, Pei J et al (2008) Lysine acetylation is a highly abundant and evolutionarily conserved modification in *Escherichia coli*. Mol Cell Proteomics 8:215–225
10. Henriksen P, Wagner SA, Weinert BT et al (2012) Proteome-wide analysis of lysine acetylation suggests its broad regulatory scope in *Saccharomyces cerevisiae*. Mol Cell Proteomics 11:1510–1522
11. Lundby A, Lage K, Weinert BT et al (2012) Proteomic analysis of lysine acetylation sites in rat tissues reveals organ specificity and subcellular patterns. Cell Rep 2:419–431

12. Weinert BT, Wagner SA, Horn H et al (2011) Proteome-wide mapping of the Drosophila acetylome demonstrates a high degree of conservation of lysine acetylation. Sci Signal 4:ra48
13. König A-C, Hartl M, Boersema PJ et al (2014) The mitochondrial lysine acetylome of Arabidopsis. Mitochondrion 19:252–260
14. Svinkina T, Gu H, Silva JC et al (2015) Deep, quantitative coverage of the lysine acetylome using novel anti-acetyl-lysine antibodies and an optimized proteomic workflow. Mol Cell Proteomics 14:2429–2440
15. Zhao S, Xu W, Jiang W et al (2010) Regulation of cellular metabolism by protein lysine acetylation. Science 327:1000–1004
16. Finkemeier I, Laxa M, Miguet L et al (2011) Proteins of diverse function and subcellular location are lysine acetylated in Arabidopsis. Plant Physiol 155:1779–1790
17. Choudhary C, Kumar C, Gnad F et al (2009) Lysine acetylation targets protein complexes and co-regulates major cellular functions. Science 325:834–840
18. Wu X, Oh M-H, Schwarz EM et al (2011) Lysine acetylation is a widespread protein modification for diverse proteins in Arabidopsis. Plant Physiol 155:1769–1778
19. Yang X-J, Seto E (2008) Lysine acetylation: codified crosstalk with other posttranslational modifications. Mol Cell 31:449–461
20. He D, Wang Q, Li M et al (2016) Global proteome analyses of lysine acetylation and succinylation reveal the widespread involvement of both modification in metabolism in the embryo of germinating rice seed. J Proteome Res 15:879–890
21. Nallamilli BRR, Edelmann MJ, Zhong X et al (2014) Global analysis of lysine acetylation suggests the involvement of protein acetylation in diverse biological processes in rice (*Oryza sativa*). PLoS One 9:e89283
22. Smith-Hammond CL, Hoyos E, Miernyk JA (2014) The pea seedling mitochondrial Nε-lysine acetylome. Mitochondrion 19:154–165
23. Xiong Y, Peng X, Cheng Z et al (2016) A comprehensive catalog of the lysine-acetylation targets in rice (*Oryza sativa*) based on proteomic analyses. J Proteomics 138:20–29
24. Zhang Y, Song L, Liang W et al (2016) Comprehensive profiling of lysine acetylproteome analysis reveals diverse functions of lysine acetylation in common wheat. Sci Rep 6:21069
25. Marchand C, Le Maréchal P, Meyer Y et al (2004) New targets of Arabidopsis thioredoxins revealed by proteomic analysis. Proteomics 4:2696–2706
26. Palmieri MC, Lindermayr C, Bauwe H et al (2010) Regulation of plant glycine decarboxylase by S-nitrosylation and glutathionylation. Plant Physiol 152:1514–1528
27. Abat JK, Mattoo AK, Deswal R (2008) S-nitrosylated proteins of a medicinal CAM plant *Kalanchoe pinnata*- ribulose-1,5-bisphosphate carboxylase/oxygenase activity targeted for inhibition. FEBS J 275:2862–2872
28. Ortega-Galisteo AP, Rodriguez-Serrano M, Pazmino DM et al (2012) S-Nitrosylated proteins in pea (*Pisum sativum* L.) leaf peroxisomes: changes under abiotic stress. J Exp Bot 63:2089–2103
29. Nakagami H, Sugiyama N, Mochida K et al (2010) Large-scale comparative phosphoproteomics identifies conserved phosphorylation sites in plants. Plant Physiol 153:1161–1174
30. Reiland S, Messerli G, Baerenfaller K et al (2009) Large-scale arabidopsis phosphoproteome profiling reveals novel chloroplast kinase substrates and phosphorylation networks. Plant Physiol 150:889–903
31. Hodges M, Jossier M, Boex-Fontvieille E et al (2013) Protein phosphorylation and photorespiration. Plant Biol 15:694–706
32. Hodges M, Dellero Y, Keech O et al (2016) Perspectives for a better understanding of the metabolic integration of photorespiration within a complex plant primary metabolism network. J Exp Bot 67:3015–3026
33. Aryal UK, Krochko JE, Ross ARS (2012) Identification of phosphoproteins in *Arabidopsis thaliana* leaves using polyethylene glycol fractionation, immobilized metal-ion affinity chromatography, two-dimensional gel electrophoresis and mass spectrometry. J Proteome Res 11:425–437
34. Sugiyama N, Nakagami H, Mochida K et al (2008) Large-scale phosphorylation mapping reveals the extent of tyrosine phosphorylation in Arabidopsis. Mol Syst Biol 4:285–286
35. Somerville CR, Ogren WL (1980) Photorespiration mutants of *Arabidopsis thaliana* deficient in serine-glyoxylate aminotransferase activity. Proc Natl Acad Sci U S A 77:2684–2687
36. Blackwell RD, Murray AJS, Lea PJ (1987) The isolation and characterisation of photorespiratory mutants of barley and pea. In: Progress in photosynthesis research. Springer, Dordrecht, The Netherlands, pp 625–628
37. McHale NA, Havir EA, Zelitch I (1988) A mutant of *Nicotiana sylvestris* deficient in serine glyoxylate aminotransferase activity. Theor Appl Genet 76:71–75

38. Heineke D, Bykova N, Gardeström P et al (2001) Metabolic response of potato plants to an antisense reduction of the P-protein of glycine decarboxylase. Planta 212:880–887
39. Xu H, Zhang J, Zeng J et al (2009) Inducible antisense suppression of glycolate oxidase reveals its strong regulation over photosynthesis in rice. J Exp Bot 60:1799–1809
40. Bauwe H, Hagemann M, Fernie AR (2010) Photorespiration: players, partners and origin. Trends Plant Sci 15:330–336
41. Garcia BA, Pesavento JJ, Mizzen CA et al (2007) Pervasive combinatorial modification of histone H3 in human cells. Nat Methods 4:487–489
42. Yates JR, Kelleher NL (2013) Top down proteomics. Anal Chem 85:6151
43. Wiśniewski JR, Zougman A, Nagaraj N et al (2009) Universal sample preparation method for proteome analysis. Nat Methods 6:359–362
44. Boersema PJ, Raijmakers R, Lemeer S et al (2009) Multiplex peptide stable isotope dimethyl labeling for quantitative proteomics. Nat Protoc 4:484–494
45. Cox J, Mann M (2008) MaxQuant enables high peptide identification rates, individualized p.p.b.-range mass accuracies and proteome-wide protein quantification. Nat Biotechnol 26:1367–1372

Chapter 6

In Vitro Analysis of Metabolite Transport Proteins

Marc-Sven Roell, Franziska Kuhnert, Shirin Zamani-Nour, and Andreas P.M. Weber

Abstract

The photorespiratory cycle is distributed over four cellular compartments, the chloroplast, peroxisomes, cytoplasm, and mitochondria. Shuttling of photorespiratory intermediates between these compartments is essential to maintain the function of photorespiration. Specific transport proteins mediate the transport across biological membranes and represent important components of the cellular metabolism. Although significant progress was made in the last years on identifying and characterizing new transport proteins, the overall picture of intracellular metabolite transporters is still rather incomplete. The photorespiratory cycle requires at least 25 transmembrane transport steps; however to date only plastidic glycolate/glycerate transporter and the accessory 2-oxoglutarate/malate and glutamate/malate transporters as well as the mitochondrial transporter BOU1 have been identified. The characterization of transport proteins and defining their substrates and kinetics are still major challenges.

Here we present a detailed set of protocols for the in vitro characterization of transport proteins. We provide protocols for the isolation of recombinant transport protein expressed in *E. coli* or *Saccharomyces cerevisiae* and the extraction of total leaf membrane protein for in vitro analysis of transporter proteins. Further we explain the process of reconstituting transport proteins in artificial lipid vesicles and elucidate the details of transport assays.

Key words Metabolite transport proteins, In vitro transport assays, Isolation of recombinant transport protein

1 Introduction

Cellular compartmentation allows for optimizing and regulating complex metabolic networks [1]. In eukaryotic plant cells, organelles (plastids and mitochondria) and compartments (endoplasmic reticulum (ER), Golgi apparatus, peroxisomes, and vacuole) are surrounded by at least one membrane. Biological membranes form a diffusion barrier for hydrophilic compounds, prohibiting the unrestricted exchange of metabolites and ions. In plants, metabolic pathways, such as photorespiration, are often interconnected between different compartments, requiring the shuttling of solutes across compartment-surrounding membranes. Membrane transport

Alisdair R Fernie et al. (eds.), *Photorespiration: Methods and Protocols*, Methods in Molecular Biology, vol. 1653, DOI 10.1007/978-1-4939-7225-8_6, © Springer Science+Business Media LLC 2017 proteins (transporters) facilitate the controlled and selective flux of metabolites and ions across cellular and intracellular membranes. This enables optimized metabolic pathways with involved enzymatic reactions taking place in the preferred environment and preventing the cellular distribution of toxic by-products [2–4].

Most transporters are integral membrane proteins, spanning the lipid bilayer by alpha-helical transmembrane domains. Pore-forming proteins, composed of beta-barrels, from the outer bacterial, mitochondrial, and plastidic membranes are a second class of transport proteins [2]. According to genome analyses, between 20 and 30% of all genes encode transmembrane proteins [5]. The prediction of transport proteins is based on their transmembrane domains. The related database for plant membrane proteins is called ARAMEMNON (http://aramemnon.botanik.uni-koeln.de) and enables access to the plant permeome of six plant species (*Arabidopsis thaliana*, *Oryza sativa*, *Zea mays*, *Brachypodium distachyon*, *Vitis vinifera*, *Populus trichocarpa*) [5, 6].

We focus here on carrier proteins mediating the passive diffusion of solutes in an electrochemical gradient-dependent manner. Membrane transport can be classified according to three different transport mechanisms: (1) uniport (transport in one direction), (2) symport (cotransport of two substrates in the same direction), and (3) antiport (transport of two substrates in opposite directions) with a net transport process that is either electrogenic or electroneutral [2].

While many solute carriers were previously identified and characterized, the understanding of plant membrane transporters is still incomplete, especially in the photorespiratory pathway. Out of nine transporters involved in the core cycle of photorespiration only the plastidic glycolate glycerate transporter (PLGG1), mediating the import of glycerate in counter-exchange with glycolate into the chloroplast, is known [7, 8]. However, including photorespiratory-associated reactions and coenzymes, a set of 25 transporters is needed for functional photorespiration [8].

Accordingly, the recombinant expression and in vitro analysis of putative transporters provide a powerful tool for investigating substrate specificities, affinities, and kinetic properties [2]. In vitro analysis of transport proteins is performed by reconstituting heterologously expressed transporters into artificial lipid vesicles (liposomes) followed by flux assays.

Choosing a suitable host for the heterologous expression of transport proteins can be challenging. The details of recombinant expression systems of transport proteins are beyond the scope of this protocol; we hence refer the reader to [2] for reviewing different expression systems suitable for functional analysis of transport proteins. Since *E. coli* and *Saccharomyces cerevisiae* represent the most frequently used expression hosts, we focus here on the purification of recombinant membrane proteins from those and further explain the technique of using total leaf membrane protein for in vitro analysis of transporters.

Purified recombinant transporters can be used for the reconstitution into liposomes (proteoliposomes). It should be taken into account that the lipid composition of liposomes should be similar to the organelle of origin of the analyzed transporter [9].

Analyzing transporter activity relies on using radioactive-labeled substrates. Transporter exchange assays are based on the uptake of radioactive-labeled substrate into the proteoliposomes in exchange for preloaded substrates (forward-exchange method) or the efflux of radioactive-labeled substrate out of the proteoliposomes (backward-exchange method). In the case of uniporter assays only the forward-exchange method can be applied [10].

Here we present three different approaches for the isolation of recombinant membrane proteins using (a) *E. coli*, (b) *Saccharomyces cerevisiae*, and (c) total leaf membrane protein. Additionally, we provide a protocol for the reconstitution of recombinant membrane proteins into liposomes using the rapid and efficient freeze-thaw-sonication procedure and a standardized protocol for transporter exchange assays. As stated above most transporters involved in photorespiration are not yet identified. Our provided protocols, schematically represented in Fig. 1, can be used for in vitro analysis of potential candidates involved in photorespiration and beyond.

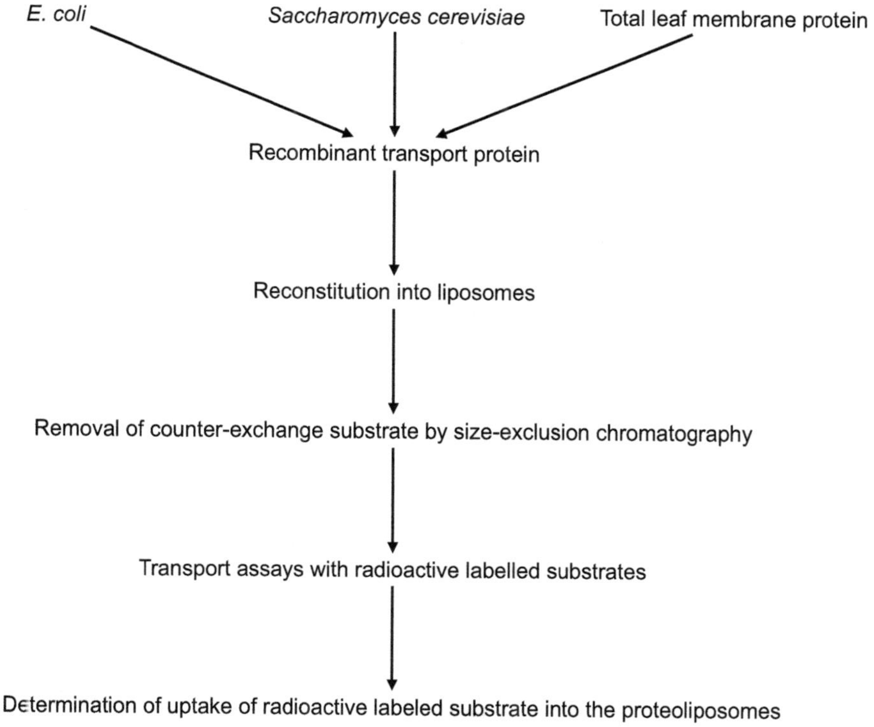

Fig. 1 Schematic overview of the in vitro analysis of metabolite transport proteins

2 Materials

2.1 Required Instrumentation

Bead mill Type MM 301 (Retsch), glass beads (0.4–0.6 mm, Sigma Aldrich), Optima™ L-80 XP Ultracentrifuge (Beckman Coulter), SW32 Ti Rotor (Beckman Coulter), Branson Sonifier 250, scintillation vial (20 ml, Sarstedt), heat block, scintillation counter LS 6000 (Beckman Coulter), ultracentrifuge vessels, mortar and pestle, magnetic stirrer, Sephadex G-25M columns (PD-10 column, GE Healthcare).

2.2 Buffers and Solution

2.2.1 Chemicals for Stock Solutions

Tris-hydroxymethyl-aminomethane (Tris) pH adjusted with hydrochloride acid (HCl), 4-(2-hydroxyethyl)-1-piperazineethanesulfonic acid (HEPES) pH adjusted with potassium hydroxide (KOH), ethylenediaminetetraacetic acid (EDTA), dithiothreitol (DTT),

4-(2-aminoethyl) benzenesulfonyl fluoride (AEBSF), pepstatin A, magnesium chloride ($MgCl_2$) stocks are prepared in MilliQ, phenylmethylsulfonyl fluoride (PMSF) is prepared in ethanol, L-α-phosphatidylcholine (Sigma, Type IV-S from soybean), 100% acetone, liquid and gas nitrogen.

2.2.2 Isolation of Recombinant Transport Protein from E. coli

1. *E. coli* lysis buffer: 50 mM Tris (HCl, pH 7.4), 1 mM EDTA, 1 mM DTT, 1 mM PMSF, 0.1 mM AEBSF, 2 μM pepstatin A (*see* **Note 1**).
2. *E. coli* membrane buffer: 50/5 buffer: 50 mM HEPES (KOH, pH 8.0), 5 mM $MgCl_2$.
3. IB wash buffer I: 10 mM Tris (HCl, pH 7.0), 0.1 mM EDTA, 1 mM DTT, 3% (v/v) Triton X-100.
4. IB wash buffer II: 10 mM Tris (HCl, pH 7.0), 0.1 mM EDTA, 1 mM DTT.
5. IB solubilization buffer I: 10 mM Tris–HCl (pH 7.0), 0.1 mM EDTA, 1 mM DTT, supplemented with non-denaturating detergent (*see* **Note 2**).
6. IB solubilization buffer II: 10 mM Tris–HCl (pH 7.0), 0.1 mM EDTA, 1 mM DTT.

2.2.3 Isolation of Saccharomyces cerevisiae Membranes

1. Yeast lysis buffer: 50 mM NaP_i (pH 7.2), 1 mM EDTA, 5% (v/v) glycerol.
2. Yeast resuspension buffer: 10 mM HEPES (KOH, pH 7.5), 10 mM $MgCl_2$.

2.2.4 Extraction of Total Membrane Proteins from Plant Leaf Materials

1. Extraction buffer: 250 mM Tris (HCl, pH 8.5), 25 mM EDTA, 30% (w/v) sucrose, 5 mM DTT, 1 mM PMSF, protease inhibitor cocktails, or combination of appropriate protease inhibitors like 0.1 mM AEBSF, 1 mM benzamidine, 1 μM pepstatin A, 2 mM leupeptin, and 2 mM antipain (*see* **Note 3**).

2. Resuspension Buffer: 50 mM HEPES (KOH, pH 7.5), 5 mM EDTA, 2 mM DTT, 1 mM PMSF and suitable protease inhibitor as in the extraction buffer (*see* **Note 4**).

2.2.5 Reconstitution of Recombinant Transport Proteins in Liposomes

1. Internal medium (without preloading substrate): 50 mM Potassium gluconate and 100 mM tricine (KOH, pH 7.6).
2. Internal medium (with preloading substrate): 20 mM Potassium gluconate, 100 mM tricine (KOH, pH 7.6), and 30 mM unlabeled counter-exchange substrate (*see* **Note 5**).
3. PD-10 buffer: 100 mM Na-gluconate, 40 mM K-gluconate, and 10 mM tricine (KOH, pH 7.6).

2.2.6 Transport Assay

1. Anion-exchange resin (acetate form, 100–200 mesh, Dowex AG1-X8 Resin Bio-Rad).
2. Cation-exchange resin (hydrogen form, 100–200 mesh, Dowex AG1-X8 Resin Bio-Rad).
3. Dowex buffer: 150 mM Sodium acetate (acetic acid, pH 7.6) *or* 150 mM lithium acetate (acetic acid, pH 7.6).
4. Rotiszint® eco plus LSC-universal cocktail (Roth).

3 Methods

3.1 Isolation of Recombinant Transport Proteins

3.1.1 Isolation of Recombinant Transport Protein from E. coli

Heterologous protein expression in *E. coli* is a well-established method. Recombinantly expressed transport proteins are either inserted into the bacterial membrane or aggregated in inclusion bodies, which is commonly the case [2]. We present a protocol starting with the preparation of an *E. coli* lysate (A). In an optimal case, recombinant protein is inserted in the bacterial membranes, which can be isolated and further used (B). Alternatively, recombinant transport proteins aggregate and are solubilized from inclusion bodies (C).

(A) Preparation of *E. coli* lysate:

If not stated otherwise all steps are carried out at 4 °C.

1. Pellet 200 ml *E. coli* expression culture by centrifugation at 3000 × *g* for 10 min and keep the cell pellet.
2. Resuspend the cell pellet in ice-cold *E. coli* lysis buffer (5 ml buffer per g wet weight) supplemented with protease inhibitors (PMSF, pepstatin A, and AEBSF).
3. Add lysozyme (1 mg/ml) and DNase I (20 μg/ml) and incubate on a shaking platform at 20 rpm for 30 min.
4. Lyse the cells by sonication using a microtip sonifier (output 4.0, duty cycle 40%) for approximately 10 min (*see*-**Note 6**).
5. Centrifuge at 12,000 × *g* for 15 min (*see* **Note 7**).

(B) Isolation of bacterial membranes:

If not stated otherwise all steps should be carried out at 4 °C.

1. Use the supernatant obtained after completing the preparation of the *E. coli* lysate.
2. For pelleting the membrane fraction, centrifuge the supernatant in an ultracentrifuge at 100,000 × *g* for 60 min (*see* **Note 8**).
3. Discard the supernatant.
4. Resuspend the membrane pellet in 50 μl cooled *E. coli* membrane buffer.
5. Isolated bacterial membranes can be directly used for reconstitution or alternatively stored at −80 °C.
6. Use between 0.5 and 5 μl of the isolated membranes for SDS-PAGE, western analysis, and determination of protein concentration.

(C) Solubilization of recombinant transport proteins from inclusion bodies:

1. Use the pellet obtained after completing preparation of *E. coli* lysate, as outlined above.
2. Resuspend the pellet in 2 ml precooled *E. coli* lysis buffer supplemented with protease inhibitors (PMSF, pepstatin A, and AEBSF).
3. Homogenize the pellet using microtip sonifier (output 3.0, duty cycle 30%) (*see* **Note 9**).
4. Centrifuge at 1000 × *g* for 10 min (*see* **Note 10**).
5. Transfer the supernatant into a new 2 ml Eppendorf tube.
6. Centrifuge at 12,000 × *g* for 15 min to pellet the inclusion bodies.
7. Resuspend the pelleted inclusion bodies in 2 ml IB wash buffer I and homogenize the pellet using microtip sonifier (output 2.0, duty cycle 20%) (*see* **Note 9**).
8. Centrifuge at 12,000 × *g* for 15 min to pellet the inclusion bodies.
9. Resuspend the pelleted inclusion bodies in 1.5 ml IB wash buffer II and homogenize the pellet using microtip sonifier (output 2.0, duty cycle 20%) (*see* **Note 9**).
10. Centrifuge at 12,000 × *g* for 15 min.
11. *Optional: Repeat the washing step 9.*
12. Solubilize the pellet in 40 μl IB solubilization buffer I (*see* **Note 2**).
13. Dilute the sample by the factor of ten with IB solubilization buffer II (*see* **Note 11**).

14. Centrifuge at 12,000 × *g* for 5 min (*see* **Note 12**).
15. Transfer the supernatant to a new tube.
16. Use the solubilized protein from inclusion bodies for reconstitution (*see* **Note 13**).
17. Use between 0.5 and 5 μl of the isolated proteins for SDS-PAGE, western analysis, and determination of protein concentration.

3.1.2 Isolation of Saccharomyces cerevisiae Membranes

If not stated otherwise all steps are carried out at 4 °C.

1. Pellet 50 ml expression culture with OD_{600} of 0.5–1.0 at room temperature.
2. Resuspend the pellet in 500 μl ice-cold yeast lysis buffer containing 1 mM PMSF, 1 μM pepstatin, and 100 μM ABSF.
3. Add one sample volume of prewashed glass beads (*see* **Note 14**).
4. Lyse the cells for 4 min at 30 Hz in a bead mill with precooled tube holders.
5. Transfer the lysed cells into a fresh 15 ml tube.
6. Pellet the glass beads for 1 min at 2200 × *g*.
7. Transfer the supernatant into an ultracentrifuge vessel.
8. Pellet the membranes for 45 min at 100,000 × *g*.
9. Discard the supernatant.
10. Resuspend the pellet (containing the isolated yeast membranes) in 50 μl yeast resuspension buffer.
11. Membrane aliquots can be directly used for reconstitution or stored at −80 °C.
12. Use between 0.5 and 5 μl of the isolated membranes for SDS-PAGE, western analysis, and determination of protein concentration.

3.1.3 Extraction of the Total Membrane Protein from Plant Leaf Materials

If not stated otherwise all steps are carried out on ice. Prior to the experiment, all equipment, tubes, and buffers to be used should be precooled.

1. Grind/homogenize the leaf materials in liquid nitrogen using mortar and pestle.
2. Between 0.1 and 1 g of leaf material is sufficient for the total membrane protein extraction. Depending on the leaf material amount, place it either in a 1.5 ml Eppendorf tube or in a 50 ml Falcon tube.
3. Add extraction buffer in a 5:1 ratio to leaf material (e.g., 5 ml extraction buffer:1 g leaf material).
4. Vortex the tube gently to get a uniform solution.

5. Incubate the tubes on ice for 10 min.
6. Centrifuge samples at 10,000 × *g* in a benchtop centrifuge for 10 min at 4 °C.
7. Carefully transfer the supernatant to the ultracentrifuge vessels without disrupting the pellets and discard the tubes with pellets.
8. Centrifuge the sample at 100,000 × *g* in an ultracentrifuge for 45 min at 4 °C.
9. Discard the supernatant.
10. Resuspend the pellet in 20–55 μl of resuspension buffer.

3.2 Reconstitution of Recombinant Transport Proteins in Liposomes

The transport activity can be measured when the transporter is in its native environment, i.e., a lipid bilayer. Transport proteins can be reconstituted into a lipid bilayer within liposomes to form proteoliposomes via the following freeze-thaw-sonication.

If not stated otherwise all steps are carried out on ice.

3.2.1 Purification of Phospholipids

An acetone-washing procedure is used to separate polar lipids from free fatty acids, which are soluble in acetone.

1. Dissolve 30 g of L-α-phosphatidylcholine in three volumes of chloroform (90 ml).
2. Add 540 ml ice-cold 100% acetone.
3. Stir the solution for 2 h at room temperature.
4. Incubate the solution overnight at 4 °C for the precipitation of phospholipids.
5. Discard the supernatant and dry the pellet using nitrogen gas.
6. The acetone-washed phospholipids can be stored at −80 °C.

3.2.2 Preparation of 3% (w/v) Phospholipid Suspension

1. Dissolve the acetone-washed phospholipids in the internal medium.
2. Sonify the solution with a sonifier (output 5.0, duty cycle 30%) for approximately 5 min until a homogenous suspension is obtained (*see* **Notes 15** and **16**).

3.2.3 The Freeze-Thaw-Sonication Reconstitution Procedure

1. Rapidly mix 950 μl of the phospholipid suspension (*see* Subheading 3.2.2) with 50 μl of the total extracted membrane protein (*see* **Note 17**).
2. Shock-freeze the mixture in liquid nitrogen. This step leads to opening of the vesicles.
3. Let the mixture thaw at room temperature. This step ensures the insertion of the proteins in the membrane.
4. Sonify the solution with a sonifier (output 4.0, duty cycle 30%) to seal the vesicles and form proteoliposomes (*see* **Note 18**).

3.2.4 Removal of the Counter-Exchange Substrate

A chemical concentration gradient between the inside and outside of the liposome is key for a successful transport assay. Hence, the unlabeled substrate outside of the liposome should be removed by size-exclusion chromatography using Sephadex G-25M columns (PD-10 column, GE Healthcare).

1. Pre-equilibrate the PD-10 columns with five column volumes of PD-10 Buffer.
2. Apply 1 ml of the proteoliposomes onto the columns.
3. Elute with one column volume of PD-10 Buffer.
4. Collect approximately 1–1.5 ml of the turbid eluate containing the proteoliposomes in a 1.5 ml Eppendorf tube.

3.3 Transport Assay

If not stated otherwise all steps are carried out at 25 °C in a temperature-controlled heat block.

1. Add 950 μl proteoliposomes to 50 μl transport medium (*see* **Note 19**) to start the reaction.
2. After the desired time (*see* **Note 20**), add 140–200 μl of the reaction mixture onto a pre-equilibrated ion-exchange column (*see* **Note 21**) to stop the reaction. The substrate will bind to the column so that no further uptake can occur.
3. Elute the proteoliposomes from the ion-exchange column by using first 350 μl and then 800 μl of 150 mM Dowex buffer (*see* **Note 22**).
4. Mix the eluted proteoliposomes with 10 ml scintillation cocktail in scintillation vials by manual shaking.
5. Add 5 μl of the transport medium to 10 ml scintillation cocktail as a standard (*see* **Note 23**).
6. The radioactivity in the eluted proteoliposomes and the standard is measured as CPM with a liquid scintillation counter.

4 Possible Results

Uptake studies into intact proteoliposomes are an established tool to determine the biochemical properties of the desired transport protein [2]. However firstly, it is important to verify that the radioactivity in the proteoliposomes is not increasing due to membrane leakiness or background activity of the isolated membranes, but rather due to an uptake event of the reconstituted transporter. Therefore, it is essential to include proper controls to the measurement to result in unambiguous interpretations of data. To exclude leakiness of the proteoliposomes and passive diffusion of the substrate inside the vesicles, samples with and without preloading of substrate should be measured. If the transporter is expressed in a recombinant expression system, isolated membranes from an empty vector control should be taken as a control to check for background activity of the isolated membrane. If the experiment is performed with isolated membranes from leaf tissues, we advise the user to use isolated membranes of wild-type plants as a control.

The liposomal uptake rate must be calculated to properly characterize the biochemical properties of the transporter. Rates are typically expressed as mol radioactive-labeled substrate per mg protein and is calculated from the measured CPM values. Thereby it is important that a defined concentration of substrate in the transport medium is measured as standard. Since the substrate concentration of the standard is known (*see* **Note 24**), the moles of substrate uptake into the proteoliposomes can be calculated. By dividing it with the protein concentration of each sample (*see* **Note 25**), the uptake rate in mol radioactive-labeled substrate per mg protein is obtained.

The amount of radioactive-labeled substrate inside the proteoliposomes increases with time until it reaches saturation. In our example, we fitted the uptake with a one-phase association kinetics model: $Y = Y0 + (\text{Plateau} - Y0) \times (1 - \exp.(-K \times x))$ (Fig. 2). Given the large variations in transporter expression and yield, we only interpret threefold differences between samples with and without preloading as significant.

The method we describe does not only characterize the activity of a transporter for a certain substrate, but also test an identified carrier's various different substrates and thereby elucidate its substrate range and its function in metabolism. Here, we describe a homo-exchange protocol for the transport protein, meaning internal and counter-exchange substrates are the same. However, with our protocol it is also possible to perform hetero-exchange experiments, where reconstituted and counter-exchange substrates differ. This might be of use for testing different substrates on one transporter. In addition to the experiments mentioned above, uptake experiments allow the determination of the biochemical properties K_m and V_{max} of a transporter. For this purpose the initial velocities have to be determined for time-kinetic measurements with different substrate concentrations, following the calculation of K_m and V_{max} with the Michaelis-Menten equation. Here, we advise the user to focus on obtaining time points in the beginning of the reaction, since it is not necessary to follow the uptake until it reaches saturation. The uptake system can also be used to perform competition experiments with different substrates in the external medium or test potential inhibitors. With regard to the latter it is possible to determine the K_i values of the inhibitors as it was described for the biochemical properties K_m and V_{max} above.

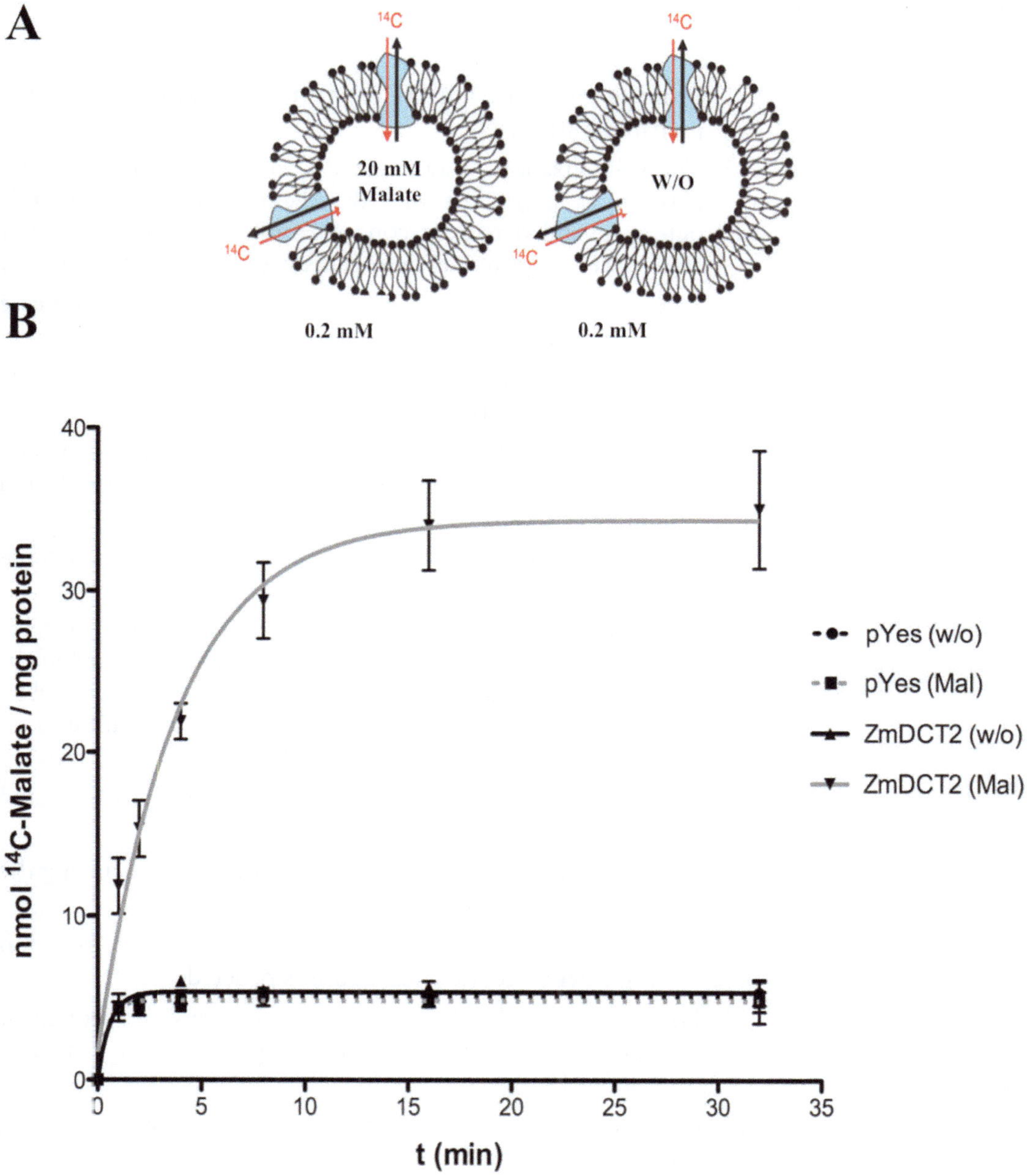

Fig. 2 Time-kinetic measurements of malate homo-exchange with reconstituted dicarboxylate transporter 2 from *Zea mays* (DCT2). Schematic picture of the proteoliposomes preloaded with (malate) and without (w/o) substrate used in the uptake experiment (**a**). Protein isolated from yeast membranes was reconstituted into liposomes with (malate) and without (w/o) preloading of substrate. Isolated membranes from the empty vector (pYes) were used as a control (**b**)

5 Notes

1. Protease inhibitors and DTT need to be added freshly to the prepared buffers.
2. The solubilization of aggregated transport proteins relies on choosing a non-denaturing detergent. Choosing a suitable detergent and detergent concentration is a critical step. For an overview of potential detergents we refer the reader to [2].
3. Besides PMSF, the combination of only two suitable protease inhibitors can be enough.
4. The method has been optimized for the rice plant and slightly modified from the original reference [11].
5. It is possible to vary the concentration of unlabeled substrate. However, the final concentration of the internal solution should be 150 mM.
6. Keep the samples on ice to prevent heat-related damages caused by prolonged sonication.
7. After centrifugation, the supernatant contains soluble proteins and *E. coli* membranes. The supernatant is further used in (b) isolation of bacterial membranes. The pellet contains cell debris and inclusion bodies and is further used in (c) solubilization of recombinant protein from inclusion bodies.
8. The obtained pellet after centrifugation contains the isolated bacterial membranes.
9. Generally, five to ten pulses are sufficient for homogenizing the pellet.
10. After centrifugation, the supernatant contains the inclusion bodies, whereas residual cell debris is pelleted.
11. For stabilizing solubilized mitochondrial transport proteins it is essential to add cardiolipin in a range of low mg/ml concentrations [9].
12. Centrifugation is used to remove unsolubilized transport proteins.
13. Solubilized transport protein can be used for direct reconstitution in liposomes. Alternatively, the detergent used for solubilization can be removed by dialysis or biobeads. It is worth mentioning that the critical micelle concentration should be calculated for the used detergent.
14. Before use, wash glass beads with 100 mM HCl for 1 h, and then wash two to three times with water, until the pH is neutral. After drying at 80 °C, the acid-washed glass beads can be stored at room temperature.

15. Regarding the reconstitution with pre-loading of the substrate, take care to add the substrate after sonication of phospholipid in the internal medium.
16. For mitochondrial membrane proteins, we recommend adding cardiolipin together with phosphatidylcholine in a ratio of 1:20. Cardiolipin is known to be essential for the structural stability and activity of mitochondrial transport proteins [9].
17. Since the detergent, which is used for the protein solubilization, could partially cause the leakage of the proteoliposomes we recommended using a ratio of 1:20 or higher for the protein/phospholipid suspension.
18. The freeze-thaw-sonication procedure we use is a very rapid and efficient method to obtain proteoliposomes with active transport proteins. However, the obtained liposomes are rather small and not equal in size. If liposomes of a consistent size are needed, it might be necessary to extrude them through membranes with distinct pore sizes. In addition, there are several other possible methods to reconstitute a transporter actively into liposomes. Therefore, we refer the reader to detailed explaining reviews [12, 13].
19. The transport medium is prepared in PD-10 buffer. Add a respective amount of unlabeled substrate to create a 100-fold concentration gradient between internal and external medium. Add radioactive-labeled substrate to have a sufficient amount of radioactivity for the measurement. Therefore, it may be necessary to check CPM values of the labeled substrate with a scintillation counter before the assay. Check the pH of the transport medium with pH paper, before adding the radioactive-labeled substrate. It should have the same pH as the PD-10 buffer.
20. The time point when a sample is taken depends on the rate of the transporter. Usually, samples are taken until 32 or 64 min to see that the uptake reaches saturation. However, when calculating the initial velocities of the uptake, we advise measuring more time points in the beginning of the reaction.
21. The ion-exchange column is prepared by dissolving the resin (Dowex AG1-X8 Resin Bio-Rad, 100–200 mesh) in 150 mM Dowex buffer. Use cation-exchange resin (hydrogen form) and lithium acetate buffer pH 7.6 for positively charged substrates and anion-exchange resin (acetate form) and sodium acetate buffer pH 7.6 for negatively charged substrates. The dissolved resin is filled into disposable glass Pasteur pipets closed with a small amount of cotton. The columns are washed five times with Dowex buffer. After the last washing step the pH of the flow through is checked. It should be the same as the Dowex buffer.

22. Use 150 mM sodium acetate buffer (acetic acid, pH 7.6) in case of a negatively charged substrate. Use 150 mM lithium acetate buffer (acetic acid, pH 7.6) in case of a positively charged substrate.
23. It is important to measure a standard for the calculation of the uptake rates from the measured CPM.
24. The addition of radioactive-labeled substrate will not change the substrate concentration of the transport medium, generated by unlabeled substrate, as the unlabeled substrate is given in high excess.
25. The protein concentration of the isolated membranes can be calculated with a common protein determination assay, like Bradford or BCA. In addition, we add a dilution factor of 2000 when the proteoliposomes are desalted with a PD-10 column.

References

1. Lunn JE (2007) Compartmentation in plant metabolism. J Exp Bot 58:35–47
2. Haferkamp I, Linka N (2012) Functional expression and characterisation of membrane transport proteins. Plant Biol (Stuttg) 14:675–690
3. Linka N, Weber APM (2010) Intracellular metabolite transporters in plants. Mol Plant 3:21–53
4. Sweetlove LJ, Fernie AR (2013) The spatial organization of metabolism within the plant cell. Annu Rev Plant Biol 64:723–746
5. Schwacke R, Flügge U-I, Kunze R (2004) Plant membrane proteome databases. Plant Physiol Biochem 42:1023–1034
6. Schwacke R, Schneider A, van der Graaff E et al (2003) ARAMEMNON, a novel database for *Arabidopsis* integral membrane proteins. Plant Physiol 131:16–26
7. Pick TR, Bräutigam A, Schulz MA et al (2013) PLGG1, a plastidic glycolate glycerate transporter, is required for photorespiration and defines a unique class of metabolite transporters. Proc Natl Acad Sci U S A 110:3185–3190
8. Eisenhut M, Pick TR, Bordych C et al (2013) Towards closing the remaining gaps in photorespiration-the essential but unexplored role of transport proteins. Plant Biol (Stuttg) 15:676–685
9. Klingenberg M (2009) Cardiolipin and mitochondrial carriers. Biochim Biophys Acta 1788:2048–2058
10. Palmieri F, Indiveri C, Bisaccia F et al (1995) Mitochondrial metabolite carrier proteins: purification, reconstitution, and transport studies. Methods Enzymol 260:349–369
11. Furbank RT, Scofield GN, Hirose T et al (2001) Cellular localisation and function of a sucrose transporter OsSUT1 developing rice grains. Aust J Plant Physiol 28:1187–1196
12. Olson F, Hunt CA, Szoka FC et al (1979) Preparation of liposomes of defined size distribution by extrusion through polycarbonate membranes. Biochim Biophys Acta 557:9–23
13. Rigaud J-L, Levy D (2003) Reconstitution of membrane proteins into liposomes. Methods Enzymol 372:65–86

Chapter 7

Quantification of Photorespiratory Intermediates by Mass Spectrometry-Based Approaches

Stéphanie Arrivault and Toshihiro Obata

Abstract

Photorespiration is an essential metabolic process in plants occurring via the oxygenase reaction of RuBisCO. In order to understand this process, it is essential to determine the amounts of intermediates involved. For this purpose we combined mass spectrometry-based approaches and the use of authentic standards for the quantification of photorespiratory intermediates. Here we describe protocols for the extraction and quantification of 2-phosphoglycolate (2PG) by LC-MS/MS and serine, glycine, glycolate, hydroxypyruvate, glyoxylate, and glycerate by GC-MS.

Key words Photorespiratory intermediates, Authentic standards, Metabolite quantification, LC-MS/MS, GC-MS

1 Introduction

Mass spectrometry-based approaches, in particular coupled with separation procedures such as liquid chromatography (LC) and gas chromatography (GC), enable accurate detection and quantification of metabolites with a wide dynamic range. For example, Refs. [1, 2] developed a targeted LC-MS/MS approach focusing on the quantification of phosphorylated intermediates from the Calvin-Benson cycle in plant material. The metabolite coverage of this method was extended to include the quantification of 2-phosphoglycolate (2-PG), the first intermediate of photorespiration. Additional photorespiratory intermediates, such as glycolate, glyoxylate, glycine, serine, hydroxypyruvate, and glycerate, can be quantified by GC-MS. Absolute amounts of metabolites are determined by comparing peak areas with those from known amount of authentic standards. The extraction protocol for GC-MS is modified from the well-established protocol described in [3], including changes in organic solvent and extraction temperature to prevent degradation of heat-sensitive metabolites (i.e., hydroxypyruvate; [4, 5]). As a result, the combination of LC-MS/MS and GC-MS

Alisdair R. Fernie et al. (eds.), *Photorespiration: Methods and Protocols*, Methods in Molecular Biology, vol. 1653, DOI 10.1007/978-1-4939-7225-8_7, © Springer Science+Business Media LLC 2017

covers major photorespiratory intermediates and therefore is a powerful approach to analyze the status of the photorespiratory pathway. Additionally these techniques are compatible with stable-isotope redistribution analyses enabling the analysis of photorespiratory flux ([6] or the Chapter 11 in this issue).

2 Materials

2.1 Reagents and Mixes

1. Water, deionized and filtered (0.2 μm filter) in Purelab-plus system, or distilled water ULC/MS grade.
2. Xylene (*see* **Note 1**).
3. Sample extraction buffer for LC-MS/MS analysis; mix chloroform and methanol (gradient grade for liquid chromatography; v:v, 3:7) and keep ice-cold (*see* **Note 1**).
4. DMF extraction buffer for GC-MS analysis; mix 25 μl of 0.2 mg/ml ribitol with 600 μl of *N,N*-dimethylformamide (*see* **Note 1**).
5. Methoxyamination reagent; dissolve methoxyamine hydrochloride at 40 mg/ml in pure pyridine (analytic grade) at 20–25 °C. This reagent needs to be prepared freshly before the experiment (*see* **Note 1**).
6. Calibration mixes for 2PG; dissolve and dilute authentic 2PG standard in water to obtain 1000, 750, 500, 375, 250, 100, and 50 nM. Solutions can be stored at −80 °C.
7. Calibration mix for GC-MS; Dissolve authentic serine, glycine, glycolate, hydroxypyruvate, glyoxylate, and glycerate standards in methanol to be 1 mg/ml. Mix the solutions and adjust the volume with methanol to obtain 20 ng/μl for each compound. This solution can be stored at −20 °C.
8. FAMEs mix; dissolve fatty acid methyl esters (FAMEs) in chloroform and mix to gain 0.8 mg/ml of methyl octanoate, methyl nonanoate, methyl decanoate, methyl laurate, methyl myristate, and methyl palmitate, and 0.4 mg/ml of methyl stearate, methyl eicosanoate, methyl docosanoate, methyl tetracosanoate, methyl hexacosanoate, methyl octacosanoate, and methyl melissate (*see* **Note 1**). This solution can be stored at 4 °C.
9. MSTFA-FAMEs mix; add 20 μl of FAMEs mix into 1 ml of *N*-methyl-*N*-(trimethylsilyl)trifluoroacetamide (MSTFA) (*see* **Note 1**).
10. HPLC elution buffers; elution buffer [A] is a 10 mM tributylamine aqueous solution, adjusted to pH 4.9 with 15 mM acetic acid (ULC/MS grade for liquid chromatography) whereas eluent [B] is methanol (*see* **Notes 1–3**).

11. Inactive gas such as argon or nitrogen.
12. Liquid nitrogen (*see* **Note 4**).

2.2 Equipment

1. Micro-vials: 1.5 ml, screw-cap, tapered bottom, 1.5 ml, safe-lock, tapered bottom, and 2.0 ml, screw-cap, round bottom (*see* **Note 5**).
2. Teflon adaptor for 1.5–2.0 ml micro-vials.
3. Oscillating ball mill.
4. 5 mm Steel balls.
5. Weighing scale.
6. Fine quality pipet tips (*see* **Note 5**).
7. Vortex.
8. Stirring rod for mixing.
9. Precooled centrifuge and bench centrifuge.
10. MultiScreen Filter Plate with Ultracel-10 membrane.
11. 96-well flat bottom microtest plates.
12. Lyophilizer.
13. Eppendorf Thermomixer Compact.
14. Rotational Vacuum Concentrator.
15. Dionex UltiMate 3000 RSLC system with Chromoleon software and ThermoFinnigan TSQ Quantum Ultra with Excalibur and TSQ Quantum software (or equivalent).
16. Analytical column Gemini C18, 2.0 mm diameter, 150 mm length, 110 Å pore size, and spherical particles of 5 μm (Phenomenex).
17. Pre-columns SecurityGuard Cartridges C18, 2.0 mm diameter, 4 mm length (Phenomenex).
18. Pre-columns Guard holder SecurityGuard Cartridges (Phenomenex).
19. Pegasus III Tof mass analyzer from LECO and corresponding software (LECO) (or equivalent).
20. Gas chromatograph (Agilent 6890 N), split/splitless injector with electronic pressure control up to 150 psi (Agilent).
21. Autosampler system (PAL Agilent). The configuration comprises an agitator-incubator oven, a 98-sample tray for 2.0-ml vials, a 32-sample tray for 10–20-ml vials, three 100-ml solvent reservoirs (i.e., a syringe wash station and a liquid version 25-ml syringe kit mounted on the robotic autosampler arm).
22. MDN-35 capillary column, 30-m length, 0.32-mm inner diameter, 0.25-mm film thickness (e.g., Macherey-Nagel or equivalent).

23. Glass vials for LC-MS/MS and GC-MS.
24. Screw cap for glass vials.
25. Microsoft operating system and Microsoft Office (Microsoft Corporation, Seattle).
26. ThermoFinnigan processing software package LCQuan-2.5.
27. Chroma TOF 1.0 software (Leco) and TagFinder software 7.

3 Methods

3.1 Plant Harvest and Aliquoting Procedures

Once harvested, make sure to keep samples at constant freezing temperature (in liquid nitrogen) to avoid degradation of metabolites. Precool all components (spatulas, vials, etc.) in liquid nitrogen before they come into contact with the sample.

1. Sample leaf materials and snap-froze in liquid nitrogen (*see* **Notes 6–8**).
2. The frozen material is transferred into micro-vial and ground into a fine powder by incorporating steel balls into the vial and using an oscillating ball mill at 25 Hz for 2 min (*see* **Note 9**).
3. Take a 15 mg aliquot of ground material into a 1.5 ml screw-cap, tapered bottom for LC-MS/MS analysis and a 50 mg aliquot of ground material into a 2 ml screw-cap, round bottom tube for GC-MS analysis (*see* **Note 9**).

3.2 Quantification of 2PG by LC-MS/MS Analysis

1. Place the aliquots in a standing position into liquid nitrogen (half the height of the tubes should be immersed, *see* **Note 10**).
2. Add 125 μl of ice-cold extraction buffer to the samples (*see* **Note 11**).
3. Thoroughly mix the powder and the extraction buffer with a stirring rod while the tubes are in liquid nitrogen. Take the tubes out of liquid nitrogen, continue to mix and place the tubes on ice.
4. Add 125 μl of ice-cold extraction buffer to clean the stirring rod.
5. Vortex for 10 s and keep on ice.
6. Store at −20 °C for 2 h.
7. Add 400 μl ice-cold water and vortex for 10 s.
8. Centrifugation at 4 °C at 22,000 × *g* for 5 min and collect the aqueous phase (upper, approximatively 570 μl) in a new 2 ml screw-cap, round bottom tube (*see* **Note 12**).
9. Repeat twice and pool with the previous aqueous phase.
10. Divide the total aqueous phase into two 2 ml screw-cap, round bottom tubes (keep the previous one and take a new one).

11. Add 400 µl ice-cold water to each tube (*see* **Note 13**).
12. Close each tube with caps punctuated with two holes, freeze the samples by immersing them into liquid nitrogen (up to half the height of the tubes) and lyophilize overnight (*see* **Note 14**).
13. Add 250 µl cold water to the samples (125 in each tube), replace punctuated caps with non-punctuated ones and shake each tube for few seconds and keep them on ice for 5 min.
14. Centrifuge at 4 °C, 22,000 × *g* for 10 min.
15. Place a 96-well plate under a MultiScreen filter plate, label the wells of the filter plates which will be used for filtration of the samples.
16. Transfer the extract from the tube into the corresponding labelled well (125 µl + 125 µl in one well for each sample).
17. Centrifuge for 2 h at 2300 × *g* (10–12 °C).
18. Transfer the filtrate from the 96-well plate into a new labelled 1.5 ml screw-cap, tapered bottom (*see* **Note 15**).

    Transfer 120 µl of standard mix or extract to a glass vial for HPLC and place into the autosampler tray precooled at 4 °C.
19. Set up the LC-MS/MS as described in [1]. For selected reaction monitoring (SRM) in negative mode the parent mass of 2PG is 155 Da and the main fragmentation product ion is 79 (with a collision energy of 10).
20. Start LC-MS/MS analysis with injection of 100 µl of a mixture of authentic standards (seven different dilutions, each measured three times), then 100 µl of extracts (randomly measured). Each tenth measurement, 100 µl of authentic standard mixture (500 nM) is injected, as well as after all extracts have been processed.
21. The obtained data (binary .RAW files) for the standards and extracts are processed with the ThermoFinnigan software package LCQuan-2.5 (*see* **Notes 16** and **17**).

3.3 Quantification of Photorespiratory Intermediates by GC-MS Analysis

Metabolites are extracted from plant leaf material by *N,N*-dimethylformamide and purified by phase separation with xylene. The samples are not heated to prevent degradation of heat-sensitive metabolites, i.e., hydroxypyruvate.

1. Add 625 µl of ice-cold DMF extraction solution and mix immediately by vortex (*see* **Note 11**).
2. Add 400 µl of water and vortex.
3. Incubate at 4 °C for 10 min with shaking at 1400 rpm.
4. Centrifuge for 8 min at 22,000 × *g* at 4 °C and transfer the supernatant into a new 2 ml screw-cap, round bottom tube.
5. Add 600 µl of xylene and mix well with vortex.
6. Incubate at 4 °C for 10 min with shaking at 1400 rpm.

7. Centrifuge for 3 min at 22,000 × *g* at 4 °C.
8. Collect two 400 μl aliquots of upper polar phase into 1.5 ml safe-lock, tapered bottom tubes. Take calibration mix for GC-MS analysis to gain appropriate amount of compounds (we usually use 0, 250, 500, and 1000 ng) into 1.5 ml tubes and mix with ribitol solution to obtain the same amount as in the samples (in this protocol 2 μg should be added). Dry them down by a vacuum concentrator overnight.
9. Immediately fill the tubes with inactive gas to prevent oxidation of metabolites. Extracts can be stored at −80 °C (*see* **Note 18**).
10. Shortly before derivatization, dry the samples for 30 min in a vacuum concentrator to prevent water condensation in the tube.
11. Add 20 μl of methoxyamine hydrochloride solution and incubate at 30 °C for 90 min with shaking at 1400 rpm.
12. Spin down the condensed solution by short centrifugation.
13. Add 180 μl of MSTFA-FAMEs mix and incubate for 30 min at 37 °C.
14. Spin down the condensed solution by short centrifugation.
15. Transfer all of the solution into a glass vial and analyze by GC-MS.
16. Set up the GC-MS as described in [3].
17. Start the GC-MS analysis with injection of 1 μl of sample.
18. The obtained data (CDF files) for the standards and extracts are processed with Chroma TOF 1.0 and TagFinder 4.0 as described in [7]. Construct a calibration curve by plotting peak intensity against amount of each authentic standard and use it for calculating metabolite contents in samples. The fragments for quantification depend on the samples but usually those are generally used: glycolate (2TMS) *m/z* 177, glyoxylate (1MEOX)(1TMS) *m/z* 160, glycine (3TMS) *m/z* 174, serine (3TMS) *m/z* 204, hydroxypyruvate (1MEOX)(2TMS) *m/z* 114, and glycerate (3TMS) *m/z* 189.

4 Notes

1. Reagents are toxic and should be handled under a fume hood.
2. Reagents are corrosive and should be handled under a fume hood.
3. If the HPLC is not equipped with degassing system in front of solvent mixer pumps of HPLC, degassing by application of a vacuum should be performed.
4. Liquid nitrogen is a low temperature refrigerant and should be handled with protective glasses and gloves. In addition, liquid nitrogen is not used in closed rooms for danger of suffocation.

5. Equipment should not be autoclaved. Minerals in water and exposure to bacterial residue within the autoclave can produce artifact peaks and unexpected ion adducts.
6. Once the plant material has been frozen in liquid nitrogen it must not thaw out even slightly before being stored at −80 °C.
7. Additional metabolites can be simultaneously quantified by LC-MS/MS, especially intermediates from the Calvin-Benson cycle, sugar and starch synthesis. However some of these metabolites have less than a second turnover time [1], so harvest has to be performed by quenching the plant material under the light field, avoiding shading [2].
8. Frozen material can be stored at −80 °C for up to 6 months.
9. Frozen homogenate can be stored at −80 °C for up to 3 months.
10. No liquid nitrogen should come into contact with the powder.
11. Extraction solutions should be precooled in −20 °C freezer.
12. To avoid collection from the chloroform phase (lower phase) leave some of the upper phase in the tube.
13. To avoid the loss of samples through the cap holes during the lyophilization process, water is added to dilute the methanol present in the aqueous phase.
14. Tubes should not be placed too close to one other in order to increase the speed of lyophilization.
15. Either use the extracts directly or store at −80 °C for up to 6 months.
16. Check manually each individual peak as the LCQuan-2.5 software sometimes does not recognize small peaks. This is time consuming but essential.
17. The potential matrix effects or loss of sensitivity during the course of the measurement are estimated and if necessary corrected. Authentic 2PG standard is analyzed several time over the entire course of the measurement and the amounts quantified (injected in between extracts and at the end of the entire measurement) are used to calculate how much percentage this represent compared to the initial standard in the calibration curve. Thus over time (more precisely over the number of runs), we can assess how the signal for 2PG was affected. If the differences along the entire measurement were ±20%, data were not corrected. If otherwise, the slope obtained based on % metabolite versus number of run was used to correct the amount accordingly.
18. **Steps 9** and **10** can be omitted in case the samples are derivatized directly after drying.

Acknowledgments

Work from S.A. and T.O. was supported by the European Commission FP7 collaborative project 3 to 4 (Contract No. 289582).

References

1. Arrivault S, Guenther M, Ivakov A et al (2009) Use of reverse-phase liquid chromatography, linked to tandem mass spectrometry, to profile the Calvin cycle and other metabolic intermediates in Arabidopsis rosettes at different carbon dioxide concentrations. Plant J 59(5):824–839
2. Arrivault S, Guenther M, Fry SC et al (2015) Synthesis and use of stable-isotope-labeled internal standards for quantification of phosphorylated metabolites by LC-MS/MS. Anal Chem 87(13):6896–6904
3. Lisec J, Schauer N, Kopka J et al (2006) Gas chromatography mass spectrometry-based metabolite profiling in plants. Nat Protoc 1 (1):387–396
4. Boldt R, Edner C, Kolukisaoglu U et al (2005) D-GLYCERATE 3-KINASE, the last unknown enzyme in the photorespiratory cycle in Arabidopsis, belongs to a novel kinase family. Plant Cell 17(8):2413–2420
5. Pick TR, Brautigam A, Schulz MA et al (2013) PLGG1, a plastidic glycolate glycerate transporter, is required for photorespiration and defines a unique class of metabolite transporters. Proc Natl Acad Sci U S A 110(8):3185–3190
6. Heise R, Arrivault S, Szecowka M et al (2014) Flux profiling of photosynthetic carbon metabolism in intact plants. Nat Protoc 9 (8):1803–1824
7. Luedemann A, von Malotky L, Erban A et al (2012) TagFinder: preprocessing software for the fingerprinting and the profiling of gas chromatography-mass spectrometry based metabolome analyses. Methods Mol Biol 860:255–286

Chapter 8

Targeted Isolation and Characterization of T-DNA Mutants Defective in Photorespiration

Stefan Timm, Katharina Modde, and Hermann Bauwe

Abstract

Transfer DNA (T-DNA) insertional lines have facilitated reverse genetic approaches in plant science and considerably accelerated the functional characterization of genes. Typically, online databases of mutant collections are searched for suitable mutant alleles of the gene of interest (GOI). Before such lines can be characterized physiologically, the genotype of the respective mutant has to be verified followed by the quantitative examination of downstream effects on the levels of the respective mRNA and the encoded protein. Here, we describe a typical workflow for the identification of photorespiratory mutants followed by phenotypic characterization according to growth under different conditions, photosynthesis on the levels of chlorophyll a fluorescence and gas exchange, and metabolite analysis.

Key words T-DNA mutant, Genotyping, Expression analysis, Phenotyping, Photosynthesis, Photorespiration, CO_2 transition, Chlorophyll a fluorescence, Gas exchange, Metabolite analysis

1 Introduction

Transfer DNA (T-DNA) insertional lines represent a valuable tool in plant science to perform reverse genetic approaches in order to unravel the role of genes and proteins for plant development, metabolism, and other processes. Initially, the term T-DNA was coined for the T-DNA fragment of *Agrobacterium tumefaciens* but is now used for any DNA segment flanked by specific 25 base pair direct repeats that are transferred by *Agrobacterium tumefaciens* into plant genomes [1]. Accordingly, a broad range of mutant collections is presently available for the model plant *Arabidopsis thaliana* (Arabidopsis) and an updated user's guide was provided recently [2]. To date more than 325,000 T-DNA insertion lines have been isolated and the T-DNA insertion site sequenced. Most of these mutants are publicly available from the Arabidopsis Biological Resource Center (ABRC) or the Nottingham Arabidopsis Stock Center (NASC) and can be found via various web portals [3]. Most of those lines (~70%) are "classic" T-DNA insertional lines

Alisdair R. Fernie et al. (eds.), *Photorespiration: Methods and Protocols*, Methods in Molecular Biology, vol. 1653, DOI 10.1007/978-1-4939-7225-8_8, © Springer Science+Business Media LLC 2017

(e.g., SALK, SAIL, WiscDSlox, and GABI collection), while others were generated by using several transposon insertions (e.g., CSHL, IMA, RIKEN, SLAT, JIC-SM collections). Because of the huge number of available mutants, there is a very good chance to find at least one but often several mutants for each of the Arabidopsis genes. If mutants for the gene of interest (GOI) cannot be spotted, they need to be created by using suppressor techniques (e.g., RNAi or antisense suppression) or the CRISPR/Cas method [4]. However, given that not all insertions fully inactivate the respective gene [5], it is essential to use at least two independent alleles and to carry out a comprehensive validation of each mutant prior in-depth physiological characterization.

To facilitate the assignment of a given mutant to a particular metabolic pathway or other features, a characteristic and easily discernible phenotype is desirable. This condition is met by genes involved in photorespiration though not by very many other genes. Generally, the respective mutants do not survive in normal air (active photorespiration) but can be grown in air enriched with high CO_2 concentrations (suppressed photorespiration). If such mutants are initially grown in high CO_2 and transferred to normal air, plants rapidly exhibit chlorotic and gradually bleaching leaves. This low- versus high-CO_2 effect has been named the "photorespiratory phenotype" [6–8]. It allows a rather simple phenotypic screening for photorespiratory mutants and was initially used to identify mutants defective in photorespiration after chemical mutagenesis [6]. Moreover, most of the so far known photorespiratory mutants share further specific responses to air CO_2 levels, such as a fast decline in photosynthetic CO_2 assimilation and specific metabolite accumulation patterns [9, 10]. In the following, we will describe how to isolate, verify, and characterize a T-DNA mutant potentially defective in photorespiration on the levels of gene expression, low-CO_2 effects, and photosynthetic-photorespiratory metabolism.

2 Materials

All solutions are prepared using ultrapure water and reagents of analytical grade. Unless stated otherwise, solutions are stored at room temperature. The mentioned equipment is only a recommendation and could be replaced by comparable devices from other manufacturers. The experimental workflow how to isolate and characterize a T-DNA mutant potentially deficient in photorespiration is schematically shown in Fig. 1, while the respective materials and methods needed to carry out each step are given within the next two sections.

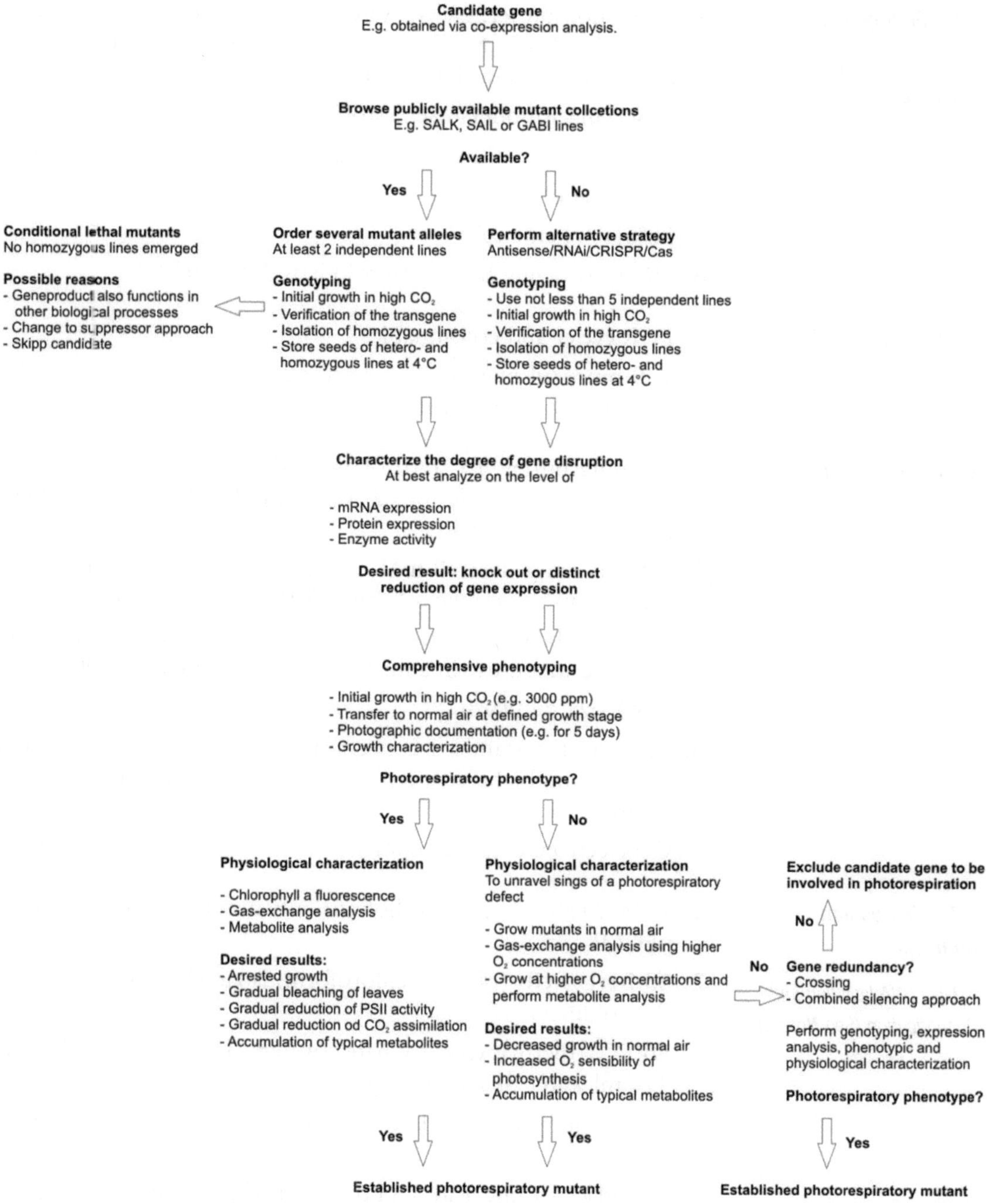

Fig. 1 Schematic overview on the isolation of a T-DNA mutant potentially deficient in photorespiration. The flow chart summarizes the steps for the basic verification and dissection of a T-DNA mutant. The working flow thereby involves genotypic verification, expression analysis on the levels of mRNA, protein abundance and enzyme activity, as well as phenotypic and physiological characterization. Following the scheme will allow distinguishing mutants defection in photorespiration or other pathways

2.1 Genotyping

2.1.1 DNA Isolation (See Note 1)

1. Extraction buffer: 2% (w/v) CTAB (cetyltrimethylammonium bromide), 1.4 M NaCl, 20 mM EDTA (ethylenediaminetetraacetic acid), 100 mM Tris–HCl, pH 8.0.
2. 90 U/mg Ribonuclease A stock solution.
3. 24:1 CI (chloroform/isoamylalcohol) mixture (by volumes).
4. Isopropanol.
5. 70% (v/v) ethanol.
6. DNA dissolving buffer: 10 mM Tris–HCl, pH 8.0.

2.1.2 PCR

1. PCR cycler and consumables.
2. 100 pmol/μl oligonucleotide stock solutions, specific for the GOI and the T-DNA-insert (*see* **Note 2**).

2.1.3 Gel Electrophoresis

1. Running buffer: 1× TAE (Tris–acetic acid–EDTA), prepare a 50× TAE stock solution, 2 M Tris, pH 8.0, 50 mM EDTA, 5.5% (v/v) acetic acid.
2. DNA loading dye: 10 mM Tris–HCl, pH 7.5, 1 mM EDTA; 0.2% (v/v) bromophenole blue, 30% (v/v) glycerol.
3. Ethidium bromide stock solution: 2.45 μM ethidium bromide in sterile water, use in a final concentration of 1:40,000.
4. DNA size standard.
5. Agarose: the standard concentration for agarose gels used is 1%, but might be adapted to the expected fragment sizes to separate.
6. Gel electrophoresis equipment.

2.2 Expression Analysis

2.2.1 RNA Isolation and Quantification (See Note 3)

1. 1 M MES (2-(*N*-morpholino) ethanesulfonic acid) buffer, pH 7.0.
2. 25:24:1 PCI (phenol/chloroform/isoamylalcohol) mixture (by volumes).
3. DEPC (diethylpyrocarbonate) water: add 1 ml DEPC to 990 ml water, shake overnight, and autoclave.
4. 3 M sodium acetate, pH 5.2: dissolve 40.8 g in 100 ml DEPC water.
5. 1 M glacial acetic acid: fill up 6.0 ml of acetic acid with DEPC water to 100 ml.
6. Z6-buffer: 8 M guanidine-HCl, 20 mM MES, 20 mM EDTA, in 100 ml sterile water, directly prior use add 350 μl of β-mercaptoethanol per 50 ml Z6-buffer.
7. 0.5 M EDTA in sterile water, pH 8.0.
8. 10× MEN-buffer: dissolve 10.3 g MOPS (morpholinopropansulfonic acid) in 200 ml 0.1 M sodium acetate, add 5 ml 0.5 M

EDTA stock solution, adjust pH to 7.0 with sodium-hydroxide and fill to 250 ml final volume with water, autoclave.

9. 70% ethanol in DEPC water.
10. 100% ethanol.
11. 5 M lithium chloride in DEPC water.
12. 1 U/μl DNaseI.
13. DNaseI reaction buffer: 100 mM Tris–HCl, pH 7.5, 25 mM $MgCl_2$, 1 mM $CaCl_2$.
14. 50 mM EDTA.

2.2.2 cDNA Synthesis

1. 2.5 μg RNA.
2. 5 μg/μl Oligo-dT or random primer, in some applications gene-specific primers are used.
3. Ribonuclease A free water.
4. 40 U/μl Ribonuclease A inhibitor.
5. 5× reverse transcriptase reaction buffer: 250 mM Tris–HCl, pH 8.3, 250 mM KCl, 20 mM $MgCl_2$, 50 mM DTT.
6. 10 mM dNTP mixture.
7. 200 U/μl Reverse transcriptase.

2.2.3 RT-PCR (See Note 4)

1. PCR cycler and consumables.
2. Gene-specific oligonucleotides that amplify a selected house-keeping gene (e.g., *S16* or Actin) and the gene of interest. Assure that the two amplicons analyzed are of similar fragment size (e.g., 350–400 bp).

2.2.4 Protein Extraction and Quantification

1. Protein extraction buffer: 75 mM Tris–HCl, 1.5 mM EDTA, pH 7.5, store at 4 °C.
2. Protease inhibitor stock solution: 60 mM $NaSO_3$ + 6 mM pefabloc, store at 4 °C.
3. Protease inhibitor stock solution: 30 mM PMSF (phenylmethylsulfonyl fluoride) in isopropanol (*see* **Note 5**), store at 4 °C.
4. Protein quantification: Roti-Nanoquant (Roth) and standard quantification assay [11].
5. Photometer.

2.2.5 Sodium Dodecyl Sulfate-Polyacrylamide Gel Electrophoresis (SDS-PAGE)

1. 0.5 M Tris–HCl (pH 6.8).
2. 1.5 M Tris–HCl (pH 8.8).
3. Acrylamide stock solution.
4. 10% (v/v) SDS, in sterile water.
5. 10% (v/v) ammonium persulfate, in sterile water.

6. TEMED (*N,N,N,N′*-tetramethyl-ethylenediamine).
7. SDS electrophoresis buffer: 1× electrophoresis buffer, prepare a 10× SDS electrophoresis buffer stock solution, 280 mM Tris, 1.92 M glycine, 1% (v/v) SDS.
8. 3× Lämmli buffer: 0.5 M Tris–HCl, pH 6.8, 30% (v/v) glycerol, 15% (v/v) β-mercaptoethanol, 0.15% (v/v) bromophenole blue, 6% (v/v) SDS, in sterile water.
9. SDS-PAGE equipment.

2.2.6 Immunoblotting

1. Methanol (≥99.9%).
2. Transfer-buffer: 1× transfer-buffer, 20% (v/v) methanol, 10% (v/v) 10× transfer-buffer, prepare a 10× transfer-buffer stock solution, 480 mM Tris, 390 mM glycine, 0.37% (v/v) SDS.
3. Skimmed milk powder.
4. 1× TBS (tris-buffered saline) buffer, prepare a 10× TBS-buffer stock solution, 200 mM Tris, pH 7.6, 1.4 M NaCl.
5. Using chemiluminescence for development: solution 1, 2.5 mM luminol in DMSO (dimethylsulfoxide), 0.4 mM p-coumaric acid in DMSO, 100 mM Tris–HCl, pH 8.5.
6. Using chemiluminescence for development: solution 2, 5.4 mM H_2O_2, 100 mM Tris–HCl, pH 8.5.
7. Developing solution (e.g., AGFA-GEVAERT G153).
8. Fixation solution (e.g., AGFA-GEVAERT G354).
9. Stripping-buffer: 47 mM Tris–HCl, pH 6.7, 2% (v/v) SDS, 1% (v/v) β-mercaptoethanol.
10. Using alkaline phosphatase for development: detection-buffer, 0.1 M Tris–HCl, pH 9.5, 0.1 M NaCl, 0.05 M $MgCl_2$.
11. Using alkaline phosphatase for development: NBT and BCIP (nitro blue tetrazolium/5-bromo-4-chloro-3-indolyl phosphate) stocks, 110 mM NBT and 160 mM BCIP, in 67% DMSO.
12. PVDF (polyvinylidene difluoride) membrane.
13. Whatman paper.
14. Radiographic films.
15. Semi-dry electroblotting device.
16. Hybridization device.

2.2.7 Enzyme Activity Measurements (Selected Example, See ***Note 6****)*

1. Protein extraction buffer: 25 mM HEPES-KOH, pH 7.6, 1 mM $MgCl_2$, 1 mM KCl, 10% glycerol (v/v), 10 mM β-mercaptoethanol (add fresh prior use), 0.1 mM PMSF (add fresh prior use), store at 4 °C.

2. Reaction buffer: 0.1 M MES, pH 6.5, 1 mg DTT/ml, add fresh prior use.
3. 3.75 mM NADH, in 0.1 M MES.
4. 3.75 mM NADPH, in 0.1 M MES.
5. 20 mM hydroxypyruvate, in 0.1 M MES.
6. 20 mM glyoxylate, in 0.1 M MES, check pH, should be 6.5.
7. Protein quantification: Roti-Nanoquant (Roth) and standard quantification assay [11].
8. Photometer.

2.3 Basic Physiological Characterization

2.3.1 Phenotyping

1. Two identical plant growth chambers (e.g., Percival AR-66C3, CLF Plant climatic, Germany, *see* **Note** 7), with built-in CO_2 concentration controlling and regulation systems (*see* **Note 8**), CO_2 gas containers, placed within a safety device.
2. Plant growth equipment and consumables.
3. Photo equipment.

2.3.2 Chlorophyll a Fluorescence

1. Imaging pulse amplitude modulated (I-PAM) fluorometer (e.g., M series, Walz, Germany).
2. Compatible laptop including software (e.g., ImagingWin V2.32).

2.3.3 Gas Exchange Analysis

1. Infrared gas analyzer (e.g., LI-6400, LI-COR, Lincoln, Nebraska, USA).
2. Compatible laptop including software: LI6400FileEx to extract data from LI-6400.
3. Gas mixing device (e.g., GMS-600, QCAL Messtechnik, Germany).
4. CO_2, O_2, and N_2 gas containers, placed within a safety device.

2.3.4 Metabolite Analysis

1. Gas chromatography, liquid chromatography, or high performance liquid chromatography coupled to mass spectrometry (GC-MS, LC-MS, or HPLC-MS) device, depending on the metabolites to be analyzed (*see* **Note 9**).

3 Methods

Unless otherwise stated, carry out all procedures at room temperature. Perform physiological characterization at a defined growth stage of Arabidopsis [12] and in the second part of the light phase if not explicitly specified.

3.1 Genotyping

This section describes the steps how to verify the T-DNA insertion in the gene of interest (GOI) on the level of genomic DNA by PCR analysis [13, 14], using gene- and insert-specific primers. Sequencing of the obtained fragments, after separation by gel electrophoresis, is highly recommended.

3.1.1 DNA Isolation

1. Harvest and freeze approximately 100 mg of leave tissue in liquid nitrogen and store at −80 °C until processing. For pulverization add a precooled (liquid nitrogen) steel ball to the 2 ml tube and grind down the leaf material using a Retsch mill.
2. Add 500 μl of 2× CTAB buffer and vortex rigorously. Incubate for 30 min at 68 °C while mixing every 5 min by vortexing.
3. Apply 3 μl ribonuclease A solution and incubate for 15 min at 37 °C. Add 500 μl CI and mix samples for 1 min thoroughly. Centrifuge for 5 min at 15,000 × *g* and transfer the aqueous phase into a new tube.
4. Add 400 μl isopropanol and mix to precipitate DNA for 15 min at −20 °C.
5. Centrifuge for 15 min at 15,000 × *g*. Discard the supernatant and wash the pellet by adding 200 μl of 70% ethanol.
6. Centrifuge for 10 min at 15,000 × *g*, discard the supernatant, and dry the pellet.
7. Dissolve the DNA in 50 μl 10 mM Tris–HCl buffer and incubate for 2 min at 68 °C.
8. Use DNA isolates for PCR analysis directly or store at −20 °C.

3.1.2 PCR to Verify the T-DNA in the GOI

1. Amplify the genomic region of the GOI, using primers spanning the predicted T-DNA impact.

   Mix 1 μl of each of the isolated DNA samples of the potential mutant with 1 μl of the forward and reverse primer derived from the genomic sequence, 7 μl of PCR grade water, and 10 μl of master mix. Simultaneously, perform positive control reactions by using wild-type DNA and negative control reactions by omitting DNA addition (*see* **Note 10**). Run a PCR program that allows the amplification of the expected fragment size according to in silico prediction.
2. Verify the T-DNA insert in the genomic region of the GOI.

   Mix 1 μl of each of the isolated DNA samples of the potential mutant with 1 μl of the forward or reverse gene-primer and 1 μl of the respective left or right border primer of the T-DNA insert, 7 μl of PCR grade water and 10 μl of master mix. Perform positive control reactions using wild-type DNA and negative control reactions by omitting DNA addition. Run a PCR program that allows the amplification of the expected fragment size according to in silico prediction.

3.1.3 Gel Electrophoresis

1. Add 2 μl of DNA loading dye to each PCR reaction.
2. Prepare an agarose-gel: e.g., dissolve and boil 1 g of agarose (1%) in 100 ml of 1× TAE-buffer (*see* **Note 11**), cool down to 50 °C, add 2.5 μl ethidium bromide and pour gel into an appropriate gel-chamber.
3. Load and separate the PCR amplicons by electrophoresis and estimate the size of the resulting fragments in comparison to a suitable DNA size standard.
4. The respective fragments of the predicted sizes should subsequently be gel-eluted (*see* **Note 12**) and sequenced in order to verify the correctness of the T-DNA insertion within the GOI (*see* **Note 13**).

3.2 Expression Analysis

This section describes the steps how to verify whether the inserted T-DNA fragment inactivates the function of the GOI. For full characterization, expression analysis should be carried out on the levels of mRNA expression, protein content, and enzymatic activity (*see* **Note 14**).

3.2.1 RNA Isolation and Quantification [15, 16]

1. Harvest and freeze 300 mg (*see* **Note 15**) of plant tissue in liquid nitrogen and store at −80 °C until extraction. For extraction use a sterile, precooled mortar and pistil and sterile, ribonuclease A free tubes. Use gloves during RNA isolation.
2. Transfer the leaf material into the mortar, add a suitable amount of liquid nitrogen, and grind down the material to a fine powder, without thawing.
3. Add 1 ml Z6-buffer to the sample and homogenize (*see* **Note 16**).
4. Transfer the homogenate into a 2 ml tube, add 500 μl PCI, and mix thoroughly. Samples can be stored on ice until all samples are processed.
5. Centrifuge 10 min at 15,000 × *g* and 4 °C. Transfer the aqueous phase (app. 750 μl) into new tube, add a 1/20 volume part (38 μl) of 1 M acetic acid and a 0.7 volume part (500 μl) of 100% ethanol. Mix and precipitate RNA 10 min at room temperature.
6. Centrifuge samples for 10 min at 15,000 × *g* and 4 °C. Carefully remove the liquid phase completely and dissolve the pellet in 500 μl DEPC-water. Subsequently, add 1 volume part (500 μl) of 5 M lithium chloride (ribonuclease A free), mix thoroughly, and incubate 30 min on ice.
7. Centrifuge for 15 min at 15,000 × *g* and 4 °C. Carefully remove the supernatant, dissolve the pellet in 1 volume part (500 μl) of 3 M sodium acetate, and mix thoroughly.

8. Centrifuge for 10 min at 15,000 × *g* and 4 °C, discard supernatant, and wash the pellet in 70% ethanol.
9. Centrifuge for 10 min at 15,000 × *g* and 4 °C, discard supernatant, and dry the pellet. Dissolve the isolate in 50 μl DEPC-water.
10. Perform DNaseI treatment: add 5 μl of DNaseI reaction buffer and 5 μl of DNaseI to each sample and incubate for 30 min at 37 °C. Following, inactivate DNaseI by adding 1 μl of 50 mM EDTA and incubating at 65 °C for 10 min.
11. Test for residual DNA contamination via PCR (*see* **Note 17**).
12. Determine RNA concentration via Nano Drop or photometrical measurements, depending of the device available in the laboratory.
13. In case isolated samples should be used for Northern blot analysis or RNA sequencing, check RNA integrity by denaturating gel electrophoresis or load and read a RNA-chip using a nucleic acid bio-analyzer device.

3.2.2 cDNA Synthesis

1. Fill up 2.5 μg RNA and 1 μl oligo-dT-primer to a total volume of 13 μl, using ribonuclease A free water, and mix thoroughly.
2. Denature RNA for 5 min at 70 °C and directly place the sample on ice for 2 min. Centrifuge for 30 s at 1000 × *g* to spin down the condensate.
3. Add 4 μl 5× reaction buffer, 2 μl dNTP mixture, 1 μl of ribonuclease A inhibitor and incubate the sample at 37 °C for 5 min.
4. Add 1 μl reverse transcriptase and incubate the sample at 42 °C for 1 h. To terminate the first-strand cDNA synthesis, place the sample at 70 °C for 15 min. Subsequently, place on ice for 2 min, centrifuge for 30 s at 1000 × *g* to spin down the condensate.
5. Dilute the synthesized cDNA samples 1:50 for initial PCR analysis.

3.2.3 Expression Analysis of the GOI via RT-PCR Analysis

1. Carry out PCR analysis in order to verify cDNA synthesis efficiency and to perform calibration thereof by using primers specific for a reliable housekeeping gene (*S16*, Actin, etc.).
2. Mix 1 μl of each of the 1:50 diluted cDNA sample with 1 μl of the forward and reverse primer, 7 μl of PCR grade water, and 10 μl of master mix. Simultaneously, perform a negative control reaction by omitting DNA addition. Initially, use a PCR program that allows the amplification of the expected fragment.
3. Determine the respective amounts of PCR cycles that allow fragment amplification in the non-saturated zone to facilitate detection of differences in the used cDNA amounts.

4. Perform further PCR reactions to calibrate the cDNA amounts of all genotypes and the wild-type control by variation of either used cDNA volume or dilution thereof.
5. Under the use of the calibrated cDNA amounts, carry out RT-PCR reactions with primers specific for the housekeeping gene, the GOI as well as a second pathway representative for comparison. Ensure that all amplicons are of a similar fragment size.
6. In case the mutant lines used are not complete knockout lines (e.g., promoter insertion line), the residual expression of the gene should be determined via quantitative real-time PCR analysis.

3.2.4 Protein Isolation and Quantification

1. Harvest and freeze 100 mg of plant tissue in liquid nitrogen and store at −80 °C until extraction. For pulverization a pre-cooled steel ball is added to the 2 ml tube containing the leaf material and grinded down to a fine powder using a Retsch mill.
2. Add 200 μl EDTA–Tris–HCl (pH 7.5), 10 μl $NaHSO_3$ + Pefabloc, and 10 μl PMSF stock solutions and mix thoroughly.
3. Centrifuge for 10 min at 15,000 × *g* and 4 °C. Transfer the supernatant into a new tube and store on ice until all samples have been processed (see **Note 18**).
4. Determine the protein concentration according to standard assays [11].

3.2.5 SDS-PAGE and Immunoblot Analysis [17]

1. To run SDS-PAGE mix 5 μg of total leaf-proteins with 0.5 volume parts of 3× Lämmli buffer (*see* **Note 19**) and boil for 10 min at 95 °C. Put samples on ice for 2 min and spin down the condensate by centrifugation for 30 s at 1000 × *g*. Use gloves for the following **steps 2–12**.
2. Prepare separating gel: mix 4 ml acrylamide, 3.35 ml water, 2.5 ml 1.5 M Tris–HCl-buffer (pH 8.8), 100 μl 10% SDS solution, 50 μl 10% ammonium persulfate, and 5 μl TEMED in a 20 ml Erlenmeyer flask (amounts sufficient for two gels) and cast gel within a 10 cm × 10 cm × 4 mm gel cassette. Allow adequate space for stacking gel and overlay with water. Wait for complete polymerization prior filling in the stacking gel.
3. Prepare stacking gel: mix 0.975 ml acrylamide, 2.725 ml water, 1.25 ml 0.5 M Tris–HCl-buffer (pH 6.8), 50 μl 10% SDS solution, 40 μl 10% ammonium persulfate, and 5 μl TEMED in 20 ml Erlenmeyer flask (amounts sufficient for two gels). Remove the water from the separation gel, fill in the stacking gel, and insert a 10-well gel comb immediately without introducing air bubbles.

4. Load samples to be analyzed, a suitable protein standard and fill 20 μl of 1× Lämmli buffer in every free lane. Ensure to leave out one lane between the protein standard and the protein samples for separating and staining of the size standard. Prepare two gels, one for Coomassie staining (calibration gel) and one for immunoblot analysis.
5. Run gel at 80 V until samples have entered the separation gel and continue at 140 V until the dye front reached the bottom of the gel.
6. Prepare two sheets of Whatman paper and one sheet of a PVDF membrane of the size, appropriate to cover the stacking gel. Saturate Whatman papers with 1× transfer-buffer. Saturate the PVDF membrane with methanol and flush with 1× transfer-buffer thereafter.
7. Build up a semi-dry electroblotting apparatus, wet the anode-plate with 1× transfer-buffer, and place one sheet of Whatman paper and the PVDF membrane thereon.
8. Turn of the power supply of the SDS PAGE, separate the gel plates using a spatula, and remove the stacking gel. Flush the separating gel with water, submit the first gel to Coomassie staining. After 30 min incubation, bleach the gel and fix it by application of 7% acetic acid. This gel serves as reference in order to exclude loading differences.
9. Put the second gel on the membrane for immunoblotting analysis. Ensure to sign the free lane between the protein standard and the protein samples.
10. Place a sheet of Whatman paper of the gel, close the blotting-device by adding the cathode-plate, and switch on the power supply (1 mA/cm^2 membrane size) for 90 min (*see* **Note 20**).
11. Wash the membrane thrice with 1× TBS-buffer for 5 min and cut of the marked protein standard from the membrane. Incubate the standard-slice for 10 min in Coomassie solution and bleach it until the protein standard is visible thereafter (*see* **Note 21**).
12. Block the membrane for 1 h in 5% skimmed milk in 1× TBS-buffer (shake at room temperature or overnight at 4 °C).
13. Wash the membrane four times in 1× TBS-buffer for 5 min. Put the membrane into 5% skimmed milk in 1× TBS-buffer, containing an appropriate dilution of the primary antibody, for 1 h (shake at room temperature or overnight at 4 °C).
14. Wash the membrane four times in 1× TBS-buffer for 5 min. Put the membrane into 5% skimmed milk in 1× TBS-buffer, containing an appropriate dilution of the secondary antibody (*see* **Note 22**), for 1 h (shake at room temperature or overnight at 4 °C).

15. Wash the membrane four times in 1× TBS-buffer for 5 min.
16. Develop the membrane according to the secondary antibody used (coupled to horse radish peroxidase or alkaline phosphatase).
17. Chemiluminescence assay:

    Prepare a mixture of solution 1 and solution 2 (1:1, by volume) and put the membrane into a transparent sheet. Take 1 ml of the solution mixture, spread it on top of the membrane and laminate transparent sheet. Perform the development of the radiographic film in the dark room. Therefore, remove the air bubbles from the membrane (*see* **Note 23**). Incubate the radiographic film with the membrane for an adequate time (e.g., 5 min) and develop the film by incubation in development solution. After getting an appropriate strength of detection signals (non-saturated area), briefly wash the film in water and fix it in the fixation solution afterwards.
    The bound antibody can be removed by using 20 ml stripping-buffer and shaking incubation for 60 min at 50 °C. After washing the membrane in water (*see* **Note 24**), start from **step 12**. With this method, the membrane can be reused to check the abundances of up to five different proteins.
18. Alkaline phosphatase assay:

    Incubate the membrane for 10 min in detection-buffer while shaking. Discard the buffer and incubate the membrane in 20 ml of a mixture of NBT/BCIP-solution in detection-buffer (use 66 μl NBT and 33 μl BCIP per 10 ml of detection-buffer). Incubate the membrane in the dark for an adequate time. After a desired strength of detection signal is visible stop the reaction by washing the membrane in water.

3.2.6 Enzyme Activity Measurements

Given that most of the enzymes involved in photorespiration and in light of the identification of novel pathway players there is no uniform protocol on how to extract and measure a particular enzyme. Therefore, we next provide one example of an experimental setup on how to determine hydroxypyruvate reductase (HPR) activity.

1. Harvest and freeze a minimum of 100 mg of leave tissue (five replicates of the wild type and the mutant potentially deficient in HPR activity) and store at −80 °C until extraction.
2. Add a precooled steel ball to the tubes containing the leaf material and grind it down to a fine powder using a Retsch mil.
3. Add 200 μl of extraction buffer to the powder and mix thoroughly. Centrifuge for 10 min at 30,000 × *g* and 4 °C. Transfer the supernatant to a new tube and store on ice until all samples were processed. Quantify the protein concentration according to standard protocols [11].

4. Have ready all buffers and chemicals for the detection of HPR activity using a coupled spectrophotometric assay: 0.1 M MES (pH 6.5), 3.75 mM NAD(P)H, 20 mM hydroxypyruvate or glyoxylate stock solutions.
5. Mix 850 μl of measuring buffer, 50 μl protein extract, and 50 μl NAD(P)H in a cuvette and run a baseline at 340 nm for at least 2 min (stable course of the curve).
6. Add 50 μl hydroxypyruvate or glyoxylate (*see* **Note 25**) to the cuvette and follow the change in absorption for another 8 min.
7. Calculate the change in absorbance of the baseline and after adding the enzyme substrate. Subtract the baseline from the actual measurement and calculate the enzyme activity according to the law of Lambert-Beer. Compare the determined activity of the mutant with the wild type.

3.3 Basic Physiological Characterization

This section should guarantee for the detection of a potential photorespiratory phenotype [6–8] and to quantify whether alterations on plant growth, photosynthesis, or metabolite abundances typical for photorespiratory mutants [10] are visible in (non)-photorespiratory conditions.

3.3.1 Phenotyping During CO_2 Transition

1. The mutant putatively defective in photorespiration should initially be grown in comparison with the respective wild type in high CO_2 conditions (suppressed photorespiration). Set a suitable plant growth chamber to 3000 ppm CO_2 (*see* **Note 26**), use a defined photoperiod (e.g., 12 h/12 h day-/night-cycle) and temperature (22/20 °C day-/night-cycle) regime, as well as controlled light qualities and intensities (120 μmol/m^2/s) and relative humidity (~70%).
2. Grow plants to a defined growth stage [12] and perform control measurements (high CO_2 control). Take representative photos of at least five individual replicates per mutant and the corresponding wild type. Quantify rosette diameters, total leaf areas, leaf number, length and width, and fresh and dry-weights.
3. Transfer the plants to normal air conditions with otherwise equal conditions. Take representative photos of the genotypes ($N = 5$) and quantify growth parameters 1, 3, and 5 days after the CO_2 shift (*see* **Note 27**).
4. In case a photorespiratory phenotype emerged transfer at least five individual plants back into a high CO_2 to allow seed production for propagation and further analysis.

3.3.2 Chlorophyll a Fluorescence

1. The mutant putatively defective in photorespiration should initially be grown in comparison with the respective wild type in high CO_2 conditions (suppressed photorespiration).

Set a suitable plant growth chamber to 3000 ppm CO_2, use defined photoperiod (e.g., 12 h/12 h day-/night-cycle) and temperature (22/20 °C day-/night-cycle) regimes and controlled light intensities (120 μmol/m^2/s) and relative humidity (~70%).

2. All genotypes should be grown in high CO_2 conditions and control measurements carried out after reaching a defined growth stage [12]. At this stage the maximum quantum yield of photosystem II (F_v/F_m), non-photochemical quenching (NPQ), and relative electron transport rates (rETR) should be determined using, for example, an imaging pulse amplitude modulations fluorometer (I-PAM). Using such a system would further substantiate the visual appearance of the mutant characterized via common photography.
3. Transfer the genotypes to normal air conditions and repeat the respective measurements (**step 2**) 1, 3, and 5 days after transferring the plants.
4. Prior the actual measurement plants are dark adapted for at least 10 min to process all of the residual photosynthetic activity. The F_v/F_m parameter is recorded after applying a saturating light pulse (3000 μmol/m^2/s) to the plants. For determination of the rETR's and NPQ varying actinic light concentrations (0–1500 μmol/m^2/s) are used.

3.3.3 Gas Exchange Analysis

1. The mutant putatively defective in photorespiration should initially be grown in comparison with the respective wild type in high CO_2 conditions (suppressed photorespiration). Set a suitable plant growth chamber to 3000 ppm CO_2, use defined photoperiod (e.g., 12 h/12 h day-/night-cycle) and temperature (22/20 °C day-/night-cycle) regimes and controlled light intensities (120 μmol/m^2/s) and relative humidity (~70%).
2. All genotypes should be grown in high CO_2 conditions and control measurements carried out after reaching a defined growth stage [12]. At this stage the net CO_2 assimilation rate (A), the CO_2 compensation point (Γ), stomatal conductance (g_s), and the ration of the internal versus external CO_2 concentration (C_i/C_a) should be determined.
3. Transfer the genotypes to normal air conditions and repeat the respective measurements (**step 2**) 1, 3, and 5 days after shifting the plants.
4. In case no clear effect on photosynthesis emerged, measurements should be carried out at higher oxygen concentrations (30% and 40% O_2 in air) to boost the phenotypic strength of the respective mutation by intensification of the oxygenation reaction of RubisCO.

5. To determine the respective parameters adapt the leaf for at least 10 min in the measuring chamber for stabilization and run standard A/C_i curves with the following parameters: 400 µl/L CO_2 in air, 300 µmol/s flow rate, 25 °C block temperature, saturating light intensity (depending on the light condition used during cultivation of the plants, needs to be determined from light response curves prior A/C_i curve measurements), vary CO_2 concentrations (400–300–200–100–50–20–0–400 ppm).

*3.3.4 Metabolite Analysis (See **Note 28**)*

1. The mutant putatively defective in photorespiration should initially be grown in comparison with the respective wild type in high CO_2 conditions (suppressed photorespiration). Set a suitable plant growth chamber to 3000 ppm CO_2, use defined photoperiod (e.g., 12 h/12 h day-/night-cycle) and temperature (22/20 °C day-/night-cycle) regimes and controlled light intensities (120 µmol/m^2/s) and relative humidity (~70%).
2. All genotypes should be grown in high CO_2 conditions and control measurements carried out after reaching a defined growth stage [12]. At this stage samples for metabolite analysis should be harvested at the mid and end of the day and night, respectively.
3. Transfer the genotypes to normal air conditions and sample leaf material at the same time points as mentioned in point two, at least 1 and 3 days after shifting the plants.

4 Notes

1. DNA isolation could also be done via commercially available plant DNA extraction kits.
2. Oligonucleotide sequences need to be derived from the T-DNA or other inserted fragments used to produce the various Arabidopsis mutant collections and the GOI analyzed. The used gene-specific primers should bind 300–400 bp up- or downstream of the insert to increase specificity.
3. RNA isolation could also be done using commercially available plant RNA extraction kits. The protocol given allows isolating high amounts of RNA for Northern blot applications or RNA-sequencing approaches.
4. Instead of semi-quantitative RT-PCR, quantitative RT-PCR should be used for absolute quantification of gene expression.
5. Suggested protease inhibitor stock solutions could be replaced by commercially available protease inhibitor cocktails.

6. Although the different (potential) photorespiratory enzymes need to be extracted and measured using different protocols an example on how to determine hydroxypyruvate reductase activity (HPR) is provided.
7. To guarantee for optimal comparison of plant growth and performance in high CO_2 (e.g., 3000 ppm) and normal air (390 ppm), it is highly recommended to keep all other parameters (light quality, light intensity, temperature cycle, photoperiodic cycle, relative humidity) similar. This is achieved best by using two identical plant growth chambers, while running the first at high CO_2 and the second with normal air.
8. The in-built CO_2 controlling and regulation system should be sufficient to apply up to 10,000 ppm CO_2 to the plant growth chamber, since photorespiratory mutants have different CO_2 requirements for phenotypic complementation [6].
9. For a general overview of metabolic alterations a GC-MS-based method detecting various metabolites involved in central carbon metabolism was proven to be appropriate [7].
10. The positive and negative controls allow checking if PCR reactions ran correctly. If there is no signal in any sample, you might have forgotten to insert a component of the PCR mix. If the agarose-gel is empty after running the electrophorese you might have forgotten to include ethidium bromide into the agarose gel or set the electric voltage the wrong way around. If there is a signal in the negative control, there is a contamination in one of the components of the PCR mix. Ensure that PCR tubes are well closed to prevent vaporizing.
11. 1% is used here, as a standard but needs to be adapted according to the fragment sizes expected. The amount of gel needs also to be adapted according to the sample amounts and gel-chambers available.
12. Gel elution is carried out using commercially available gel-extraction kits. For some fragments (suboptimal primer-binding efficiency, long PCR amplicons) it is recommended to sub-clone them to facilitate sequencing.
13. The orientation of sequencing and the respective results can be compared to the initial analysis done by each stock center deposited on their own website or the TAIR data base.
14. It is recommended to analyze both the protein abundance and the respective activity thereof, respectively. However, given that a well working antibody is probably unavailable or the enzymatic activity difficult to detect, one of both methods is sufficient for the first time.
15. For fast analysis of the expression of the GOI it might be more feasible to use smaller amounts of plant tissue and a commercial

available plant RNA extraction kit. The given procedure is on a larger scale to also allow Northern blot or RNA-sequencing analysis.

16. The amount of Z6-buffer used needs to be adapted if larger amounts of leaf tissue is used.
17. To test for residual DNA contamination 1 μl of the isolated RNA is used for PCR analysis by checking the amplification of a selected gene. It is recommended to perform a positive control using wild-type DNA and a negative control using water to exclude PCR mistakes.
18. Store the protein samples on ice until usage. Otherwise, the proteins will denature and enzymes activity is lost.
19. 5 μg protein is recommended for highly abundant photorespiratory proteins. In some cases, the amount of protein loaded needs to be adapted.
20. It is important to remove possible air bubbles between the membrane and the separating gel. Otherwise, the proteins cannot be transferred to the membrane.
21. Do not forget to cut off the protein standard. Once the protein standard is blocked with 5% skimmed milk, Coomassie solution will color the whole membrane stripe.
22. The use of the secondary antibody depends on the primary antibody raised (rabbit, donkey, goat, etc.) and the detection method elected (chemiluminescence, alkaline phosphatase).
23. It is important to remove the air bubbles and surplus solution mixture from the top of the membrane by using a piece of paper and swiping the solution mixture to the borders of the transparent sheet. Otherwise, the air bubbles could disturb the signal detection. Too much solution mixture on top of the membrane could lead to unnecessary background signals.
24. Wash the membrane with water until it no longer smells like β-mercaptoethanol.
25. A "classic" hydroxypyruvate reductase prefers 3HP reduction with NADH as cofactor but also accepts glyoxylate as substrate and NADPH as a cofactor though to a much lower efficiency [18].
26. In some cases, photorespiratory mutants have different demands on CO_2 supplementation. Therefore, a given mutant needs to be grown at concentrations up to 10,000 ppm [19].
27. In some cases the experiments need to be carried out for more than 5 days, for instance if an intermediate photorespiratory mutant is used and the phenotypic or physiological responses emerge only if the mutant is exposed for more than 5 days to normal air.

28. Here, we only provide the basic overview of the work flow to be carried out if samples of a photorespiratory mutant shall be analyzed. For recommendations on the chromatography method, technical details or comprehensive information on extraction protocols and data handling please see previous reports on this topic [20–22].

Acknowledgments

Work in the author's laboratory is funded by the German research foundation (DFG), especially in frame of the research unit on photorespiration (PROMICS—FOR1186, to H.B.). We wish to thank various Bachelor and Master Students involved in the isolation and characterization of photorespiratory mutants, namely: Franziska Flügel, Silja Frankenbach, Nina Gehm, Jonas Giese, and Julia Walter.

References

1. Koncz C, Nemeth K, Redei GP, Schell J (1992) T-DNA insertional mutagenesis in Arabidopsis. Plant Mol Biol 5:963–976
2. O'Malley RC, Barragan CC, Ecker JR (2015) A user's guide to the Arabidopsis T-DNA insertion mutant collections. Methods Mol Biol 1284:323–342
3. O'Malley RC, Ecker JR (2010) Linking genotype to phenotype using the Arabidopsis unimutant collection. Plant J 6:928–940
4. Puchta H (2015) Using CRISPR/Cas in three dimensions: towards synthetic plant genomes, transcriptomes and epigenomes. Plant J. doi:10.1111/tpj.13100
5. Krysan PJ, Young JC, Sussman MR (1999) T-DNA as an insertional mutagen in Arabidopsis. Plant Cell 12:2283–2290
6. Timm S, Bauwe H (2013) The variety of photorespiratory phenotypes – employing the current status for future research directions on photorespiration. Plant Biol (Stuttg) 4:737–747
7. Timm S, Mielewczik M, Florian A, Frankenbach S, Dreissen A, Hocken N et al (2012) High-to-low CO_2 acclimation reveals plasticity of the photorespiratory pathway and indicates regulatory links to cellular metabolism of Arabidopsis. PLoS One 8:e42809
8. Somerville CR (2001) An early Arabidopsis demonstration. Resolving a few issues concerning photorespiration. Plant Physiol 1:20–24
9. Orf I, Timm S, Bauwe H, Fernie AR, Hagemann M, Kopka J, Nikoloski Z (2016) Can cyanobacteria serve as a model of plant photorespiration?—a comparative meta-analysis of metabolite profiles. J Exp Bot 10:2941–2952
10. Timm S, Florian A, Fernie AR, Bauwe H (2016) The regulatory interplay between photorespiration and photosynthesis. J Exp Bot 10:2923–2929
11. Bradford MM (1976) A rapid and sensitive method for the quantitation of microgram quantities of protein utilizing the principle of protein-dye binding. Anal Biochem 72:248–254
12. Boyes DC, Zayed AM, Ascenzi R, McCaskill AJ, Hoffman NE, Davis KR, Gorlach J (2001) Growth stage-based phenotypic analysis of Arabidopsis: a model for high throughput functional genomics in plants. Plant Cell 7:1499–1510
13. Mullis K, Faloona F, Scharf S, Saiki R, Horn G, Erlich H (1986) Specific enzymatic amplification of DNA in vitro: the polymerase chain reaction. Cold Spring Harb Symp Quant Biol 51:263–273
14. Murray MG, Thompson WF (1980) Rapid isolation of high molecular weight plant DNA. Nucleic Acids Res 19:4321–4325
15. Logemann J, Schell J, Willmitzer L (1987) Improved method for the isolation of RNA from plant tissues. Anal Biochem 1:16–20

16. Chirgwin JM, Przybyla AE, MacDonald RJ, Rutter WJ (1979) Isolation of biologically active ribonucleic acid from sources enriched in ribonuclease. Biochemistry 24:5294–5299
17. Towbin H, Staehelin T, Gordon J (1979) Electrophoretic transfer of proteins from polyacrylamide gels to nitrocellulose sheets: procedure and some applications. Proc Natl Acad Sci U S A 9:4350–4354
18. Timm S, Florian A, Jahnke K, Nunes-Nesi A, Fernie AR, Bauwe H (2011) The hydroxypyruvate-reducing system in Arabidopsis: multiple enzymes for the same end. Plant Physiol 2:694–705
19. Eisenhut M, Bräutigam A, Timm S, Florian A, Tohge T, Fernie AR, Bauwe H, Weber APM (2016) Photorespiration is crucial to the dynamic response of photosynthetic metabolism and stomatal movement to altered CO_2 availability. Mol Plant. doi:10.1016/j.molp.2016.09.011
20. Fernie AR, Aharoni A, Willmitzer L, Stitt M, Tohge T, Kopka J et al (2011) Recommendations for reporting metabolite data. Plant Cell 7:2477–2482
21. Arrivault S, Guenther M, Ivakov A, Feil R, Vosloh D, van Dongen JT et al (2009) Use of reverse-phase liquid chromatography, linked to tandem mass spectrometry, to profile the Calvin cycle and other metabolic intermediates in Arabidopsis rosettes at different carbon dioxide concentrations. Plant J 5:826–839
22. Lisec J, Schauer N, Kopka J, Willmitzer L, Fernie AR (2006) Gas chromatography mass spectrometry-based metabolite profiling in plants. Nat Protoc 1:387–396

Chapter 9

Exploiting Natural Variation to Discover Candidate Genes Involved in Photosynthesis-Related Traits

Franklin Magnum de Oliveira Silva, Lucas de Ávila Silva, Wagner L. Araújo, Agustin Zsögön, and Adriano Nunes-Nesi

Abstract

Naturally occurring genetic variation in plants can be very useful to dissect the complex regulation of primary metabolism as well as of physiological traits such as photosynthesis and photorespiration. The physiological and genetic mechanisms underlying natural variation in closely related species or accessions may provide important information that can be used to improve crop yield. In this chapter we describe in detail the use of a population of introgression lines (ILs), with the *Solanum pennellii* IL population as a study case, as a tool for the identification of genomic regions involved in the control of photosynthetic efficiency.

Key words Introgression lines population, Fluorescence, Photosynthesis, Photorespiration, Natural variation

1 Introduction

Reducing yield losses caused by photorespiration is a promising breeding goal [1]. Plants have developed an array of different strategies, involving anatomic and metabolic traits, to cope with photorespiration and reduce its detrimental effects [2]. Natural genetic variation in plant photorespiration and photosynthesis can thus be used as a tool to identify genetic determinants which can be exploited for further improvement of crop yield [3–5]. In spite of the growing number of studies on this topic, the genetic factors responsible for variation in photosynthesis, photorespiration, and related parameters, such as respiration, still remain largely unknown [3, 4]. Identification of genes encoding auxiliary factors or modifiers of photorespiration and photosynthesis could provide further insight into the regulation of yield and thus information of immediate practical value. Furthermore, such factors have not been identified in previous mutant screenings due to potential

Alisdair R. Fernie et al. (eds.), *Photorespiration: Methods and Protocols*, Methods in Molecular Biology, vol. 1653, DOI 10.1007/978-1-4939-7225-8_9, © Springer Science+Business Media LLC 2017

redundancy and numerous genes involved with small effects on the observed phenotypes [3].

Quantitative Trait Locus (QTL) mapping is a powerful technique for dissecting the genetic basis of phenotypic differences between species and cultivars [4, 6, 7]. The creation of populations of introgression lines (ILs), whereby the genome of a species is broken down in discrete, overlapping chromosomal segments, boosted significant progress in QTL mapping in both monocots [8–10] and eudicots [11–13]. A particularly useful collection of lines is the *Solanum pennellii* introgression population, in which defined genomic segments of the *S. pennellii* genome replaced homologous regions in the tomato (*S. lycopersicum* cv M82) genomic background (Fig. 1). Originally mapped by Restriction Fragment Length Polymorphism (RFLP) [11], and more recently resequenced thoroughly [14], the *S. pennellii* IL population is very useful to map QTL, given that any phenotypic difference between an IL and the recurrent parental line is attributed solely to donor parent genes within the introgressed chromosomal segment [15]. An added advantage is the existence of overlapping segments of *S. pennellii* genome in different ILs, which allows fine-mapping of candidate genes. The assessment of ILs thus represents a powerful tool for identifying and understanding the genetic determinants involved in quantitative traits linked to processes such as photorespiration, photosynthesis, and underlying traits (*see* **Note 1**).

For QTL mapping of photosynthesis-related traits, a detailed characterization of the experimental material is a critical requirement for a meaningful analysis by bioinformatics tools and for publication needs. While information directly related to the experimental approach (*e.g.*, daylength and species in a diurnal cycle experiment) is often recorded in a standardized way along with the results in spreadsheet formats, other relevant information (*e.g.*, air temperature, light intensity, plant age at sampling) is often left out. It should also be mentioned that performing chlorophyll fluorescence and gas exchange measurements on the same fully expanded source leaves is important for an accurate and full understanding of the response of the photosynthetic apparatus [19, 20].

In this chapter we describe the use of an IL population in tomato as a powerful tool to identify genetic factors involved in the regulation of photorespiratory and photosynthetic related parameters.

2 Materials

1. The *S. pennellii* population of ILs is composed of 76 distinct lines curated and distributed by The C.M. Rick Tomato Genetics Resource Center, University of California Davis (http://tgrc.ucdavis.edu/).

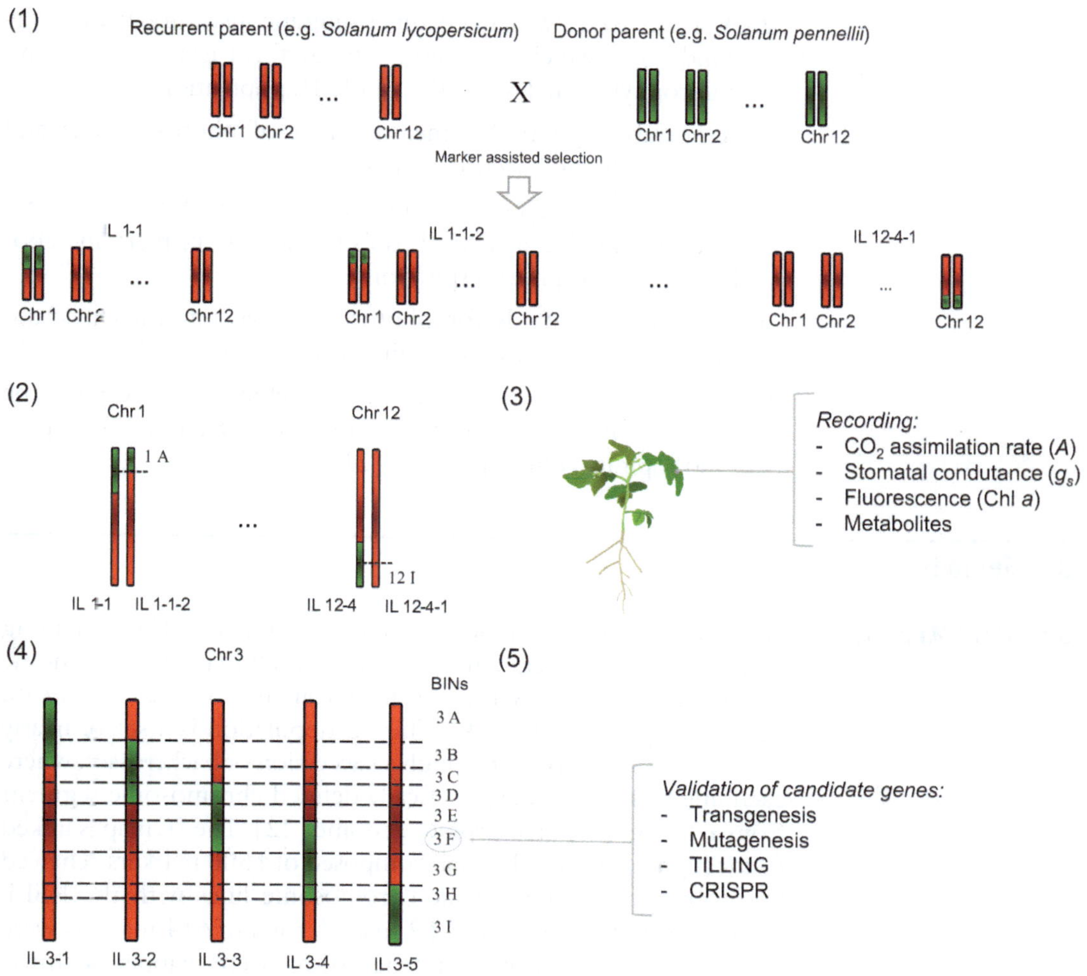

Fig. 1 Schematic representation of a pipeline designed to detect genomic regions and identify candidate genes controlling photosynthesis-related traits. (*1*) Cultivation of an IL population covering the entire genome (*2*) enables mapping traits to specific bins. The IL population was previously produced through controlled introgression and marker-assisted selection generating a set of recurrent parent lines with single introgressed segments [16]. (*3*) Evaluation of gas exchange, fluorescence (Chl *a*), metabolites [17], and other methods [18] should be carried out twice or more in the whole population. (*4*) ILs showing significant differences in the trait of interest (ILs 3-3 and IL 3-4, in the example) indicate genomic regions (chromosome position using the bin mapping method) with significant variation in the parameter evaluated relative to recurrent parent. Precise genome localization is possible due to the existence of overlapping introgressed regions (Bin 3F, in the example). (*5*) The identification of genes potentially related to trait of interest then requires creating a list of the genes between the flanking markers that delimit the bin. Subsequent sequence analysis using the NCBI protein database and the Blast algorithm (Blastx, Blast2GO and Blastp) results in a catalogue of candidate genes with sequence homology to previously characterized expressed sequences. The validation of candidate genes can be carried out by transgenesis, mutagenesis, TILLING, and CRISPR. Allele comparison is finally performed through identification of nucleotide and amino-acid polymorphisms

2. A facility to grow many plants simultaneously, under the assay conditions, which is necessary to perform gas exchange and chlorophyll *a* analysis of the whole IL population.
3. A portable system for measurement of photosynthesis and chlorophyll fluorescence parameters.
4. Gas chromatography mass spectrometry (GC-MS) systems are recommended as analytical tools to measure primary metabolites related to photorespiration.
5. Authentic standards for photorespiratory intermediates and other metabolites for determination of absolute values of each.
6. Reference library mass spectra and retention indices can be obtained from the Golm Metabolome Database (<http://gmd.mpimp-golm.mpg.de>) [21].

3 Methods

3.1 Plant Materials

Solanum pennellii ILs population is composed of 76 ILs containing a defined genomic segment of the *S. pennellii* (LA0716) genome replacing homologous regions in a tomato (*S. lycopersicum* cv. M82) background [11, 14]. This IL population is a set of nearly isogenic lines developed through a succession of backcrosses, where each line carries a single genetically defined chromosome segment from a divergent (polymorphic) genome [22]. The IL map is linked to the high-resolution F2 map composed of 1500 markers achieved by probing all of the specific chromosome lines with the RFLP markers from the framework F2 map. A total of 614 markers were probed and the ends of the introgressions were mapped with the resolution of the F2 map [23]. In addition, a set of 285 smaller sublines with marker-defined introgressions smaller than the original set of the 37 largest ILs of the initial population and large quantities of F2 seeds derived from selfing of the hybrids of the ILs with M82 were recently generated [24]. Thus, it is expected that this new population will speed up the discovery of genes underlying traits of interest.

3.2 Growth Conditions

1. Seeds can be germinated in peat and subsequently rooted plantlets are acclimatized and transferred to 1 L pots containing 8 g of commercial substrate supplemented with chemical fertilizer (N: 4; P_2O_5: 14; K_2O: 8).
2. The population is grown in a randomized block design for minimizing the effect of variability when it is associated with discrete units (e.g., location, operator, plant, batch, time). The usual case is to randomize one plant of the 76 ILs and at least one M82 control plant within each block (*see* **Note 2**).

3. Environmental conditions within the glasshouse can be semi-controlled: natural light with temperature maintained within the range of 15 °C (night) to 25 °C (day) and relative humidity of 45–50% (*see* **Note 2–4**).

3.3 Photosynthesis and Chlorophyll Fluorescence Measurements

1. Measurement of photosynthesis via CO_2 uptake is carried out with an Infra-Red Gas Analyzer (IRGA). The IRGA compares the CO_2 concentration in gas passing through a chamber containing a leaf/plant and the concentration of CO_2 leaving the chamber.
2. Since maximal photosynthetic rate is observed in the morning, gas exchange and fluorescence measurements are performed at the beginning of the light period. Thus, CO_2 assimilation rate (A), stomatal conductance (g_s), internal CO_2 concentration (C_i), and transpiration (E) are measured between 1 and 6 h after dawn.
3. Measurements are performed on the second terminal leaflet of the third fully expanded leaf from the apex of 4-week-old nonflowering plants (*see* **Note 5**). The IRGA is set to an external CO_2 concentration of 400 μmol/mol air and flow rate of 300 μmol/s. The air temperature and VPD of 25 °C and 1.2 kPa, respectively, and artificial saturating light of 1300 μmol photons/m^2/s with 10% blue light in order to maximize stomatal opening.
4. Chlorophyll *a* fluorescence parameters are determined simultaneously with gas exchange measurements using an open-flow infrared gas exchange analyzer system equipped with an integrated fluorescence chamber.
5. For analytical methods used for generation of dataset the chlorophyll fluorescence parameters are measured in the same leaf used for gas exchange analysis. After 1 h of dark adaptation, the leaf should be illuminated with weak modulated measuring beams (0.03 μmol/m^2/s) to obtain the initial fluorescence (F_0). Subsequently, saturating white light pulses of 8.000 μmol photons/m^2/s is applied for 0.8 s to ensure maximum fluorescence emissions (F_m).
6. In light-adapted leaves, steady-state fluorescence yield (F_s) is obtained following a saturating white light pulse (8000 μmol/m^2/s, 0.8 s) which is applied to achieve light-adapted maximum fluorescence (F_m'). Thereafter, actinic light is turned off and far-red illumination is applied (2 μmol/m^2/s) to measure light-adapted initial fluorescence (F_0') [25].
7. The chlorophyll fluorescence dataset obtained is used to determine actual quantum yield of PSII (1), coefficients of photochemical quenching (2) and non-photochemical quenching (3), and electron transport rate (4), where PPFD is

photosynthetic photon flux density; α is leaf absorptance and β reflects the partitioning of absorbed quanta between PSII and PSI, and the product $\alpha.\beta$ is adopted as described in the literature for tomato as equal to 0.43 [25, 26].

$$\delta\text{PSII} = (F_m{'} - F_s)/F_m{'} \tag{1}$$

$$\text{qP} = (F_m{'} - F_s)/(F_m{'} - F_0{'}) \tag{2}$$

$$\text{NPQ} = (F_m - F_m{'})/F_m{'} \tag{3}$$

$$\text{ETR} = \delta\text{PSII}.\alpha.\beta.\text{PPFD} \tag{4}$$

3.4 Estimation of Photorespiration

A demand exists for protocols allowing rapid screening of photorespiration that provide high-resolution information regarding a plant's physiological status, often in relation to plant-based growth/developmental parameters and yield or stress, [27, 28] and where a high number of measurements need to be carried out in a short time period. Determination of photosynthetic and chlorophyll *a* fluorescence parameters is an efficient tool to derive values of photorespiration [29]. From this relationship, it is possible to estimate the rate of total electron flow through PSII by means of in vivo coupled measurements of those parameters on intact leaves, to partition the electron flow between assimilative and non-assimilative processes, and to derive values of photorespiration [29, 30].

1. To estimate the photorespiration (Pr), as previously described by [29], the electron transport rate (ETR) should be calculated on the basis of Eq. 5.
2. Once ETR is known, it is possible to estimate photorespiration according to Eq. 6. Dark respiration (R_d) can be measured using the same gas exchange system as described above after at least 1 h in the dark period and after that divided by two ($R_d/2$) to estimate the mitochondrial respiration rate in the light (R_L) [31–33].

$$\text{ETR} = \delta\text{PSII}.\alpha.\beta.\text{PPFD} \tag{5}$$

$$\text{Pr} = 1/12(\text{ETR} - 4.(A + R_L)) \tag{6}$$

3.5 Quantification of Metabolic Intermediates of the Photorespiratory Pathway

The integration of molecular and genetic approaches with biochemical ones, such as metabolite profiling, is now driving rapid advances in knowledge of the key metabolic roles and connections of the enzymes and genes of the photorespiratory pathway [34]. For photorespiratory metabolism, eight intermediates, *glutamate*, *glutamine*, *glycine*, *serine*, *glycerate*, *2-oxoglutarate*, *GABA*, and *glycolate*, have been reported on the interplay between photorespiration and other metabolic pathways [17].

These targeted metabolites can be determined using an established GC-MS-based protocol as described previously [35]. For this analysis, approximately 25 mg of leaf samples are collected in the middle of the light period and stored at −80 °C. Chromatograms and mass spectra can be evaluated using computational tools, such as Chroma TOF® 4.2 (Leco, St Joseph, MI) and TagFinder 4.0 [36]. Metabolites are manually identified using TagFinder by comparing to the reference library mass spectra and retention indices in the Golm Metabolome Database [21]. Quantities of metabolites are expressed as relative intensity, based on peak integration, and relative to the internal standard. Metabolite profiling data should be reported following recent recommendations [37].

3.6 Genomic Localization and Analyses of the Identified QTL

The *S. pennellii* ILs have been extensively phenotyped for numerous traits, such as morphology, yield, fruit quality, and fruit primary and secondary metabolites for identification of QTL [38]. Chromosomal location of putative QTL is determined first by two-way analysis of variance (ANOVA). Thereafter, QTL location is determined using Dunnett's test to compare statistically each IL with the control tomato cultivar M82. Once QTL are identified, bin mapping can be a qualitative endeavor. The criteria to select the QTL-associated bins for further comparative genomic analyses can be: (1) regions covered by at least one IL displaying significant effect when tested by Dunnett's post-hoc test (Fig. 2a) and (2) selection of regions spanned by more than two ILs (overlapping region), which showed the same phenotypic direction (Fig. 2b).

Following screening of the IL population for photosynthesis-related parameters in which chromosome regions are identified as containing the genetic determinants responsible for a specific phenotype, subsequent genomic analyses (described below) are necessary to further resolve the gene or genes involved in determination of the trait of interest.

3.7 Identification of Candidate Genes for Photorespiration

The candidate gene approach, which consists of looking for genes segregating around a locus putatively responsible for the variation of a trait, has been proposed as a way to start QTL characterization [39, 40]. Once the bin (defined genomic region between two known genetic markers) is identified, sequences of the selected region can be retrieved from both the M82 and *S. pennellii* genomes [41] by probing the flanking markers (according to the IL genetic map deposited in the Sol Genomics Network website) in both genome sequences to identify polymorphisms between the parental lines. Thus, all polymorphisms in the region of interest are evaluated in more detail to select the most likely gene(s) affecting the analyzed trait and bioinformatically predict the potential functional effects of each identified polymorphism [42].

One approach to reducing the number of candidate genes is to conduct further experiments using larger numbers of segregating

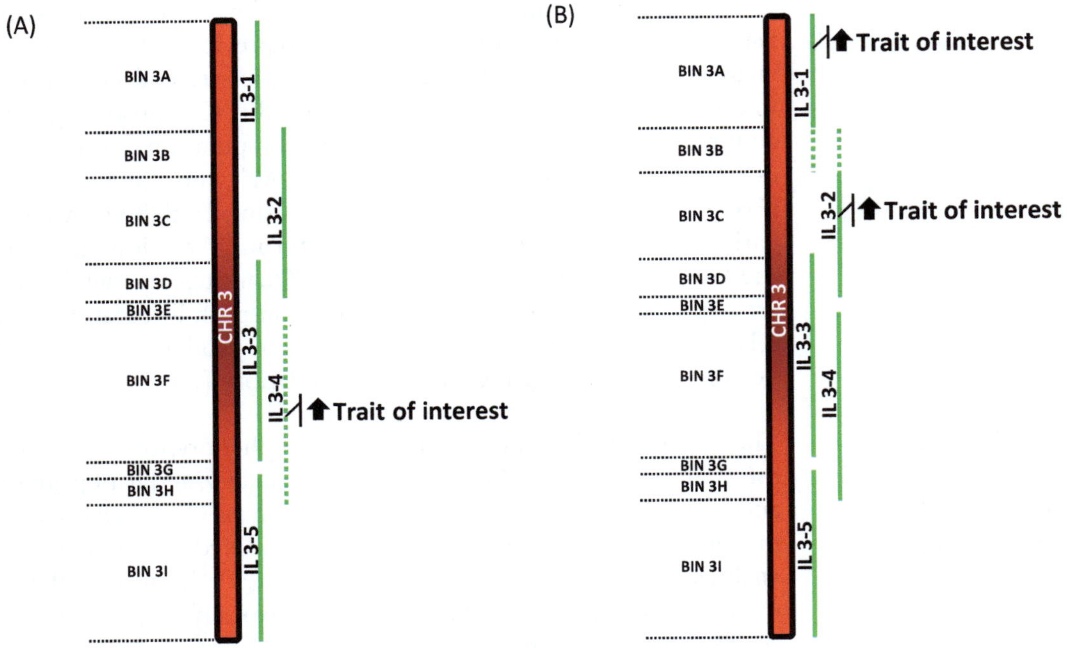

Fig. 2 A random hypothetical example of putative QTL detected in chromosome *3* of the population of 76 introgression lines (ILs) of *Solanum pennellii* (*green lines*) into a genetic background of *Solanum lycopersicum* (M82). Composite interval mapping was implemented using IL cartographer from www.sgn.cornell.edu. *Black arrows* indicate positive changes relative to the trait of interest in independent experiments and *dashed lines* represent the chromosome fragments to putative QTL. Candidate QTL in the region covered by one IL displaying significant effect between IL and its recurrant parent (M82) (**a**). Candidate QTL in an overlapping region (*dashed line*) between two ILs that show the same phenotypic direction (**b**)

progeny to reduce the size of the QTL-containing interval (*see* **Note 6**). Then, classical methods such as positional cloning [43], insertional mutagenesis [44], and reverse genetics (TILLING, CRISPR) should be used to identify the relevant genes. A complementary approach would be to use bioinformatics tools and genome and "Omics" information to assign genes in the QTL interval to bins of higher probability than other candidates [45].

4 Notes

1. In the same way as for the *Solanum pennellii* ILs population, the power of the approach for QTL detection in germplasm development and extraction of valuable and novel alleles from exotic germplasm can be realized in other model systems, including major food and biofuel crops such as pepper, rice, barley, wheat, maize, soybean, and Arabidopsis.
2. Grow more than four M82 control plant per block. In case of photosynthesis and chlorophyll fluorescence parameters, it is

important to collect data set along the day due to slight fluctuations in environmental conditions.

3. High variability due to changes in environmental conditions. Therefore, it is highly recommended to repeat the experiment and analysis in two or more independent experiments and select the ILs with conserved changes.
4. The majority of research on natural genetic variation on ILs population must be carried under controlled conditions to establish the plant genetic effect (*G*) in the phenotypic trait. However, because of the strong effect of $G \times E$ on QTL mapping, evaluating genotypes in field conditions is essential to gain insight into the extent of $G \times E$ into its underlying genetic causes.
5. High density of plants can cause competition between plants per light and thus affecting photosynthetic performance. In the case of 4 weeks, old tomato plants 30 cm between plants and 35 cm between rows is recommended.
6. Identification of loci and specific alleles underlying variation in quantitative traits is an important goal for dissecting complex traits. Despite major advancements in genomics technology, moving from QTL to causal alleles remains a major challenge in genetics research. Then, use of sub ILs to reduce the genomic regions and reduce the number of candidate genes are the ideal raw material for QTL validation, refinement of QTL location, and, ultimately, gene discovery.

Acknowledgments

Financial support was provided by Conselho Nacional de Desenvolvimento Científico e Tecnológico (CNPq), Fundação de Amparo à Pesquisa do Estado de Minas Gerais (FAPEMIG) (grant numbers APQ-00688-12, CRA BDS-00040-14 and BPD-00019-16), Max Planck Society to A.N.N. and W.L.A. Research fellowships granted by CNPq to A.N.N. and W.L.A. are also gratefully acknowledged.

References

1. Betti M, Bauwe H, Busch FA et al (2016) Manipulating photorespiration to increase plant productivity: recent advances and perspectives for crop improvement. J Exp Bot 67:2977–2988
2. Sage RF (2013) Photorespiratory compensation: a driver for biological diversity. Plant Biol 15:624–638
3. Flood PJ, Harbinson J, Aarts MGM (2011) Natural genetic variation in plant photosynthesis. Trends Plant Sci 16:327–335
4. Nunes-Nesi A, Nascimento VL, Silva FMO, Zsögön A, Araújo WL, Sulpice R (2016) Natural genetic variation for morphological and molecular determinants of plant growth and yield. J Exp Bot 67:2989–3001
5. Lawson T, Kramer DM, Raines CA (2012) Improving yield by exploiting mechanisms underlying natural variation of photosynthesis. Curr Opin Biotechnol 23:215–220
6. Tanksley SD (1997) Seed banks and molecular maps: unlocking genetic potential from the wild. Science 277:1063–1066
7. Collard BCY, Jahufer MZZ, Brouwer JB, Pang ECK (2005) An introduction to markers, quantitative trait loci (QTL) mapping and

marker-assisted selection for crop improvement: the basic concepts. Euphytica 142:169–196
8. Tian F, Li DJ, Fu Q et al (2006) Construction of introgression lines carrying wild rice (*Oryza rufipogon* Griff.) segments in cultivated rice (*Oryza sativa L.*) background and characterization of introgressed segments associated with yield-related traits. Theor Appl Genet 112:570–580
9. Li Z, Fu BY, Gao YM, Xu JL, Ali J, Lafitte HR et al (2005) Genome-wide introgression lines and their use in genetic and molecular dissection of complex phenotypes in rice (*Oryza sativa* L.) Plant Mol Biol 59:33–52
10. Jena KK, Kochert G, Khush GS (1992) RFLP analysis of rice (*Oryza sativa* L.) introgression lines. Theor Appl Genet 84:608–616
11. Eshed Y, Zamir D (1994) A genomic library of *Lycopersicon pennellii* in *Lycopersicon esculentum* – a tool for fine mapping of genes. Euphytica 79:175–179
12. Canady MA, Meglic V, Chetelat RT (2005) A library of *Solanum lycopersicoides* introgression lines in cultivated tomato. Genome 48:685–697
13. Do PT, Prudent M, Sulpice R, Causse M, Fernie AR (2010) The influence of fruit load on the tomato pericarp metabolome in a *Solanum chmielewskii* introgression line population. Plant Physiol 154:1128–1142
14. Chitwood DH, Kumar R, Headland LR, Ranjan A, Covigton MF, Ichihashi Y et al (2013) A quantitative genetic basis for leaf morphology in a set of precisely defined tomato introgression lines. Plant Cell 25:2465–2481
15. Fulop D, Ranjan A, Ofner I et al (2016) A new advanced backcross tomato population enables high resolution leaf QTL mapping and gene identification. G3 (Bethesda) 6 (10):3169–3184
16. Eshed Y, Zamir D (1995) An introgression line population of *Lycopersicon pennellii* in the cultivated tomato enables the identification and fine mapping of yield-associated QTL. Genetics 141:1147–1162
17. Obata T, Florian A, Timm S, Bauwe H, Fernie AR (2016) On the metabolic interactions of (photo)respiration. J Exp Bot 67:3003–3014
18. Busch FA (2013) Current methods for estimating the rate of photorespiration in leaves. Plant Biol 15:648–655
19. Gu J, Yin X, Struik PC, Stomph TJ, Wang H (2012) Using chromosome introgression lines to map quantitative trait loci for photosynthesis parameters in rice (*Oryza sativa* L.) leaves under drought and well-watered field conditions. J Exp Bot 63:455–469
20. Gu J, Yin X, Stomph TJ, Struik PC (2014) Can exploiting natural genetic variation in leaf photosynthesis contribute to increasing rice productivity? A simulation analysis. Plant Cell Environ 37:22–34
21. Kopka J, Schauer N, Krueger S et al (2005) GMD@CSB.DB: the Golm metabolome database. Bioinformatics 21:1635–1638
22. Zamir D (2001) Improving plant breeding with exotic genetic libraries. Nat Rev Genet 2:983–989
23. Eshed Y, Abu-Abied M, Saranga Y, Zamir D (1992) *Lycopersicon esculentum* lines containing small overlapping introgressions from *L. pennellii*. Theor Appl Genet 83:1027–1034
24. Alseekh S, Ofner I, Pleban T, Tripodi P, Di Dato F, Cammareri M et al (2013) Resolution by recombination: breaking up *Solanum pennellii* introgressions. Trends Plant Sci 18:536–538
25. Aucique-Perez CE, Rodrigues FÁ, Moreira WR, DaMatta FM (2014) Leaf gas exchange and chlorophyll a fluorescence in wheat plants supplied with silicon and infected with *Pyricularia oryzae*. Phytopathology 104:143–149
26. Maxwell K, Johnson GN (2000) Chlorophyll fluorescence – a practical guide. J Exp Bot 51:659–668
27. Montes JM, Melchinger AE, Reif JC (2007) Novel throughput phenotyping platforms in plant genetic studies. Trends Plant Sci 12:433–436
28. Furbank RT, Von Caemmerer S, Sheehy J, Edwards G (2009) C4 rice: a challenge for plant phenomics. Funct Plant Biol 36:845–856
29. Valentini R, Epron D, Angelis P, Matteucci G, Dreyer E (1995) In situ estimation of net CO_2 assimilation, photosynthetic electron flow and photorespiration in Turkey oak (*Q. cerris* L.) leaves: diurnal cycles under different levels of water supply. Plant Cell Environ 18:631–640
30. Peterson RB (1990) Effects of irradiance on the in vivo CO_2: O_2 specificity factor in tobacco using simultaneous gas exchange and fluorescence techniques. Plant Physiol 94:892–898
31. Niinemets Ü, Sack L (2006) Structural determinants of leaf light-harvesting capacity and photosynthetic potentials. In: Esser K, Lüttge U, Beyschlag W, Murata J (eds) Progress in botany, 67th edn. Springer, Berlin
32. Niinemets Ü, Cescatti A, Rodeghiero M, Tosens T (2005) Leaf internal diffusion conductance limits photosynthesis more strongly in older leaves of Mediterranean evergreen

broad-leaved species. Plant Cell Environ 28:1552–1566

33. Niinemets Ü, Díaz-Espejo A, Flexas J, Galmés J, Warren CR (2009) Role of mesophyll diffusion conductance in constraining potential photosynthetic productivity in the field. J Exp Bot 60:2249–2270
34. Fernie AR, Bauwe H, Eisenhut M, Florian A, Hanson DT, Hagemann M et al (2013) Perspectives on plant photorespiratory metabolism. Plant Biol 15:748–753
35. Osorio S, Alba R, Nikoloski Z, Kochevenko A, Fernie AR, Giovannoni JJ (2012) Integrative comparative analyses of transcript and metabolite profiles from pepper and tomato ripening and development stages uncovers species-specific patterns of network regulatory behavior. Plant Physiol 159:1713–1729
36. Luedemann A, Von Malotky L, Erban A, Kopka J (2012) TagFinder: preprocessing software for the fingerprinting and the profiling of gas chromatography-mass spectrometry based metabolome analyses. Methods Mol Biol 860:255–286
37. Fernie AR, Klee HJ (2011) The use of natural genetic diversity in the understanding of metabolic organization and regulation. Front Plant Sci 2:1–10
38. Schauer N, Semel Y, Balbo I, Steinfath M, Repsilber D, Selbig J et al (2008) Mode of inheritance of primary metabolic traits in tomato. Plant Cell 20:509–523
39. Pflieger S, Lefebvre V, Causse M (2001) The candidate gene approach in plant genetics: a review. Mol Breed 7:275–291
40. Etienne C, Rothan C, Moing A, Plomion C, Bodénès C, Svanella-Dumas L et al (2002) Candidate genes and QTLs for sugar and organic acid content in peach [*Prunus persica* (L.) Batsch]. Theor Appl Genet 105:145–159
41. Bolger A, Scossa F, Bolger ME, Lanz C, Maumus F, Tohge T et al (2014) The genome of the stress-tolerant wild tomato species *Solanum pennellii*. Nat Genet 46:1034–1038
42. Ahsan M, Lundberg AE, Kierczak M, Siegel PB, Carlborg O, Marklund S (2013) Identification of candidate genes and mutations in QTL regions for chicken growth using bioinformatic analysis of NGS and SNP-chip data. Front Genet 4:1–8
43. Chen C, DeClerck G, Tian F, Spooner W, McCouch S, Buckler E (2012) PICARA, an analytical pipeline providing probabilistic inference about a priori candidates genes underlying genome-wide association QTL in plants. PLoS One 7:e46596
44. Ikeda M, Miura K, Aya K, Kitano H, Matsuoka M (2013) Genes offering the potential for designing yield-related traits in rice. Curr Opin Plant Biol 16:213–220
45. Schauer N, Semel Y, Roessner U, Gur A, Balbo I, Carrari F et al (2006) Comprehensive metabolic profiling and phenotyping of interspecific introgression lines for tomato improvement. Nat Biotechnol 24:447–454

Chapter 10

Metabolic Engineering of Photorespiration

Martin K.M. Engqvist and Veronica G. Maurino

Abstract

The introduction of two alternative glycolate catabolic pathways in the chloroplasts of *Arabidopsis thaliana* rendered plants with increased biomass. To introduce these synthetic pathways, the selected genes were stepwise integrated in the nuclear genome of wild-type plants. These plants were transformed by *Agrobacterium tumefaciens* carrying the binary vectors using the floral dip method. Selection of transformants was conducted using different selection agents and the expression of the transgenes was confirmed by PCR and enzyme activity measurements.

Key words PCR amplification, Binary vectors, Arabidopsis transformation, *Agrobacterium tumefaciens*

1 Introduction

The prime function of the C_2-pathway (photorespiratory pathway) is to salvage glycolate 2-P by conversion to glycerate 3-P, which reenters the C_3-reductive cycle [1]. During photorespiration both, CO_2 and NH_3, are released within the mitochondria, resulting in a loss of at least 25% of the CO_2 fixed in ambient air in C_3-plants. Reduction of flux through photorespiration has been viewed as a major way to improve plant carbon fixation and yield. Introduction of two alternative chloroplastic bypasses to photorespiration in *Arabidopsis thaliana*, which divert part of the glycolate away from the photorespiratory pathway and metabolize it in the chloroplasts, enabled the production of plants with increased biomass and apparent rate of CO_2 assimilation when growing under short day conditions [2, 3]. Here we describe the methods used to produce *A. thaliana* plants with functional chloroplastic glycolate catabolic cycles: the GMK pathway (Subheadings 2.1 and 3.1): Co-expression of glycolate oxidase, malate synthase, and catalase in *A. thaliana* chloroplasts [2, 4]; and (Subheadings 2.2 and 3.2) the bacterial pathway: expression of the bacterial glycolate catabolic pathway in *A. thaliana* chloroplasts [3].

Alisdair R. Fernie et al. (eds.), *Photorespiration: Methods and Protocols*, Methods in Molecular Biology, vol. 1653, DOI 10.1007/978-1-4939-7225-8_10,

Both of these approaches use stepwise integration of the transgenes in the nuclear genome of *A. thaliana*. It is however feasible to construct a single vector carrying all necessary genes. The cloning methods for both approaches are described using restriction enzymes. This could be replaced by the use of more recent, scarless, cloning techniques such as Gibson assembly [5], ligase cycling reaction [6], CPEC [7], or SLiCE [8].

2 Materials

2.1 GMK Pathway: Co-expression of Glycolate Oxidase, Malate Synthase, and Catalase in A. thaliana Chloroplasts

2.1.1 Cloning of Glycolate Oxidase, Malate Synthase and Catalase

1. Cloning vector: pCR-Blunt II-TOPO (Invitrogen).
2. Organisms: *Arabidopsis thaliana* (Arabidopsis) seedlings, *Cucurbita maxima* (pumpkin) seedlings, *Escherichia coli*, *Agrobacterium tumefaciens*.
3. *Lycopersicon esculentum* (tomato) genomic DNA.
4. TRIzol reagent kit (Gibco-BRL).
5. SuperScript II reverse transcriptase kit (Invitrogen).
6. PCR primers 10 pmol/μl stock solutions in water: GO-fow (5′-TACAATTGGAGATCACTAACGTTACCGAGT-3′), GO-rev1 (5′-TGGGACACTCCACGTCCTTAGTCTAGACTAG-TA-3′), MS-fow (5′-ACACCGGTCGCTGGGAATGTATTCT-GAATCGGCA-3′), MS-rev1 (5′-CACATAGGCATACATCAT-CCCAGGTGAGTCGACGTT-3′), KatE-fow1 (5′-ACACCG-GTCGCAACATAACGAAAAGAACCCA-3′), KatE-rev1 (5′-ACGTCGACTCAGGCAGGAATTTTGTCAATCT-3′), PGM-fow (5′-TAGGTACCCAATCAACAATGACGTCGACCTAC-3′), PGM-rev (5′-GAGATTAAATCGTTGCCGACGAAGCA-ATTGTA-3′), rbcS3C-fow (5′-ACGAGCTCATCCAGAAT-TGGCGTTGGATTA-3′) and rbcS3C-rev (5′-AGCAACGGT-GGAAGAGTCAGTTGCAACCGGTAT-3′).
7. Platinum Pfx DNA polymerase PCR kit (Invitrogen).
8. Restriction enzymes: *Age*I, *Kpn*I, *Mun*I, *Sac*I, and *Sal*I.
9. T4 DNA ligase.
10. Gel extraction kit: QIAprep Spin Miniprep Kit (Qiagen).
11. Plasmid miniprep kit: QIAprep Spin Miniprep Kit (Qiagen).
12. Thermal cycler.
13. Temperature-controlled water bath.
14. Liquid LB medium: 10 g/L tryptone, 5 g/L yeast extract, 10 g/L NaCl.
15. Solid LB medium in Petri dishes, 10 g/L tryptone, 5 g/L yeast extract, 10 g/L NaCl, 15 g/L agar.
16. Antibiotics: kanamycin.

17. Temperature-controlled growth cabinet at 37 °C.
18. Agarose gel electrophoresis chambers, power supply, agarose, and TAE (Tris-acetate-EDTA) buffer.
19. Competent *E. coli*.

2.1.2 Generation of Plant Binary Vectors

1. Modified pGreenII plant binary vectors [9]: pGreenII 35S-Kan vector (modified pGreenII plasmid carrying the CaMV35S promoter and the kanamycin resistance gene); pGreenII 35S-Hyg vector (modified pGreenII plasmid carrying the CaMV35S promoter and the hygromycin resistance gene); and pGreenII 35S-BASTA vector (modified pGreenII plasmid carrying the CaMV35S promoter and the phosphinothricin resistance gene).
2. Restriction enzymes: *Kpn*I, *Sac*I, *Sal*I, and *Spe*I.
3. T4 DNA ligase.
4. Gel extraction kit: QIAquick Gel Extraction Kit (Qiagen).
5. Plasmid miniprep kit: QIAprep Spin Miniprep Kit (Qiagen).
6. Temperature-controlled growth cabinet at 37 °C.
7. Agarose gel electrophoresis chambers, power supply, agarose, and buffer.
8. Competent *E. coli* cells.
9. Liquid LB medium: 10 g/L tryptone, 5 g/L yeast extract, 10 g/L NaCl.
10. Solid LB medium in Petri dishes, 10 g/L tryptone, 5 g/L yeast extract, 10 g/L NaCl, 15 g/L agar.
11. Antibiotics and selection compounds: kanamycin, hygromycin, and BASTA.

2.1.3 Generation of Primary Transgenic Lines

1. YEB medium: Nutrient Broth or Beef Extract 5 g/L, Yeast Extract 1 g/L, Peptone 5 g/L, Sucrose 5 g/L, MgSO4 2 mM, pH 7.4.
2. Silwet L-77.
3. Electroporator.
4. Spectrophotometer.

2.1.4 Propagation and Selection of Transgenes

1. MS Medium Murashige and Skoog Basal medium (Duchefa).
2. Antibiotics and selection compounds: rifampicin, kanamycin, carbenicillin, sulfadiazine, phosphinothricin.
3. Sucrose.
4. Earth growth substrate: three parts soil (Gebr. Patzer), one part vermiculite (Basalt Feuerfest).
5. Temperature-controlled, illuminated, plant growth chamber.
6. Ethanol.

2.1.5 Confirmation of Transgene Expression by PCR

1. Leaf material from transgenic plants.
2. TRIzol reagent kit (Gibco-BRL).
3. SuperScript II reverse transcriptase kit (Invitrogen).
4. PCR kit with Taq DNA polymerase (Qiagen).
5. Primers 10 pmol/μl stock solutions in water: PGM-fow (5′-TAGGTACCCAATCAACAATGACGTCGACCTAC-3′), GO-rev1 (5′- TGGGACACTCCACGTCCTTAGTCTAG-ACTAGTA-3′), rbcS3C-fow2 (5′-ATGGCTTCTTCAGTAA-TGTCCTCAGCAGCT-3′), MS-rev1 (5′- CACATAG-GCATACATCATCCCAGGTGAGTCGACGTT-3′), KatE-rev1 (5′-ACGTCGACTCAGGCAGGAATTTTGTCAA TCT-3′), Actin2S-for (5′-TGTACGCCAGTGGTCCTACAACC-3′) and Actin2B-rev (5′-GAAGCAAGAATGGAACCACCG-3′).
6. Thermal cycler.
7. Agarose gel electrophoresis chambers, power supply, agarose, and TAE (Tris-acetate-EDTA) buffer.

2.1.6 Enzyme Activity Confirmation

Measurement of Glycolate Oxidase in Protein Extracts of Isolated Chloroplasts

1. Chloroplast isolation buffer: 50 mM HEPES/KOH (pH 7.5), 300 mM Mannitol, 1 mM EDTA, 1 mM $MgCl_2$, 0.1% (w/v) BSA, 0.02% (w/v) Na-ascorbate, 0.5% (w/v) polyvinyl-pyrrolidone.
2. SH-buffer: 50 mM HEPES/KOH (pH 7.5), 333 mM mannitol.
3. Protein extraction buffer: 100 mM HEPES/KOH (pH 7.2), 1 mM EDTA, and 10 mM 2-mercaptoethanol.
4. Glycolate oxidase assay buffer: 100 mM triethanolamine (pH 7.8), 3 mM EDTA, 0.75 mM oxidized glutathione, and 4 mM phenylhydrazine (modified from [10]).
5. Sodium glycolate solution in H_2O.
6. Centrifuge.
7. Spectrophotometer with selectable wavelength.
8. Liquid nitrogen.
9. Mortar and pestle.
10. Miracloth.
11. Percoll.
12. Quartz cuvette.

Measurement of Malate Synthase in Protein Extracts of Isolated Chloroplasts

1. Chloroplast isolation buffer: 50 mM HEPES/KOH (pH 7.5), 300 mM Mannitol, 1 mM EDTA, 1 mM $MgCl_2$, 0.1% (w/v) BSA, 0.02% (w/v) Na-ascorbate, 0.5% (w/v) polyvinylpyrrolidone.
2. SH-buffer: 50 mM HEPES/KOH (pH 7.5), 333 mM mannitol.

3. Protein extraction buffer: 50 mM Tris–HCl (pH 8.0) and 1 mM $MgCl_2$.
4. Malate synthase assay buffer: 50 mM Tris–HCl (pH 8.0), 5 mM $MgCl_2$, 3 mM acetyl-CoA, and 1.6 mM 5,5-dithiobis 2-nitrobenzoic acid (modified from [11]).
5. Glyoxylate solution in H_2O, pH-adjusted to ~6.0 with NaOH.
6. Centrifuge.
7. Spectrophotometer with selectable wavelength.
8. Liquid nitrogen.
9. Mortar and pestle.
10. Miracloth.
11. Percoll.
12. Quartz cuvettes.

Measurement of Catalase in Protein Extracts of Isolated Chloroplasts

1. Chloroplast isolation buffer: 50 mM HEPES/KOH (pH 7.5), 300 mM Mannitol, 1 mM EDTA, 1 mM $MgCl_2$, 0.1% (w/v) BSA, 0.02% (w/v) Na-ascorbate, 0.5% (w/v) polyvinyl-pyrrolidone.
2. SH-buffer: 50 mM HEPES/KOH (pH 7.5), 333 mM mannitol.
3. Protein extraction buffer: 50 mM KH_2PO_4 (pH 7.0), 1% (v/v) polyvinylpyrrolidone (PVP)-40 and 0.1% (v/v) Triton X-100.
4. Catalase assay buffer: 50 mM KH_2PO_4 (pH 7.0).
5. H_2O_2 in water.
6. Centrifuge.
7. Spectrophotometer with selectable wavelength.
8. Liquid nitrogen.
9. Mortar and pestle.
10. Miracloth.
11. Percoll.
12. Quartz cuvette.

2.2 Bacterial Pathway: Expression of the Bacterial Glycolate Catabolic Pathway in *A. thaliana* Chloroplasts

2.2.1 Plasmid Constructs

1. Plant binary expression vectors containing the enhanced CaMV35S promoter and the chloroplast targeting peptide of the potato rbcS1 gene (Thomas Rademacher, Institute for Biology I, RWTH-Aachen): pTRA-K-rbcS1-cTP, containing the nptII cassette of pPCV002 for selection of transgenic plants on kanamycin; pTRA-Sul-rbcS1-cTP and pSuper-PAM-Sul-rbcS1-cTP carrying the sulfadiazine resistance cassette; and pTRA-Hyg-rbcS1-cTP carrying the hygromycin resistance cassette. pTRA-Hyg-rbcS1-cTP-PCK and pTRA-PT-rbcS1-cTP-gfp containing the phosphinothricin cassette for selection of

transgenic plants on BASTA (Teresa Sikora; Institute for Biology I; RWTH-Aachen).

2. Plasmids used as templates: pET-TSR and pET-GCL (Rafijul Bari, Institute for Biology I, RWTH-Aachen),
3. Primers 10 pmol/µl stock solutions in water: TSR-Mlu-fow (5′-TCGAACGCGTTAGGTGCATGAAACTGGGATT-TATTG-3′), T7-rev (5′-GCTAGTTATTGCTCAGCGG-3′), GLC-Mlu-fow (5′-TCGAACGCGTTAGGTGCATGGGGA-TGGCAAAAAT-3′), GlcD-Asc-fow (5′-TCGAGGCGCGC-CAGGTGCATGAGCATCTTGTACGAAGAG-3′) and GlcD-Xba-rev (5′-TCGATCTAGATCAGAAACGCTCCAGTTCA-GGG-3′), GlcE-Asc-fow (5′-TCGAGGCGCGCCAGGTG-CATGCTACGCGAGTGTGATTAC-3′), GlcE-Xba-rev (5′-TCGATCTAGATCAAAGTTCCGCGTACATGCGAC-3′), GlcF-Asc-fow (5′-TCGAGGCGCGCCAGGTGCATGCAA-ACCCAATTAACTGAAG-3′), GlcF-Xba-rev (5′-TCGATC-TAGATTATTCCTTTTCAAGGGC-3′).
4. Restriction enzymes: *Asc*I, *Bpu*1102I, *Ecl*136II, *Eco*RI, *Mlu*I, *Pme*I, *Sgr*AI, *Xba*I.
5. Taq-DNA polymerase.
6. T4-DNA-Ligase.
7. QIAquick Gel Extraction Kit (Qiagen).

2.2.2 Generation and Selection of Primary Transgenic Lines

1. MS Medium Murashige and Skoog Basal medium (Duchefa).
2. YEB medium: Nutrient Broth or Beef Extract 5 g/L, Yeast Extract 1 g/L, Peptone 5 g/L, Sucrose 5 g/L, MgSO4 2 mM, pH 7.4.
3. Antibiotics and selection compounds: rifampicin, kanamycin, carbenicillin, sulfadiazine, phosphinothricin.
4. Sucrose.
5. Silwet L-77.
6. Ethanol.
7. Electroporator.

2.2.3 Multiplex PCR

1. Thermal cycler.
2. 0.2 ml PCR reaction tubes.
3. FastStart High Fidelity enzyme (Roche Applied Science, Mannheim, Germany).
4. dNTP mix: dATP, dCTP, dGTP, and dTTP each 2 mM.
5. Multiplex PCR buffer D, 10×: Tris–HCl, pH 8.3 100 mM, KCl 500 mM, $MgCl_2$, 25 mM.
6. DMSO 100%.

7. Primers, 10 pmol/μl stock solutions in water: GL-FW1 (5′-CGCACCGACTGAAACCTGCTTC-3′), TSR-FW1 (5′-CTCAACCTGGCACTGCAAAGTGCG-3′), GlcD-FW1 (5′-GGCGGGCTGGAGATGATGGATAAC-3′), GlcE-FW1 (5′-TCGGAAGCTACGGTTGTCTTGGCG-3′), and GlcF-FW1 (5′-GCCGAT AAAGCACGTCAGGTCAGT-3′), and GT-DEF-Rev1 (5′-GCTCAACACATGAGCGAAACC-3′).

3 Methods

3.1 GMK Pathway: Co-expression of Glycolate Oxidase, Malate Synthase, and Catalase in A. thaliana Chloroplasts

3.1.1 Plasmid Constructs

Cloning of Phosphoglucomutase (PGM) Transit Peptide

1. Isolate *A. thaliana* RNA from 100 mg of leaves using the TRIzol reagent.
2. Convert RNA to first-strand cDNA using SuperScript II reverse transcriptase.
3. Amplify the 225 bp which represent the transit peptide from the *A. thaliana* phosphoglucomutase gene (At5g51820) in a PCR reaction with the PGM-fow and PGM-rev primers, Platinum Pfx DNA polymerase and using *A. thaliana* first-strand cDNA as template. The oligonucleotides were designed to introduce unique *Kpn*I and *Mun*I restriction sites at the 5′ and 3′ ends of the PCR fragment.
4. Run the PCR product on an agarose gel.
5. Cut out the PCR product from the gel on a UV table and isolate DNA using the gel extraction kit.
6. Store DNA at −20 °C for later use.

Cloning of Rubisco Small Subunit Promoter and Transit Peptide

1. Amplify the promoter (715 bp) and transit peptide (172 bp) from the tomato Rubisco small subunit (X66072) in a PCR reaction with the rbcS3C-fow and rbcS3C-rev primers, Platinum Pfx DNA polymerase and using tomato genomic DNA as template. The oligonucleotides were designed to introduce unique *Sac*I and *Age*I restriction sites at the 5′ and 3′ ends of the PCR fragment.
2. Run the PCR product on an agarose gel.
3. Cut out the PCR product from the gel on a UV table and isolate DNA using the gel extraction kit.
4. Store DNA at −20 °C for later use.

Cloning of Glycolate Oxidase (GC) cDNA

1. Isolate *A. thaliana* RNA from 100 mg of leaves using the TRIzol reagent.
2. Convert RNA to first-strand cDNA using SuperScript II reverse transcriptase.

   Amplify the cDNA corresponding to GO from *A. thaliana* (At3g14420) in a PCR reaction with the GO-fow and GO-rev1 primers and using the Platinum Pfx DNA polymerase PCR kit. The oligonucleotides were designed to introduce unique *Mun*I and *Spe*I restriction sites at the 5′ and 3′ ends of the PCR fragment (*see* **Note 1**).

3. Use the PCR product in a reaction with pCR-Blunt II-TOPO according to the manufacturer's instructions to create the GO-TOPO plasmid.
4. Transform competent *E. coli* cells with the GO-TOPO plasmid.
5. Select resistant colonies and grow *E. coli* cultures.
6. Isolate plasmid from the *E. coli* cultures using a miniprep kit.
7. Sequence the DNA inserted into the plasmid to ensure that no mutations are present.

Cloning of Malate Synthase (MS) cDNA

1. Isolate *C. maxima* RNA from 100 mg of leave using the TRIzol reagent.
2. Convert RNA to first-strand cDNA using SuperScript II reverse transcriptase.

   Amplify the cDNA corresponding to MS from *C. maxima* cotyledons (X56948) in a PCR reaction with the MS-fow and MS-rev1 primers and using the Platinum Pfx DNA polymerase PCR kit. The oligonucleotides were designed to introduce unique *Age*I and *Sal*I restriction sites at the 5′ and 3′ ends of the PCR fragment (*see* **Note 1**).
3. Use the PCR product in a reaction with pCR-Blunt II-TOPO according to the manufacturer's instructions to create the MS-TOPO plasmid.
4. Transform competent *E. coli* cells with the MS-TOPO plasmid.
5. Select resistant colonies and grow *E. coli* cultures.
6. Isolate plasmid from the *E. coli* cultures using the miniprep kit.
7. Sequence the DNA inserted into the plasmid to ensure that no mutations are present.

Cloning of Catalase (KatE) Gene

1. Amplify the cDNA corresponding to KatE from *E. coli* (M55161) using *E. coli* DNA in a PCR reaction with the KatE-fow1 and KatE-rev1 primers and using the Platinum Pfx DNA polymerase PCR kit. The oligonucleotides were designed to introduce unique *Age*I and *Sal*I restriction sites at the 5′ and 3′ ends of the PCR fragment.
2. Use the PCR product in a reaction with pCR-Blunt II-TOPO according to the manufacturer's instructions to create the KatE-TOPO plasmid.
3. Transform competent *E. coli* cells with the KatE-TOPO plasmid.

4. Select resistant colonies and grow *E. coli* cultures.
5. Isolate plasmid from the *E. coli* cultures using the miniprep kit.
6. Sequence the DNA inserted into the plasmid to ensure that no mutations are present.

3.1.2 Generation of Binary Vectors

Construction of the 35S:GO Vector

1. Digest the GO-TOPO plasmid and the PGM transit peptide with *Kpn*I and *Mun*I.
2. Insert the transit peptide upstream of the GO cDNA through ligation to create the PGM-GO-TOPO plasmid.
3. Repeat Subheading "Cloning of Glycolate Oxidase (GO) cDNA", **steps 4–7** with the newly ligated plasmid.
4. Digest the pGreenII 35S-Kan vector and the PGM-GO-TOPO plasmid with *Kpn*I and *Spe*I and subsequently ligate the fragments to create the 35S:GO binary vector [4].
5. Repeat Subheading "Cloning of Glycolate Oxidase (GO) cDNA", **steps 4–7** with the 35S:GO binary vector.

Construction of the rbcS3C:MS Vector

1. Digest the MS-TOPO plasmid and the Rubisco small subunit promoter and transit peptide fragment with *Sac*I and *Age*I.
2. Insert the fragment upstream of the MS cDNA in the MS-TOPO plasmid through ligation to create rbcS3C-MS-TOPO plasmid.
3. Digest the pGreenII 35S-Hyg vector and the rbcS3C-MS-TOPO plasmid with *Sac*I and *Sal*I and subsequently ligate the fragments to create the rbcS3C:MS binary vector [4].
4. Repeat Subheading "Cloning of Glycolate Oxidase (GO) cDNA", **steps 4–7** with the rbcS3C:MS binary vector.

Construction of the rbcS3C:KatE Vector

1. Digest the KatE-TOPO plasmid and the Rubisco small subunit promoter and transit peptide with *Sac*I and *Age*I.
2. Insert the Rubisco small subunit promoter and transit peptide upstream of the KatE gene through ligation to create rbcS3C-KatE-TOPO plasmid.
3. Digest the pGreenII 35S-BASTA vector and the rbcS3C-KatE-TOPO plasmid with *Sac*I and *Sal*I and subsequently ligate the fragments to create the rbcS3C:KatE binary vector [2].
4. Repeat Subheading "Cloning of Glycolate Oxidase (GO) cDNA", **steps 4–7** with the rbcS3C:KatE binary vector.

3.1.3 Generation of Transgenic Lines

Preparation of *A. tumefaciens*

1. Transform *A. tumefaciens* GV3101 carrying the pSoup helper plasmid (Rif^r, Tet^r) by electroporation with the binary vectors produced in Subheading 3.1.2.
2. Plate transformed *A. tumefaciens* on solid YEB medium containing 100 μg/ml rifampicin, 10 μg/ml tetracycline, and

either of 25 μg/ml kanamycin, 25 μg/ml hygromycin, or 10 μg/ml phosphinothricin for 35S:GO, rbcS3C:MS, and rbcS3C:KatE, respectively. NOTE without pSoup helper plasmid the pGreen vector will not replicate.

3. Prepare a pre-culture by inoculating a single colony of *A. tumefaciens* carrying a plasmid into 5 ml YEB medium containing 100 μg/ml rifampicin, 10 μg/ml tetracycline, and either of 25 μg/ml kanamycin, 25 μg/ml hygromycin, or 10 μg/ml phosphinothricin, for 35S:GO, rbcS3C:MS, and rbcS3C:KatE, respectively.
4. Incubate at 28 °C for 2 days with shaking at 150 rpm.
5. Inoculate the pre-culture into 200 ml YEB medium containing the same antibiotics as the pre-culture.
6. Incubate for further 2 days at 28 °C with shaking at 200 rpm.
7. Spin down cells at 5000 × *g* at 4 °C for 20 min.
8. Resuspend the pellet in 5% sucrose till OD_{600} reaches 0.8.
9. Add 0.04% Silwet L-77 (400 μl/1000 ml).
10. Use the cell suspension directly for floral dip transformation of *A. thaliana*.

Floral Dip Transformation of A. thaliana

1. Grow *A. thaliana* plants ecotype Columbia-0 at 8 h light, 16 h dark at 20 °C and a flux density of 100 mmol/m^2/s for 4–6 weeks.
2. Transfer the plants to 16 h light and 8 h dark at 23–25 °C to enhance flower production.
3. Clip the first bolts to induce the proliferation of many secondary inflorescences.
4. Dip the reproductive tissues of the plants in the Agrobacterium solution three times for 3–10 min each [5].
5. Place the plants under a dome or cover for 16–24 h in order to maintain high humidity. Transfer the plants to 16 h light and 8 h dark at 23–25 °C and a flux density of 100 mmol/m^2/s.
6. Tie the loose bolts with wax paper.
7. Stop watering the plants after seed maturation.
8. Harvest the dry seeds for further selection.

3.1.4 Selection and Propagation of the Transgenic Lines

Generation of Plants Expressing GO (GO-Lines, Fig. 1)

1. Sterilize seeds harvested from plants transformed with 35S:GO by submerging in 70% (v/v) ethanol for 15 min.
2. Wash seeds three times with 98% (v/v) ethanol.
3. Dry seeds on a sterile Whatman paper.
4. Sow seeds on MS-medium containing 50 μg/ml kanamycin and keep in darkness for 4 days at 4 °C to synchronize germination.

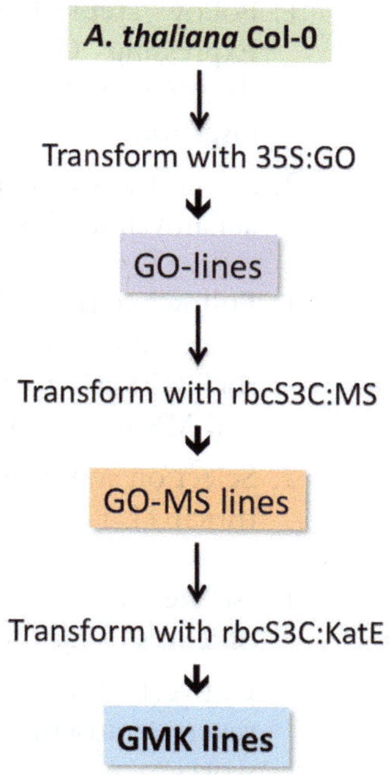

Fig. 1 Simplified representation of the generation of GMK lines as in [2]. GO-lines = plants expressing glycolate oxidase in chloroplasts; GO-MS lines = plants expressing GO and MS in chloroplasts; GMK lines = plants expressing GO, MS, and KatE in chloroplasts

5. Grow the selected plants at 8 h light, 16 h dark at 20 °C and a flux density of 100 mmol/m^2/s for 10–15 days.
6. After 14 days of growth, transfer surviving seedlings to soil.
7. Grow plants for a further 10–20 days in the same conditions.
8. Confirm genomic insertions of transgene by PCR on genomic DNA (Subheading 3.1.5) and confirm activity of transgenes in enzymatic assays (Subheading 3.1.6).
9. Transfer the plants to 16 h light and 8 h dark at 23–25 °C and a flux density of 100 mmol/m^2/s to enhance flowering and seed production.
10. Allow plants expressing transgenes to self-pollinate.
11. Allow plants to set seed and harvest them, keeping the seeds from each plant separate.
12. Repeat **steps 1–11** for two more rounds.
13. Any plant yielding offspring which all have the insertion is homozygous for it.

Generation of Plants Expressing GO and MS (GO-MS Lines, Fig. 1)

1. Follow **steps 1–13** in Subheading "Generation of Plants Expressing GO (GO-Lines, Fig. 1)" using the plasmid rbcS3C:MS to transform homozygous lines for the GO-insertion.
2. Select transformants on MS-medium containing 50 μg/ml kanamycin and 30 mg/ml hygromycin.

Generation of Plants Expressing GO, MS, and KatE (GMK Lines, Fig. 1)

1. Follow **steps 4–13** in Subheading "Generation of Plants Expressing GO (GO-Lines, Fig. 1)" using the plasmid rbcS3C:KatE to transform homozygous plants for the GO- and MS-insertions.
2. Select on soil irrigated with water containing 1 ml phosphinothricin/L H_2O.

3.1.5 Confirmation of Transgenes Expression

1. Isolate RNA from the transgenic plants by extracting 100 mg of leaves using the TRIzol reagent.
2. Convert RNA to first-strand cDNA using SuperScript II reverse transcriptase. NOTE make sure the DNase digestion works.
3. Amplify the full-length genes using the generated first-strand cDNA as template and the Taq DNA polymerase PCR kit. Primers PGM-fow and GO-rev1 for GO (1.3 kb PCR product), rbcS3C-fow2 and MS-rev1 for MS (1.8 kb PCR product), and rbcS3C-fow2 and KatE-rev1 for KatE (2.4 kb PCR product). As a control, amplify the actin2 gene using primers Actin2S-for and Actin2B-rev.
4. Run the PCR product on an agarose gel and verify that the expected band sizes are present.

3.1.6 Enzyme Activities in Extracts of Isolated Chloroplasts

Measurement of Glycolate Oxidase in Protein Extracts of Isolated Chloroplasts

1. Collect 5 g of leaf material from selected transgenic plants.
2. Freeze material in liquid nitrogen and quickly grind with mortar and pestle.
3. Add 50 ml of chloroplast isolation buffer.
4. Filter solution through three layers of Miracloth (cheese cloth) to remove large debris.
5. Centrifuge extracts at 1000 × *g* for 10 min at 4 °C to pellet chloroplasts.
6. Remove supernatant and resuspend pellet in 1 ml SH-buffer.
7. Add 0.5 ml of the extract to 1 ml 35% Percoll gradient (35% Percoll, 65% SH-buffer).
8. Centrifuge at 500 × *g* for 5 min at 4 °C to pellet chloroplasts.
9. Wash pellet in 1 ml of SH-buffer.
10. Centrifuge at 1000 × *g* for 10 min at 4 °C to pellet chloroplasts.

11. Add 1 ml glycolate oxidase extraction buffer to pellet to extract enzymes.
12. Spin extracts at 20,000 × *g* for 5 min at 4 °C to clarify extract.
13. Transfer the supernatant to a new tube without disturbing pellet.
14. Store extract on ice.
15. Perform enzyme assays in 1 ml reactions, in glycolate oxidase assay buffer and using 50 µl enzyme extract.
16. Place cuvette in spectrophotometer (with a temperature of 25 °C) and record readings at 320 nm for 4 min to establish a baseline.
17. Add sodium glycolate to a concentration of 2.3 mM in the reaction and mix by covering cuvette with parafilm and inverting.
18. Replace cuvette in spectrophotometer and record readings at 320 nm for 10 min.
19. Calculate enzyme activity using the extinction coefficient for phenylhydrazone at 324 nm of 16.8 mM^{-1} cm^{-1}.

Measurement of Malate Synthase in Protein Extracts of solated Chloroplasts

1. Perform **steps 1–11** in Subheading "Measurement of Glycolate Oxidase in Protein Extracts of Isolated Chloroplasts".
2. Add 1 ml malate synthase extraction buffer to pellet to extract enzymes.
3. Spin extracts at 20,000 × *g* for 5 min at 4 °C to clarify extract.
4. Transfer the supernatant to a new tube without disturbing pellet.
5. Store extract on ice.
6. Perform enzyme assays in 1 ml reactions, in malate synthase assay buffer and using 50 µl enzyme extract.
7. Place cuvette in spectrophotometer (with a temperature of 25 °C) and record readings at 320 nm for 4 min to establish a baseline.
8. Add glyoxylate solution to a concentration of 4 mM in the reaction and mix by covering cuvette with parafilm and inverting.
9. Replace cuvette in spectrophotometer and record readings at 410 nm for 10 min.
10. Calculate enzyme activity using the extinction coefficient for 4-nitrothiolate at 410 nm of 13.7 M^{-1} cm^{-1}.

Measurement of Catalase in Protein Extracts of Isolated Chloroplasts

1. Prepare materials listed in **steps 1–11** in Subheading "Measurement of Glycolate Oxidase in Protein Extracts of Isolated Chloroplasts."
2. Add 1 ml catalase extraction buffer to pellet to extract enzymes.
3. Spin extracts at 20,000 × *g* for 5 min at 4 °C to clarify extract.
4. Transfer the supernatant to a new tube without disturbing pellet.
5. Store extract on ice.
6. Perform enzyme assays in 1 ml reactions, in catalase assay buffer, and using 50 μl enzyme extract.
7. Place cuvette in spectrophotometer (with a temperature of 25 °C) and record readings at 240 nm for 4 min to establish a baseline.
8. Add H_2O_2 solution to a concentration of 10 mM in the reaction and mix by covering cuvette with parafilm and inverting.
9. Replace cuvette in spectrophotometer and record readings at 410 nm for 10 min.
10. Calculate enzyme activity using the extinction coefficient for H_2O_2 at 240 nm of 43.6 M^{-1} cm^{-1}.
11. Normalize catalase activity measurements to the glyceraldehyde 3-phosphate dehydrogenase activity, determined in chloroplast extracts as previously described [12].

3.2 Bacterial Pathway: Expression of the Bacterial Glycolate Catabolic Pathway in *A. thaliana* Chloroplasts [3]

3.2.1 Plasmid Constructs

Construction of the Plasmid pTRA-K-rbcS1-cTP-TSR-His, GCL-His

1. Amplify by PCR the coding sequences of tartronate semialdehyde reductase (TSR) using the primers TSR-Mlu-fow and T7-rev and the plasmid pET-TSR as template (*see* **Note 2**).
2. Amplify by PCR the coding sequences of glyoxylate carboligase (GCL) using the primers GCL-Mlu I-fow and T7-rev and the plasmid pET-GCL as template.
3. Isolate the PCR products with QIAquick Gel Extraction Kit (Qiagen).
4. Digest the TSR-His and GLC-His PCR products with *Mlu*I and *Bpu*1102I at 37 °C.
5. Digest pTRA-K-rbcS1-cTP with *Mlu*I and *Ecl*136II at 37 °C.
6. Clone TSR-His and GLC-His into pTRA-K-rbcS1-cTP to produce the plasmids pTRA-K-rbcS1-cTP-TSR-His and pTRA-K-rbcS1-cTP-GCL-His.
7. Digest pTRA-K-rbcS1-cTP-GCL-His with *Pme*I and *Bpu*1102I at 37 °C and isolate the GCL-His insert.
8. Digest pTRA-K-rbcS1-cTP-TSR-His with *Asc*I at 37 °C, fill with Klenow fragment, digest with *Bpu*1102I at 37 °C, and isolate the pTRA-K-rbcS1-cTP-TSR-His vector backbone.

9. Ligate the GCL-His insert with pTRA-K-rbcS1-cTP-TSR-His to produce the vector pTRA-K-rbcS1-cTP-TSR-His,GCL-His (*see* **Note 3**).

Construction of the Plasmids pTRA-Hyg-rbcS1-cTP-glcD,glcE and pSuper-PAM-Sul-rbcS1-cTPGlcD, GlcE

1. Amplify the coding sequences of glcD and glcE subunit of *E. coli* GlcDH using the primer pairs GlcD-Asc-fow and GlcD-Xba-rev and GlcE-Asc-fow and GlcE-Xba-rev, respectively, and bacterial genomic DNA as template.
2. Digest the glcD and glcE PCR products with *Asc*I and *Xba*I at 37 °C.
3. Digest pTRA-K-rbcS1-cTP with *Mlu*I and *Xba*I at 37 °C.
4. Clone glcD and glcE into pTRA-K-rbcS1-cTP to produce the plasmids pTRA-K-rbcS1-cTP-glcD and pTRA-K-rbcS1-cTP-glcE.
5. Digest pTRA-K-rbcS1-cTP-glcE with *Asc*I and *Pme*I at 37 °C.
6. Isolate the glcE insert.
7. Digest pTRA-Hyg-rbcS1-cTP-PCK with *Asc*I and *Pme*I at 37 °C.
8. Clone glcE into pTRA-Hyg-rbcS1-cTP-PCK to produce the plasmids pTRA-Hyg-rbcS1-cTP-glcE.
9. Digest pTRA-K-rbcS1-cTP-glcD with *Asc*I and *Sgr*AI at 37 °C and isolate the glcD insert.
10. Digest pTRA-Hyg-rbcS1-cTP-glcE with *Pme*I and *Sgr*AI at 37 °C and isolate the pTRA-Hyg-rbcS1-cTP-glcE vector backbone.
11. Ligate glcD with pTRA-Hyg-rbcS1-cTP-glcE to produce the vector pTRA-Hyg-rbcS1-cTP-glcD,glcE (*see* **Note 4**).
12. Digest pTRA-hyg-rbcS1-cTP-GlcD, glcE and pSuper-PAM-SulrbcS1-cTP with *Asc*I and *Pme*I.
13. Isolate the glcD,glcE insert and the vector backbone.
14. Ligate glcD,glcE with pSuper-PAM-SulrbcS1-cTP to produce the vector pSuper-PAM-Sul-rbcS1-cTP-glcD,glcE.

Construction of the Plasmid pTRA-PT-rbcS1-cTP-glcF

1. Amplify the coding sequences of glcF subunit of *E. coli* GlcDH using the primers GlcF-Asc-fow and GlcF-Xba-rev and bacterial genomic DNA as template.
2. Digest glcF PCR product with *Asc*I and *Xba*I.
3. Digest pTRA-K-rbcS1-cTP vector with *Mlu*I and *Xba*I.
4. Ligate glcF with pTRA-K-rbcS1-cTP to produce the vector pTRA-K-rbcS1-cTP-glcF.
5. Digest pTRA-K-rbcS1-cTP-glcF with *Eco*RI and *Xba*I.
6. Isolate the glcF insert.

7. Digest pTRA-PT-rbcS1-cTP-gfp with *Eco*RI and *Xba*I.
8. Isolate the vector backbone.
9. Ligate glcF with pTRA-PT-rbcS1-cTP to produce the vector pTRA-PT-rbcS1-cTP-glcF.

3.2.2 Generation and Selection of Primary Transgenic Lines

Preparation of *Agrobacterium tumefaciens*

1. Transform by electroporation *A. tumefaciens* GV3101 (pMP90RK): Gm^r, Km^r, Rif^r [13] with the binary vectors produced.
2. Prepare a pre-culture by inoculating a single colony of *A. tumefaciens* carrying a plasmid into 5 ml YEB medium containing 100 μg/ml rifampicin, 25 μg/ml kanamycin, 50 μg/ml carbenicillin.
3. Incubate at 28 °C for 2 days with shaking at 150 × *g*.
4. Inoculate the pre-culture into 200 ml YEB medium containing 100 μg/ml rifampicin, 25 μg/ml kanamycin, 50 μg/ml carbenicillin.
5. Incubate for further 2 days at 28 °C with shaking at 200 rpm.
6. Spin down cells at 5000 × *g* at 4 °C for 20 min.
7. Resuspend the pellet in 5% sucrose till OD_{600} reaches 0.8.
8. Add 0.04% Silwet L-77 (400 μl/1000 ml).
9. Use the cell suspension directly for floral dip transformation of *A. thaliana*.

Floral Dip Transformation of *A. thaliana*

1. Grow *A. thaliana* plants ecotype Columbia-0 at 8 h light, 16 h dark at 20 °C and a flux density of 100 mmol/m^2/s.
2. Transfer the plants to 16 h light and 8 h dark at 23–25 °C to enhance flower production.
3. Clip the first bolts to induce the proliferation of many secondary inflorescences.
4. Dip the reproductive tissues of the plants in the Agrobacterium solution three times for 3–10 min each [14].
5. Place the plants under a dome or cover for 16–24 h in order to maintain high humidity. Transfer the plants to 16 h light and 8 h dark at 23–25 °C and a flux density of 100 mmol/m^2/s.
6. Tie the loosed bolts with wax paper.
7. Stop watering the plants after seed maturation.
8. Harvest the dry seeds for further selection.

Selection of Primary Transgenic Plants

1. Sterilize the harvested seeds by submerging in 70% (v/v) ethanol for 15 min.
2. Wash three times with 98% (v/v) ethanol.
3. Dry seeds on a sterile Whatman paper.

4. Sow seeds of plants transformed with pTRA-K-rbcS1-cTP-TSR-His,GCL-His on MS-medium containing 50 μg/ml kanamycin.
5. Sow seeds of plants transformed with pSuper-PAM-Sul-rbcS1-cTPGlcD,GlcE on MS-medium containing 20 μg/ml sulfadiazine.
6. Sow seeds of plants transformed with pTRA-PT-rbcS1-cTP-glcF on soil and irrigate with water containing 1 ml phosphinothricin/L H_2O.
7. Grow the selected plants (GT, DE, and F plants) at 8 h light, 16 h dark at 20 °C and a flux density of 100 mmol/m^2/s for 10–15 days.
8. Grow plants that survive for further 10–20 days in the same conditions.
9. Transfer the plants to 16 h light and 8 h dark at 23–25 °C and a flux density of 100 mmol/m^2/s to enhance flowering and seed production.
10. Analyze individual selected GT, DE, and F plants for gene expression by Real-Time RT-PCR.
11. Propagate segregants of high-expressing plants by antibiotic selection as described above to obtain homozygous lines.

3.2.3 Production of Plants Expressing the Whole Pathway

Production of Plants Expressing glcD, glcE, and glcF (DEF Lines, Fig. 2)

1. Transform T3 or T4 homozygous and highly expressing F plants with pTRA-Hyg-rbcS1-cTP-glcD,glcE and pSuper-PAM-Sul-rbcS1-cTPglcD,glcE as described in Subheading "Floral Dip Transformation of *A. thaliana*."
2. Select individual T1 plants on phosphinothricin and sulfadiazine as described in Subheading "Selection of Primary Transgenic Plants."
3. Analyze selected DEF plants for gene expression by Real-Time RT-PCR (*see* **Note 5**).
4. Select high-expressing DEF plants for genetic crosses.

Production of Plants Expressing GCL, TSR, glcD, glcE, and glcF (GT-DEF Lines, Fig. 2)

1. Cross T3 segregants of high-expressing GT plants with selected high-expressing DEF plants.
2. Select individual T1 plants by multiplex PCR (*see* Subheading 3.2.4).
3. Analyze selected GT-DEF plants for gene expression by Real-Time RT-PCR (*see* **Note 5**).

3.2.4 Confirmation of Transgenic Plants by Multiplex PCR

1. Extract genomic DNA from GT-DEF segregants.
2. Perform PCR with 50–100 ng genomic DNA, 2.5 units of FastStart High Fidelity, 0.5 μl of the forward primers, and

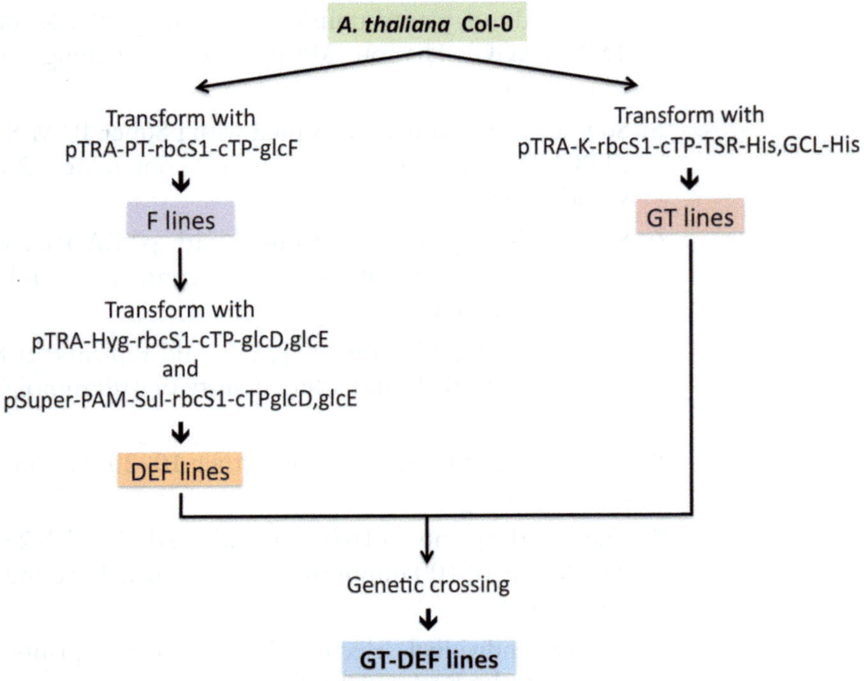

Fig. 2 Simplified representation of the generation of GT-DEF lines as in [3]. GT lines = plants expressing GCL and TSR in chloroplasts; F lines = plants expressing glcF in chloroplasts; DEF lines = plants expressing glcD, glcE, and glcF in chloroplasts; GT-DEF lines = plants expressing GCL, TSR, and DEF in chloroplasts

2.5 μl of the reverse primer, 2.5 μl 10× multiplex PCR buffer D, 1 μl DMSO (4% in the reaction mixture), and H_2O_2 until 25 μl.

3. Perform PCR in a thermocycler as follows: initialization at 95 °C for 2 min, 35 cycles of (1) denaturation at 95 °C for 20 s, (2) annealing at 60 °C for 20 s, and (3) extension at 72 °C for 1 min, final termination at 72 °C for 5 min, hold at 4 °C forever.

4 Notes

1. The nucleotides encoding the last amino acids (ARL), which represent peroxisomal targeting signal, should be better omitted.
2. Bacterial genomic DNA can also be used as template.
3. GCL and TSR are separated by a scaffold attachment region (SAR) as a transcription separator.
4. Low number of transformants may be obtained.
5. DEF plants show enhanced growth compared to the wild-type and flat leaves with elongated petioles.

Acknowledgements

This work was supported by grants of the Deutsche Forschungsgemeinschaft, MA2379/4-1, FOR 1186, and EXC 1028 to V.G.M.

References

1. Maurino VG, Peterhansel C (2010) Photorespiration: current status and approaches for metabolic engineering. Curr Opin Plant Biol 13:249–256
2. Maier A, Fahnenstich H, von Caemmerer S, Engqvist MKM, Weber APM, Flugge UI, Maurino VG (2012) Transgenic introduction of a glycolate oxidative cycle into A-thaliana chloroplasts leads to growth improvement. Front Plant Sci 3:38
3. Kebeish R, Niessen M, Thiruveedhi K, Bari R, Hirsch HJ, Rosenkranz R, Stabler N, Schonfeld B, Kreuzaler F, Peterhansel C (2007) Chloroplastic photorespiratory bypass increases photosynthesis and biomass production in *Arabidopsis thaliana*. Nat Biotechnol 25:593–599
4. Fahnenstich H, Scarpeci TE, Valle EM, Flugge UI, Maurino VG (2008) Generation of hydrogen peroxide in chloroplasts of Arabidopsis overexpressing glycolate oxidase as an inducible system to study oxidative stress. Plant Physiol 148:719–729
5. Gibson DG, Young L, Chuang RY, Venter JC, Hutchison CA 3rd, Smith HO (2009) Enzymatic assembly of DNA molecules up to several hundred kilobases. Nat Methods 6:343–345
6. de Kok S, Stanton LH, Slaby T, Durot M, Holmes VF, Patel KG, Platt D, Shapland EB, Serber Z, Dean J, Newman JD, Chandran SS (2014) Rapid and reliable DNA assembly via ligase cycling reaction. ACS Synth Biol 3:97–106
7. Quan J, Tian J (2009) Circular polymerase extension cloning of complex gene libraries and pathways. PLoS One 4:e6441
8. Zhang Y, Werling U, Edelmann W (2012) SLiCE: a novel bacterial cell extract-based DNA cloning method. Nucleic Acids Res 40:e55
9. Fahnenstich H, Saigo M, Niessen M, Zanor MI, Andreo CS, Fernie AR, Drincovich MF, Flugge UI, Maurino VG (2007) Alteration of organic acid metabolism in Arabidopsis overexpressing the maize C(4)NADP-malic enzyme causes accelerated senescence during extended darkness. Plant Physiol 145:640–652
10. Yamaguchi K, Nishimura M (2000) Reduction to below threshold levels of glycolate oxidase activities in transgenic tobacco enhances photoinhibition during irradiation. Plant Cell Physiol 41:1397–1406
11. Smith CV, Huang CC, Miczak A, Russell DG, Sacchettini JC, Honer zu Bentrup K (2003) Biochemical and structural studies of malate synthase from Mycobacterium tuberculosis. J Biol Chem 278:1735–1743
12. Ferri G, Comerio G, Iadarola P, Zapponi MC, Speranza ML (1978) Subunit structure and activity of glyceraldehyde-3-phosphate dehydrogenase from spinach chloroplasts. Biochim Biophys Acta 522:19–31
13. Koncz C, Schell J (1986) The promoter of Tl-DNA gene 5 controls the tissue-specific expression of chimeric genes carried by a novel type of Agrobacterium binary vector. Mol Gen Genet 204:383–396
14. Clough SJ, Bent AF (1998) Floral dip: a simplified method for Agrobacterium-mediated transformation of *Arabidopsis thaliana*. Plant J 16:735–743

Chapter 11

$^{13}CO_2$ Labeling and Mass Spectral Analysis of Photorespiration

Leonardo Perez de Souza, Marek Szecówka, Alisdair R. Fernie, and Takayuki Tohge

Abstract

Photorespiratory metabolism is compartmented over the chloroplast, peroxisome, cytosol, and mitochondria, and due to its complex structure it is often the case that metabolite levels alone are not able to fully describe photorespiration. Metabolic fluxes represent a more meaningful biological description of metabolism, adding to metabolite levels and often revealing different aspects of the system such as the presence of inactive metabolic pools of photorespiratory intermediates. We describe here a protocol for the $^{13}CO_2$ feeding of *Arabidopsis* and tracing of ^{13}C enriched metabolites for metabolic fluxes estimation, which allows high throughput analysis of labeling pattern on different metabolites involved in photorespiration and downstream processes.

Key words Metabolic flux, Mass spectrometry, $^{13}CO_2$ labeling, Photorespiratory metabolism

1 Introduction

Photorespiration is an essential pathway for the detoxification of 2-phosphoglycolate (2PG) resulting from the oxygenation of Rubisco and recovery of its substrate 3-phosphoglycerate (3PGA); the whole process requires reactions in different subcellular compartments, namely the chloroplast, peroxisome, cytosol, and mitochondria [1]. Photorespiratory flux is considered to be one of the most important biological processes in terms of mass flow only behind photosynthesis [2, 3]. As documented in several recent works, photorespiratory flux was suggested to be strongly connected not only to photosynthesis but also to other important metabolic pathways like the TCA cycle, 1-C metabolism, redox signaling, the GABA shunt, and nitrogen assimilation [1–5]. Most of the works regarding photorespiratory metabolism relies on the metabolite profiling of different mutants and conditions, and draw conclusions from changes in total levels of metabolites. Whilst this is a common and powerful approach it has, by itself, the

Alisdair R. Fernie et al. (eds.), *Photorespiration: Methods and Protocols*, Methods in Molecular Biology, vol. 1653,
DOI 10.1007/978-1-4939-7225-8_11,

limitation of missing information regarding metabolite compartmentation and possible changes in metabolic fluxes, which are ultimately the most representative description of metabolism having a great potential to reveal regulatory aspects and possible interactions between different pathways [6].

Isotopic labeling techniques are an essential part of all these experiments [7] and have been an important step in the determination of metabolic fluxes for long time; nevertheless recent advances in analytical tools such as mass spectrometry had a significant impact on these approaches allowing for much more comprehensive experiments using stable isotopes [8]. In this context, the use of $^{13}CO_2$ feeding for labeling of photosynthetic tissue has been suggested as a standard for metabolomics analysis, and even though this approach imposes many challenges for experimental setup and data interpretation, it has a great potential for the analysis of the metabolic state of an organism under physiological conditions [9]. One of the critical differences of $^{13}CO_2$ labeling in comparison with the use of other substrates is that labeling cannot reach steady state, an important assumption for further interpretation of results and modeling of fluxes [7]. The recent development of modeling based in nonsteady state isotopic labeling strategies such as Kinetic flux profiling (KFP) [6] and Instationary metabolic flux analysis (INST-MFA) [10], in combination with the use of high throughput mass spectrometry for metabolite profiling allowed to fully explore the benefits of $^{13}CO_2$ labeling. We describe here a protocol for the $^{13}CO_2$ feeding of Arabidopsis followed by data treatment for the estimation of metabolic fluxes that is suitable for the analysis of most photorespiratory intermediates as well as many metabolites involved in related pathways. We introduce the procedures for $^{13}CO_2$ labeling and data treatment once data acquisition and mathematical modeling are going to be discussed in other chapters.

2 Materials

Use ultrapure water and high quality of gas and chemicals for the experiments. Transparent chamber should be able to tightly close in order to avoid dilution of $^{13}CO_2$.

2.1 Plant Cultivation and $^{13}CO_2$ Feeding

1. 6 cm diameter plant cultivation pots.
2. Soil.
3. Water/gas-permeable plastic foil.
4. Elastic bands.
5. Gas flow controllers calibrated for N_2, O_2, and CO_2 and smart control software (Brooks Instrument).
6. Nitrogen gas.

7. Oxygen gas.
8. $^{12}CO_2$ gas.
9. $^{13}CO_2$ (isotopic purity 99 atom percentage) gas.
10. PVC tube PVC Rauclair-E tubes (diameter/thickness 4/1).
11. Transparent labeling boxes (Magenta™ GA-7 plant culture box, Sigma, cat. no. V8505).
12. Gas washing bottle for humidifier.
13. Gas washing bottle for CO_2 trap.
14. Plastic funnel (diameter 65 mm, length 68 mm, stem diameter 15 mm).
15. Liquid nitrogen.
16. Water, deionized and filtered (0.2-μm filter), or distilled water ULC/MS grade.
17. Soda lime.

2.2 Data Evaluation

1. CORRECTOR 1.91 (http://www-en.mpimp-golm.mpg.de/03-research/researchGroups/01-dept1/Root_Metabolism/smp/CORRECTOR/index.html).

3 Methods

3.1 Plant Cultivation

The procedure presented in this chapter is optimized for *Arabidopsis thaliana* (Arabidopsis) under standard growth conditions. But, it can be adapted to other plant species in case those plants are of suitable size for the labeling chamber.

1. Cultivate Arabidopsis plants in 8 h day/16 h light/dark cycles for 2 weeks. We used an average irradiance of 115 μmol/m^2/s, temperatures of 22 °C day/20 °C night, and 50% relative humidity previously [6].
2. Prepare 6 cm diameter pots filled with watered soil covered with water/gas-permeable plastic foil fixed by elastic bands (*see* **Note 1**).
3. Make a small hole in the top of the plastic foil using a scalpel.
4. Make a small hole in the soil, with the flaps of the foil held back with a tweezer.
5. Transfer 2-week-old Arabidopsis plants carefully with its root covered by soil (*see* **Note 2**).
6. Grow the plants in 8 h day/16 h light/dark cycles for 3 weeks in individual pots (*see* **Note 3**).
7. Plants with fully expanded leaves at stage 3.90 [11] (fresh weight of plant about 300 mg) are normally used for $^{13}CO_2$ feeding experiments (*see* **Note 4**).

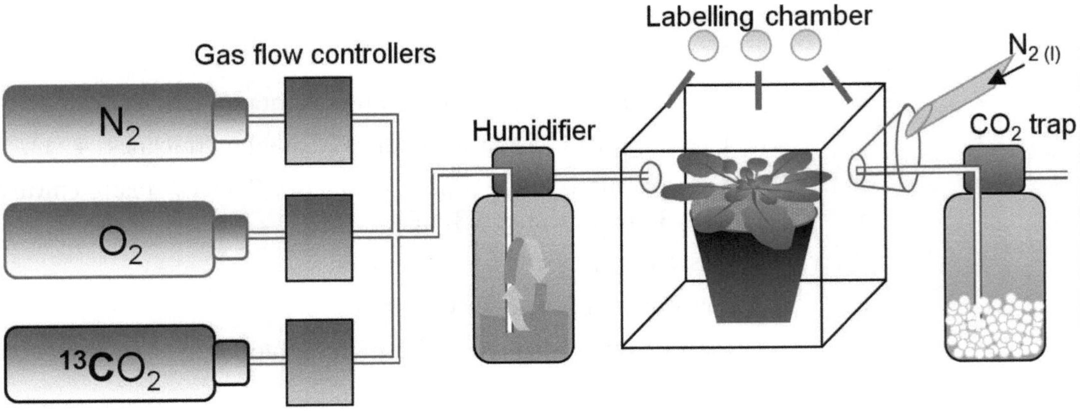

Fig. 1 $^{13}CO_2$ labeling system used for Arabidopsis. System used for $^{13}CO_2$ gas labeling experiment using gas flow controllers, humidifier, labeling chamber, and CO_2 trap is presented. The funnel is only opened when it is used to pour liquid N_2 onto the plant for harvesting

3.2 Preparation of Labeling System

The experimental setup for dynamic $^{13}CO_2$-labeling of plant leaves should be designed to achieve several important points: (1) *Air tightness*: the gas labeling chamber should be tightly closed in order to avoid dilution of $^{13}CO_2$, (2) *Chamber size*: the volume of chamber should be minimized to avoid the time delay between introduction and full equilibration of $^{13}CO_2$ in the chamber, (3) *Leaves*: the leaf should be fully exposed to the light and not damaged by closing the chamber, (4) *Temperature*: the temperature in the chamber should be constant, (5) *Harvest*: metabolism should be quenched immediately, under the light and $^{13}CO_2$ regime, and without any exposure to unlabeled CO_2 before quenching. Model of gas labeling system is presented in Fig. 1.

1. Prepare the gas flow controllers for each gas type. Each gas type should be supplied separately at a fixed flow rate and mixed in a single PVC tube. This procedure results in a final concentration of 78% N_2, 21% O_2, 400 ppm $^{12}CO_2$ or $^{13}CO_2$, and a final flow of 5 L/min (*see* **Note 5**).
2. Connect the three PVC pipe lines (N_2, O_2, and CO_2) into one.
3. Prepare a humidifier after passing through gas flow controllers, the gas mixture flows into a transparent labeling chamber.
4. Prepare PVC pipe line splitting to three PVC pipe lines.
5. Assemble the chamber using a 380 mL Magenta GA-7 plant culture box with three transparent PVC pipes inserted through tight holes in the box for entry of the gas mixture (*see* **Note 6**).
6. Insert a funnel through a tight hole directed towards the plants in the Magenta GA-7 plant culture box (*see* **Note 7**).
7. Block the funnel, except for a PVC pipe tube for exit of the gas.
8. Install CO_2 trap leading to a bottle containing soda lime to capture the $^{13}CO_2$ (*see* **Note 8**).

9. Insert a funnel through a tight hole, with the end directed towards the plant. This funnel is initially blocked, except for a tube for exit of the gas mixture leading to a bottle containing soda lime to capture the $^{13}CO_2$. The funnel will be further used to pour liquid N_2 onto the plant to quench metabolism at harvest. Secure all inserted tubes in an airtight manner (Fig. 1).
10. Check all inserted tubes in an airtight manner.
11. Check system by testing the level of $^{12}CO_2$ in the chamber using an infrared CO_2 gas analyzer (Li Cor LI 800) (*see* **Note 9**).
12. Check the temperature in the labeling chamber with thermocouples (*see* **Note 10**).

3.3 $^{13}CO_2$ Feeding

1. Set up the onset of light in the growth chamber at 2 h before starting the experiments.
2. Move individual plants quickly from the growth chamber to the labeling chamber in their pots (<5 s) (*see* **Note 11**).
3. Harvest the plant material at designed time points, and freeze immediately to quench the metabolism (*see* **Note 12**).
4. Repeat steps more than three times for biological replicates for each labeling condition and control.

3.4 Measurement of ^{13}C Labeled Photorespiration Intermediates by GC-MS

1. Methods and procedures of the measurement of photorespiration intermediates are described in Chapter 2.

3.5 Evaluation of Isotope Redistribution and ^{13}C–Enrichment of PR Intermediates

After proceeding with sample preparation and data acquisition for GC-MS, process the data obtained for standards and extracts of metabolites involved in photorespiration using Chroma TOF 1.0 and TagFinder 4.0 as previously described [7, 12, 13]. Since MS detects ionized compounds separated by their mass-to-charge ratio (m/z), the m/z of ^{13}C-labeled compounds is increasing by an amount which equals the number of incorporated stable isotopes (*see* Fig. 2a and b). Therefore, by determining the ratio of intensity of the monoisotopic ion and its isotopic ions, the ratio of stable isotope labeling can be quantified [14].

1. Generate a peak intensity matrix containing all available mass isotopomers of characteristic mass fragments that represent the photorespiratory intermediates (Table 1).
2. Create an input file .txt according to the indications provided in the documentation file of the CORRECTOR software tool (http://www-en.mpimp-golm.mpg.de/03-research/researchGroups/01-dept1/Root_Metabolism/smp/CORRECTOR/index.html) [15].

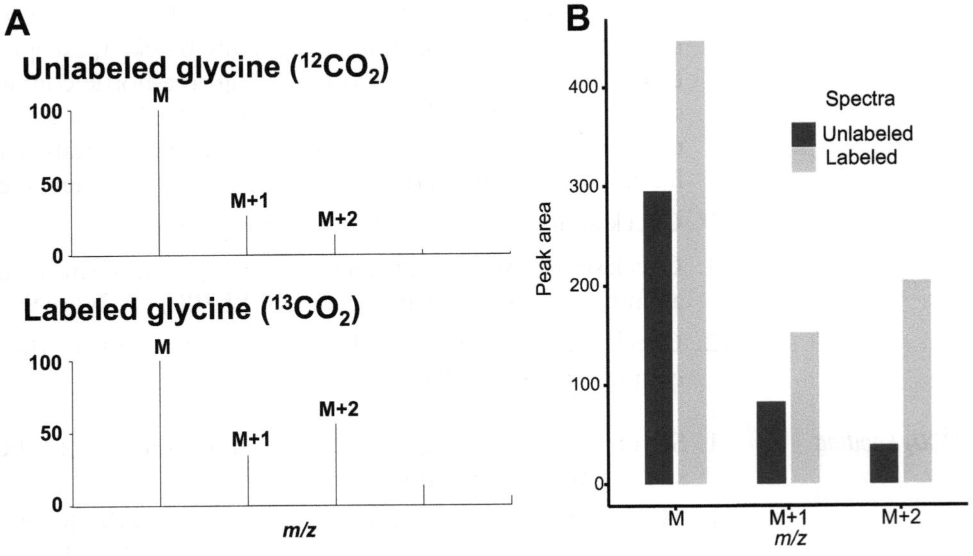

Fig. 2 Mass spectra (**a**) and chromatographic peak areas (**b**) for unlabeled and labeled glycine showing the mass shift towards heavier mass isotopomers

Table 1
Major mass fragments of photorespiratory intermediates detected by GC-MS in $^{13}CO_2$ labeling experiments

Metabolite	Chemical structure	Analyte	Fragment structure	Labeled carbons	Lowest fragment mass	Highest fragment mass
Glycolic acid	$C_2H_4O_3$	Glycolic acid_2TMS	$C_6H_{17}O_2Si_2$	1	177	178
Glycolic acid	$C_2H_4O_3$	Glycolic acid_2TMS	$C_7H_{17}O_3Si_2$	2	205	207
Glycine	$C_2H_5NO_2$	Glycine_3TMS	$C_7H_{20}N_1Si_2$	1	174	175
Glycine	$C_2H_5NO_2$	Glycine_3TMS	$C_9H_{26}O_1N_1Si_3$	1	248	249
Glycine	$C_2H_5NO_2$	Glycine_3TMS	$C_{10}H_{26}O_2N_1Si_3$	2	276	278
Serine	$C_3H_7NO_3$	Serine_3TMS	$C_8H_{22}O_1N_1Si_2$	2	204	206
Serine	$C_3H_7NO_3$	Serine_3TMS	$C_8H_{20}O_2N_1Si_2$	2	218	220
Serine	$C_3H_7NO_3$	Serine_3TMS	$C_{11}H_{28}O_3N_1Si_3$	3	306	309
Glyceric acid	$C_3H_6O_4$	Glyceric acid_3TMS	$C_7H_{17}O_2Si_2$	2	189	191
Glyceric acid	$C_3H_6O_4$	Glyceric acid_3TMS	$C_{11}H_{27}O_4Si_3$	3	307	310

3. Estimate the metabolic pools which are assumed to be unlabeled at the time point before starting the labeling.
4. Process the data using the CORRECTOR software tool implemented with an isoconfig.cfg file containing analytes of interest with respective fragment information (Table 1). For each input file three output files are obtained, output 1 gives the corrected MS data, output 2 the relative mass isotopomer abundance for each mass isotopomer (Table 1), and output 3 the ^{13}C enrichment (*see* **Note 13**).
5. Evaluate data for reproducibility and variation between biological replicates, and consecutive time points by observing the relative mass isotopomer abundance of each mass isotopomer over time (Fig. 3) (*see* **Note 14**) before further analysis.
6. Calculate the enrichment and the s.d. of the unlabeled and labeled fractions for each metabolite m and each time point t from corresponding samples.

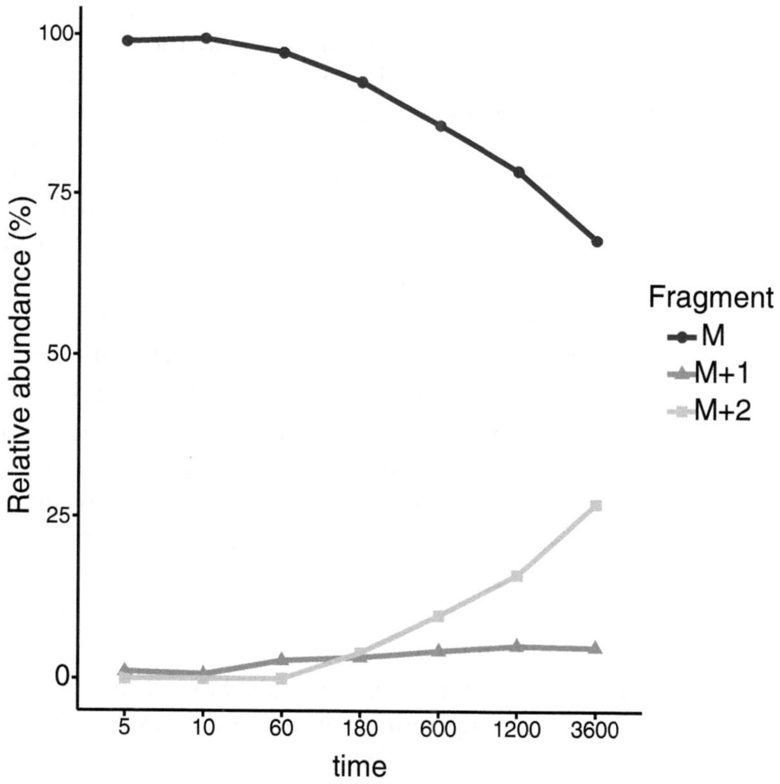

Fig. 3 Relative abundance of different glycine mass isotopomers over time

4 Notes

1. Covering with water/gas-permeable plastic foil fixed by elastic bands between the rosette and the soil allows the collection of the rosette clean of soil.
2. Check the leaves are above the plastic foil.
3. Short day light growth condition makes much easier handling plants to avoid induction of flowering before the rosette reaches a suitable size.
4. Plant watering should be carefully controlled to keep the pot moist but not waterlogging.
5. It is important to use a steady high gas flow rate for rapid replacement of $^{12}CO_2$ by $^{13}CO_2$ during the labeling treatment. It can also minimize the possibility of mixture of $^{13}CO_2$ with ambient $^{12}CO_2$ during the brief period when box is opened for quenching metabolism by liquid N_2 at harvest.
6. The boxes should be highly transparent in order to allow the transmission of the irradiance. It is necessary to measure the light intensity (within ~97%) in the box before starting any experiments. If the irradiance is transmitted a lower intensity amount, an extra light supply can be adopted for providing the sufficient intensity of light irradiance to the plant in the labeling chamber. It is essential to make sure the boxes are resistant to liquid nitrogen before starting the experiment. In case the box lid presents any cracks, after being subjected to liquid nitrogen, it must be replaced by a new one.
7. Since this funnel will be used to pour liquid N_2 into the labeling chamber to quench metabolism at plant harvest, all tubes inserted to the chamber should be tightly closed.
8. The $^{13}CO_2$ which is passing through the labeling chamber should be trapped by a "CO_2 trap" (Fig. 1) with soda lime to avoid distribution of $^{13}CO_2$ to the other plants.
9. Infrared CO_2 gas analyzer (Li-Cor LI-800) can be used for tracking the $^{12}CO_2$ concentration. For example, after 3–5 s, the $^{12}CO_2$ concentration inside the chamber should be approximating 0 ppm.
10. The plant temperature can differ from the air temperature. Since temperature measuring devices placed in direct contact under light irradiance can overestimate temperatures, the temperature in the labeling chamber should be carefully checked with thermocouples. The temperature of the flowing gas should also be checked and controlled to avoid affecting the temperature in the labeling chamber.

11. One type of control should contain plants that are placed in the chamber, but without exposure to $^{13}CO_2$ at different time points of the experiment.
12. Check in preliminary experiments the time points in which photorespiration rate and metabolite levels are constant in your plant system and growth conditions, since labeling needs to be performed in metabolic steady state. The frozen rosette is extremely fragile and some leaves or leaf parts may break and fall on the working surface during the handling.
13. Due to fragmentation of the analyte by GC-TOFMS, detectable peaks are not the molecular parent ion (or non-fragmented derivate). Since it is usually assumed that the labeling within the fragment is representative of the labeling of the entire molecule, it is important to note how many carbons are included in those fragments. Relative mass isotopomer abundance $z^{(i)}$ for each mass isotopomer with N carbon atoms in which i ^{13}C atoms are incorporated is calculated by corrector following the equation:

$$z^{(i)} = \frac{m_i}{\sum\limits_{j=0}^{N} m_j}. \quad (1)$$

where m_j represents the mass isotopomer abundance for each metabolite. Note that the relative abundance of the unlabeled isotopomer is termed the unlabeled content. For GC-TOFMS data the mass isotopomer abundance (m_j) is given by its peak intensity corrected for natural abundance by the CORRECTOR software. ^{13}C enrichments of metabolites are calculated by CORRECTOR from the relative mass isotopomer abundances following the equation:

$$^{13}\text{C enrichment} = \frac{\sum_{i=0}^{N} i\ m_i}{N\sum_{j=0}^{N} m_j} = \frac{\sum_{i=0}^{N} i\ z^{(i)}}{N}. \quad (2)$$

14. The factors such as gas leakage in the labeling chamber or the PVC tubes, incorrect illumination, and poor watering of an individual pot leading to water deficit and stomatal leakage can lead to large variation and slower labeling of individual samples. For early time points, $^{13}CO_2$ from earlier experiments may lead to artifactually rapid labeling, due to the plants having assimilated some $^{13}CO_2$ before the start of the of the labeling treatment.

Acknowledgement

This work was financially supported by the Max Planck Society. We thank National Council for Scientific and Technological Development CNPq-Brazil for financially supporting LPS.

References

1. Bauwe H, Hagemann M, Kern R, Timm S (2012) Photorespiration has a dual origin and manifold links to central metabolism. Curr Opin Plant Biol 15(3):269–275. doi:10.1016/j.pbi.2012.01.008
2. Bauwe H, Hagemann M, Fernie AR (2010) Photorespiration: players, partners and origin. Trends Plant Sci 15(6):330–336. doi:10.1016/j.tplants.2010.03.006
3. Florian A, Araújo WL, Fernie AR (2013) New insights into photorespiration obtained from metabolomics. Plant Biol 15(4):656–666. doi:10.1111/j.1438-8677.2012.00704.x
4. Foyer CH, Bloom AJ, Queval G, Noctor G (2009) Photorespiratory metabolism: genes, mutants, energetics, and redox signaling. Ann Rev Plant Biol 60(1):455–484. doi:10.1146/annurev.Arplant.043008.091948
5. Obata T, Florian A, Timm S, Bauwe H, Fernie AR (2016) On the metabolic interactions of (photo)respiration. J Exp Bot 67 (10):3003–3014. doi:10.1093/jxb/erw128
6. Szecowka M, Heise R, Tohge T, Nunes-Nesi A, Vosloh D, Huege J, Feil R, Lunn J, Nikoloski Z, Stitt M, Fernie AR, Arrivault S (2013) Metabolic fluxes in an illuminated Arabidopsis rosette. Plant Cell 25(2):694–714. doi:10.1105/tpc.112.106989
7. Heise R, Arrivault S, Szecowka M, Tohge T, Nunes-Nesi A, Stitt M, Nikoloski Z, Fernie AR (2014) Flux profiling of photosynthetic carbon metabolism in intact plants. Nat Protoc 9 (8):1803–1824. doi:10.1038/nprot.2014.115
8. Romisch-Margl W, Schramek N, Radykewicz T, Ettenhuber C, Eylert E, Huber C, Romisch-Margl L, Schwarz C, Dobner M, Demmel N, Winzenhorlein B, Bacher A, Eisenreich W (2007) $^{13}CO_2$ as a universal metabolic tracer in isotopologue perturbation experiments. Phytochemistry 68 (16–18):2273–2289. doi:10.1016/j.phytochem.2007.03.034
9. Eisenreich W, Bacher A (2007) Advances of high-resolution NMR techniques in the structural and metabolic analysis of plant biochemistry. Phytochemistry 68(22–24):2799–2815. doi:10.1016/j.phytochem.2007.09.028
10. Ma F, Jazmin LJ, Young JD, Allen DK (2014) Isotopically nonstationary 13C flux analysis of changes in *Arabidopsis thaliana* leaf metabolism due to high light acclimation. Proc Natl Acad Sci U S A 111(47):16967–16972. doi:10.1073/pnas.1319485111
11. Boyes DC, Zayed AM, Ascenzi R, McCaskill AJ, Hoffman NE, Davis KR, Gorlach J (2001) Growth stage-based phenotypic analysis of Arabidopsis: a model for high throughput functional genomics in plants. Plant Cell 13 (7):1499–1510. doi:10.1105/tpc.13.7.1499
12. Lisec J, Schauer N, Kopka J, Willmitzer L, Fernie AR (2006) Gas chromatography mass spectrometry-based metabolite profiling in plants. Nat Protoc 1(1):387–396. doi:10.1038/nprot.2006.59
13. Luedemann A, Malotky L, Erban A, Kopka J (2012) TagFinder: preprocessing software for the fingerprinting and the profiling of gas chromatography–mass spectrometry based metabolome analyses. Methods Mol Biol 860:255–286. doi:10.1007/978-1-61779-594-7_16
14. Hasunuma T, Harada K, Miyazawa S-I, Kondo A, Fukusaki E, Miyake C (2010) Metabolic turnover analysis by a combination of in vivo ^{13}C-labelling from $^{13}CO_2$ and metabolic profiling with CE-MS/MS reveals rate-limiting steps of the C3 photosynthetic pathway in *Nicotiana tabacum* leaves. J Exp Bot 61 (4):1041–1051. doi:10.1093/jxb/erp374
15. Huege J, Goetze J, Dethloff F, Junker B, Kopka J (2014) Quantification of stable isotope label in metabolites via mass spectrometry. Methods Mol Biol 1056:213–223. doi:10.1007/978-1-62703-592-7_20

Chapter 12

Isotopically Nonstationary Metabolic Flux Analysis (INST-MFA) of Photosynthesis and Photorespiration in Plants

Fangfang Ma, Lara J. Jazmin, Jamey D. Young, and Doug K. Allen

Abstract

Photorespiration is a central component of photosynthesis; however to better understand its role it should be viewed in the context of an integrated metabolic network rather than a series of individual reactions that operate independently. Isotopically nonstationary ^{13}C metabolic flux analysis (INST-MFA), which is based on transient labeling studies at metabolic steady state, offers a comprehensive platform to quantify plant central metabolism. In this chapter, we describe the application of INST-MFA to investigate metabolism in leaves. Leaves are an autotrophic tissue, assimilating CO_2 over a diurnal period implying that the metabolic steady state is limited to less than 12 h and thus requiring an INST-MFA approach. This strategy results in a comprehensive unified description of photorespiration, Calvin cycle, sucrose and starch synthesis, tricarboxylic acid (TCA) cycle, and amino acid biosynthetic fluxes. We present protocols of the experimental aspects for labeling studies: transient $^{13}CO_2$ labeling of leaf tissue, sample quenching and extraction, mass spectrometry (MS) analysis of isotopic labeling data, measurement of sucrose and amino acids in vascular exudates, and provide details on the computational flux estimation using INST-MFA.

Key words Transient isotopic labeling, Metabolic flux analysis, Carbon-13 tracer, Photosynthesis, Photorespiration, Carbon assimilation, Primary metabolism

1 Introduction

Photorespiration competes with photosynthetic carbon assimilation due to the bifunctional substrate specificity (CO_2 and O_2) of RuBisCO [1, 2]. When the pathway is considered strictly for its primary function, to convert 2-phosphoglycolate (2PG) into 3-phosphoglycerate, energy and CO_2 are released. Thus the operation of the pathway is considered to be costly to the cell [3, 4]. Engineering RuBisCO to favor carboxylation over oxygenation has remained a major target for improving crop productivity; though altering the substrate specificity of RuBisco has remained challenging [5, 6]. Alternatively, manipulating or bypassing photorespiration has produced mixed results sometimes leading to more

Alisdair R. Fernie et al. (eds.), *Photorespiration: Methods and Protocols*, Methods in Molecular Biology, vol. 1653,
DOI 10.1007/978-1-4939-7225-8_12,

biomass and other times not, possibly because the role of photorespiration is more central to metabolism than considered in most descriptions. In fact, photorespiration is intimately tied not only to the Calvin cycle but also to C_1 metabolism, amino acid biosynthesis, and tricarboxylic acid (TCA) cycle metabolism [7, 8]. In light of the other roles it is less surprising that reduced photorespiration can impair photosynthetic capacity [9–11] and negatively impact other pathways [12, 13]. Given this complexity, it is difficult to decipher the exact roles of photorespiration under different conditions, and novel methods are needed to examine the interactions between photorespiration and other central metabolic pathways.

There are a number of approaches used to measure photorespiration currently. As summarized by Sharkey [14] and Busch [15] methods including post-illumination burst of CO_2, inhibition of photorespiration at low O_2 concentration, CO_2 influx into CO_2-free air, CO_2 assimilation in ^{14}C labeled air, biochemical photosynthetic models based on RuBisCO kinetics, CO_2 release in mitochondria, ammonia production, O_2 consumption, and estimation with ^{18}O or ^{13}C are each with their own limitations. Some are qualitative, others rely on assumptions and parameters that would require reconsideration each time that the environment is perturbed. Importantly, these methods are not inherently designed to investigate photorespiration as part of an integrated metabolic network.

Techniques to quantitatively assess intracellular fluxes using isotopic tracers and MFA have developed rapidly. Labeling studies at steady state are suitable to map metabolic fluxes in heterotrophic systems. By feeding the biological systems of interest, the isotope tracers (i.e., ^{13}C) incorporate into various metabolic pathways and pools yielding different labeling patterns that can be computationally analyzed by MFA to determine fluxes. ^{13}C is the preferred isotope tracer for quantifying cell physiology in non-photosynthetic tissues. However conventional ^{13}C steady state MFA is ineffective for evaluating autotrophic metabolism in leaves that assimilate carbon from CO_2. As indicated in Fig. 1, once steady state labeling is achieved in leaves, all metabolites are uniformly ^{13}C-labeled irrespective of fluxes and pool sizes. Therefore transient measurements of isotope incorporation following a step change from $^{12}CO_2$ to $^{13}CO_2$ can provide estimates for fluxes in leaves with the application of INST-MFA. Figure 2 indicates an overview of the MFA process. Leaf metabolism is kept at a pseudo-steady state while ^{13}C is introduced into central metabolic pathways such that the intermediate pools become labeled over time. Labeled leaf samples are collected at multiple time points during the pre-steady state labeling period (i.e., 15 min for C_3 plants) and the isotopic labeling data are analyzed by mass spectrometry (MS). The relative abundances of different isotope isomers (i.e., "isotopomers") associated with each measured metabolite are plotted

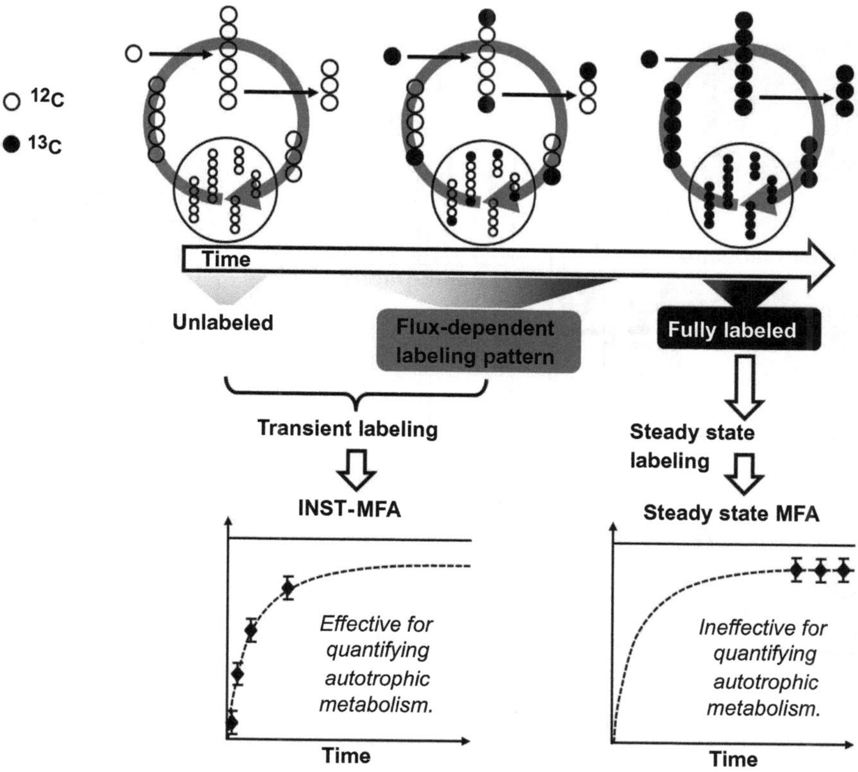

Fig. 1 Evaluating autotrophic metabolism requires transient labeling and INST-MFA. The upper part of figure indicates the incorporation of ^{13}C into an autotrophic system, such as leaves. Because of a single carbon source from CO_2, Calvin cycle intermediates become fully labeled over time, thus making the conventional steady state MFA ineffective for quantifying autotrophic metabolism. Labeling data collected during the isotopically transient period can be computationally analyzed to estimate fluxes via INST-MFA

and input into publically available software packages [16] such as Isotopomer Network Compartmental Analysis (INCA; [17]) and OpenMebius [18]. Labeling patterns are the result of flux through metabolic pathways; therefore the computational estimate of flux occurs through minimizing the lack of fit between simulated and measured labeling patterns (i.e., an inverse problem). Flux parameters in isotopomer models are iteratively adjusted to converge to an optimal solution. An acceptable model is based upon the high quality isotopic labeling data, and the development of robust algorithms for the derivation of global solutions to nonlinear minimization problems. Other methods that are aimed at modeling specifically the unlabeled fraction of the isotopomer pool, referred to as kinetic flux profiling (KFP), are methodologically described elsewhere [19, 20]. By applying KFP, Szecowka et al. [21] successfully determined central metabolic fluxes in Arabidopsis rosettes related to photosynthesis, photorespiration, and end-product synthesis.

Performing INST-MFA is technically challenging at several levels. Experimental design must consider the following aspects:

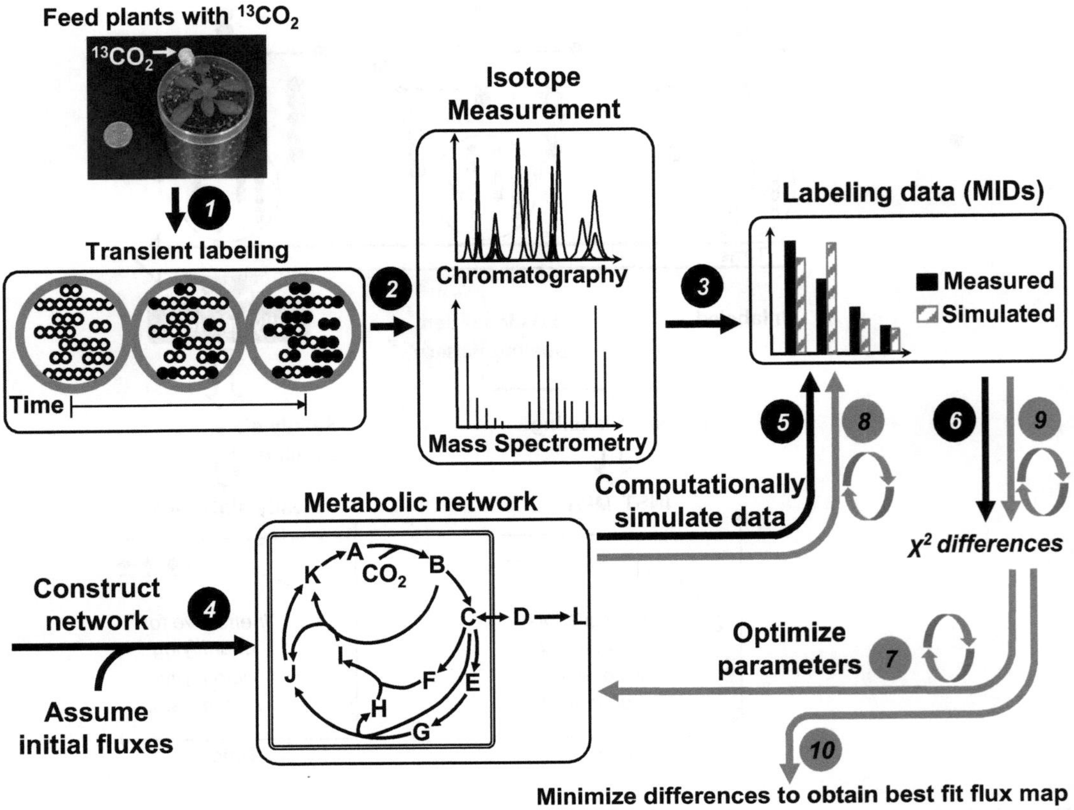

Fig. 2 An overview of the INST-MFA process. $^{13}CO_2$ is fed into leaves through central metabolism. Transient labeling data are collected (①), analyzed via mass spectrometry (②) to generate the mass isotopomer distributions (MIDs) data (③), which are input into INCA. A computational model is applied to simulate the labeling data (④ and ⑤) and determine fluxes by minimizing the lack of fit between simulated and measured labeling patterns (⑥–⑩). Flux parameters in the model are iteratively adjusted until the optimization converges. Once the best fit is obtained, a flux map can be presented

(1) plant metabolism maintained in a pseudo-steady state during the entire ^{13}C labeling period; (2) carefully chosen time points to provide the best description of the dynamic labeling process; (3) a $^{13}CO_2$ concentration that is precisely controlled; and (4) capacity to quench tissues rapidly to minimize metabolic artifacts of the labeling process. The measurement of isotopic composition in intermediates takes advantage of liquid chromatography with tandem mass spectrometry (LC-MS/MS) and utilizes electrospray ionization (ESI). ESI that is regarded as a "soft" ionization technique desolvates molecules keeping otherwise labile bonds (such as phosphate bonds) largely intact and is therefore crucial to the INST-MFA approach. In addition, finding the solution of large-scale ordinary differential equation (ODE) models is computationally burdensome but has been in part aided by the application of elementary metabolite unit (EMU) decomposition algorithm [22] to INST-MFA [23]. Overcoming these challenges has led to

quantification of central metabolism in cyanobacteria, algae, and plants [24–28] with INST-MFA.

In this chapter, we describe detailed protocols for isotopic labeling investigations and INST-MFA of photorespiratory fluxes as a component of the larger network. To cope with cellular heterogeneity and subcellular compartments in leaves, dilution parameters to accommodate photosynthetic inactive pools are introduced into the model, which proves to be essential for obtaining the statistically acceptable fits to labeling data. Compared with other methods for accessing photorespiration, INST-MFA provides an estimate of photorespiration based on multiple measurements through an integrated approach. The model presents photorespiration as a component of central metabolism that is tied to: Calvin cycle, sucrose and starch synthesis, a bifurcated TCA cycle, and amino acid biosynthesis. This quantitative strategy is consistent with the known complex features of photosynthetic carbon assimilation and therefore will provide insights on plant growth and response to many environmental perturbations not currently studied or understood.

2 Materials

2.1 Transient $^{13}CO_2$ Labeling Experiment and Quenching

1. Mature Arabidopsis plants at metabolic steady state.
2. $^{13}CO_2$ gas (premixed or pure; Sigma).
3. Ultra-zero air (CO_2 free; Airgas).
4. Labeling chambers for individual plants.
5. Tygon tubes.
6. Plastic manifold.
7. Benchtop atmospheric pressure gas analyzer (RGA).
8. Liquid N_2.

2.2 Net CO_2 Assimilation Measurement

1. Li-Cor 6400 portable photosynthesis system.
2. ImageJ (for leaf area measurement).
3. Light intensity meter.

2.3 Chlorophyll Content Measurement

1. 80% (v/v) acetone.
2. SpectraMax microplate reader.

2.4 Total RuBisCO Protein Measurement

1. Novex Tris-Glycine SDS sample buffer containing 5% (vol/vol) β-mercaptoethanol.
2. RuBisCO standard protein (Sigma).
3. Molecular weight markers (Invitrogen).
4. Novex 4–12% (w/v) Bis-Tris 10-well gels.

5. Invitrogen Simply Blue safe stain.
6. ImageJ.

2.5 Extraction of Metabolites and Sample Preparation

1. Methanol:Chloroform (7:3, v:v) eventually combined with double distilled water (ddH_2O).
2. 1,4-Piperazinediethanesulphonic acid (PIPES) stock solution (300 μM).
3. Bead mill homogenizer.
4. Benchtop centrifuge.
5. 0.45 μm cellulose acetate centrifuge tube filters.
6. SpeedVac concentrator.
7. Ion exchange columns (Dowex 50 and Dowex 1, Sigma).
8. *N*-Tert-butyldimethylsilyl-*N*-methyltrifluoroacetamide (MTBSTFA) + 1% tert-butyldimethylchlorosilane (TBDMCS) with acetonitrile (*see* **Note 1**) (1:1, v:v).
9. LC and GC injection vials.

2.6 Sucrose Concentration Measurement

1. 75% (v/v) methanol to be combined with chloroform and ribitol—an internal standard.
2. Methoxyamine hydrochloride (MOX) and *N*-methyl-*N*-trimethylsilyl-trifluoroacetamide (MSTFA) to be combined with pyridine for derivatization.
3. Thermomixer.

2.7 Starch Concentration Measurement

1. 80% (v/v) ethanol.
2. Dimethyl sulfoxide (DMSO) with α-amylase and amyloglucosidase to hydrolyze starch.
3. Total starch assay Kit (Megazyme).
4. SpectraMax M2 microplate reader (Molecular Devices).

2.8 Collection of Vascular Exudates

1. Glass plates.
2. Razors.
3. Glass pipettes.
4. Calibrated capillary pipets (5 μL).

2.9 Amino Acid Concentration Measurement

1. Dowex 50 ion exchange columns.
2. Norvaline (internal standard).

2.10 Mass Spectrometry Analysis and Data Processing

1. AB Sciex QTRAP4000 linked to Shimadzu HPLC.
2. Thermo Scientific ISQ single quadrupole GC-MS.
3. Synergi 4-μm Hydro-RP column (150 mm × 2 mm; Phenomenex).

4. DB5 silica column (30 m length, 0.25 mm diameter, 0.25 μm film thickness).
5. Running buffers comprising: tributylamine (TBA, 10 mM) plus 11 mM acetic acid (Buffer A), Buffer B is methanol.
6. Computers equipped with Analyst® software and Thermo Scientific™ Xcalibur™ software.

3 Methods

3.1 Measure Net CO_2 Assimilation Rate

1. Measure net CO_2 assimilation rate of Arabidopsis rosettes via a Li-COR 6400XT portable photosynthesis system (*see* **Note 2**).
2. Setup of Li-COR 6400 system:
   (a) Leaf chamber: 2 × 3 cm standard.
   (b) Light source: red/blue.
   (c) Light intensity: 200 μmol/m^2/s.
   (d) CO_2 concentration in reference chamber: 400 ppm.
   (e) Air temperature: 22 ± 1 °C.

3.2 Measure Chlorophyll Content

1. Chlorophyll is extracted and quantified according to [29] (Arnon, 1949). Extract chlorophyll from 10 mg leaf fresh weight (FW) with 80% (v/v) acetone (10 mL) in the dark for 24 h.
2. After centrifugation at 1500 × *g*, load 200 μL supernatant to flat bottom 96-well plates. 80% (v/v) acetone is used as the blank.
3. Chlorophyll concentrations are determined spectrophotometrically at 663 and 647 nm with a SpectraMax M2 microplate reader.
4. Readings are converted to chlorophyll amount using published equations in [29].

3.3 Measure RuBisCO Protein Content

1. Protein in 10 mg (FW) leaf tissues is extracted with 1 mL Novex Tris-Glycine SDS sample buffer containing 5% (v/v) β-mercaptoethanol heated to 100 °C for 10 min. 1 mg RuBisCO standard is prepared using the same procedure as a positive control.
2. Protein extracts, SeeBlue_2 molecular weight markers, and the RuBisCO standard are loaded to Novex 4–12% (w/v) Bis-Tris 10-well gels and processed with NuPAGE SDS running buffer at a constant voltage of 200 V.
3. Gels are rinsed with double distilled water (ddH_2O) and stained with coommassie blue stain for 1 h and detained in ddH_2O overnight at 4 °C.

4. Gels are photographed, scanned, and the RuBisCO band thickness of samples is calculated with imaging software.
5. Total RuBisCO concentration in samples is quantified through comparison with external standard curves (*see* **Note 3**).

3.4 Measure Sucrose and Starch Contents in Unlabeled Leaves

1. Weigh out 100 mg (FW) frozen leaf powder in 2 mL centrifuge tubes.
2. Add 1.4 mL 75% methanol (−20 °C), 24 μg ribitol (internal standard) to the tube, and one metal bead into the tube.
3. Macerate the leaf tissues with a bead mill homogenizer for 5 min at 30 reps/s, shake the tubes in a thermomixer at 70 °C and 950 rpm for 30 min.
4. Let sit 10 min, then centrifuge at 16,000 × *g*, transfer the supernatant to a glass vial, add 750 μL chloroform (−20 °C) and 1.5 mL ice-cold ddH_2O, and vortex briefly.
5. Centrifuge at 2200 × *g* for 15 min to fractionate the polar and nonpolar phases. Transfer the upper methanol-H_2O phase (containing polar metabolites) to a new vial and this is the extract needed for sucrose determination. The lower chloroform phase with nonpolar metabolites is kept for starch analysis.
6. Dry a small amount of the extract from **step 5** by lyophilization (5 μL).
7. Dried extracts are derivatized with 40 μL of methoxyamine hydrochloride (MOX, 20 mg/mL in pyridine) at 37 °C for 2 h, then with 60 μL of MSTFA for another 30 min.
8. Derivatized samples are analyzed on the Thermo Scientific ISQ single quadrupole GC-MS system with the same settings as described above. Full scan mode is used. Sucrose content can be determined by interpolating from the linear relationship between the peak area and the concentrations of sucrose standard.
9. Re-extract the remaining chloroform phase from **step 5** with 1.5 mL 80% (v/v) ethanol three times in a thermomixer (30 min at 80 °C and 950 rpm). The supernatant is discarded and contains water-soluble metabolites while the remaining pellet contains starch.
10. Dry down the pellets using vacuum centrifugation (i.e., speed vac).
11. Add DMSO to aid starch hydration, extraction, and digestion.
12. Digest starch with α-amylase and amyloglucosidase.
13. Starch concentration as glucose equivalents is measured spectrophotometrically at 510 nm and quantified in comparison with the standard curve plotting with glucose standard.

3.5 Determine the Appropriate Plant Developmental Stage for Labeling Study

1. Net CO_2 assimilation rate, chlorophyll content, and RuBisCO protein contents in leaves of 3- to 5-week-old Arabidopsis plants are measured as described above.
2. In our hands, 4-week-old plants grown under 16/8 h cycles have significant biomass but without generation of reproductive tissues. The plants have comparable chlorophyll content, RuBisCO levels, and net CO_2 assimilation rates. Plants at this stage are used for the labeling experiments (*see* **Note 4**).

3.6 Determine the Appropriate Labeling Time Window During the Day

1. Leaf sucrose and starch content in 4-week-old plants are monitored from 7 am to noon.
2. At midday, sucrose content in leaves does not change whereas starch consistently increases linearly indicating that plants are in a metabolic steady state (*see* **Note 5**). Plants can then labeled during this time (Fig. 3a).

3.7 Construct Labeling Chambers for Individual Arabidopsis Plants

1. A 330 mL plant culture cylinder can be modified for labeling individual Arabidopsis plants that have a planar leaf structure (Fig. 3b).
2. With the cap on, the headspace in each cylinder after including a small potted Arabidopsis plant is about 60 mL. A small headspace is essential for rapid gas turnover. Caps are highly transparent and only block the light intensity by 5%.

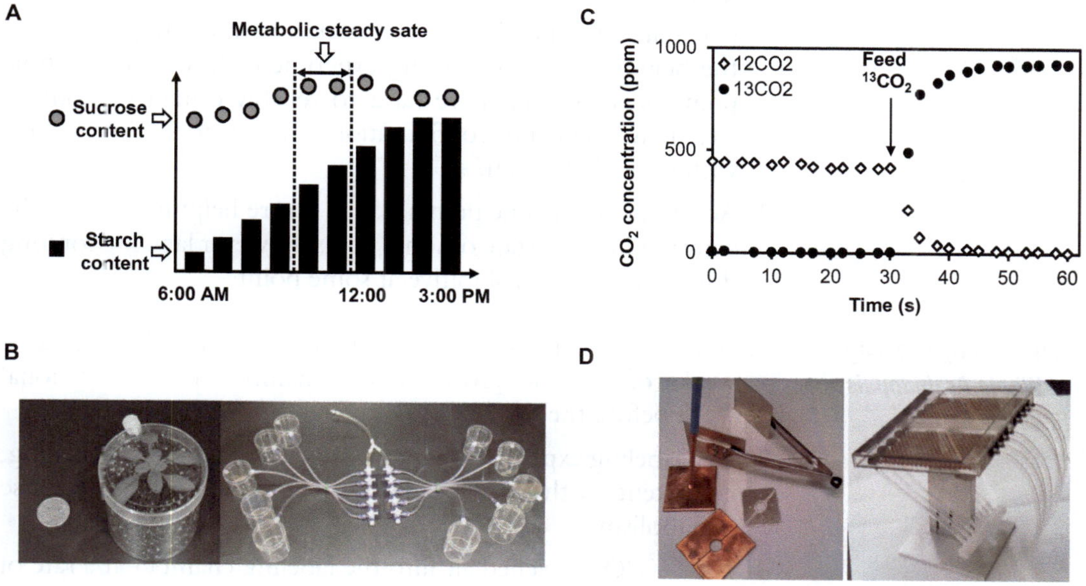

Fig. 3 $^{13}CO_2$ labeling of plant tissue. (**a**) Determine the appropriate labeling time window by monitoring semi-diurnal changes in leaf sucrose and starch contents. (**b**) Labeling chamber for individual Arabidopsis plants. (**c**) The precise control of $^{13}CO_2$ concentration in labeling chambers via RGA. (**d**) Custom-made aluminum or copper freeze clamps (*left*) are constructed to efficiently quench labeled plant tissues. Other labeling device (*right*) is constructed to label leaves with different canopy structures (i.e., broad leaves)

3. Gas is introduced into the labeling chamber through a small hole in the cap that is connected to the gas supply by Tygon tubing. Another small hole is drilled in the wall of cap for exhaust.
4. A manifold can be used to connect multiple labeling chambers with Tygon tubing and label several plants at the same time (Fig. 3b).

3.8 Prepare $^{13}CO_2$ Gas for Labeling Experiments

1. Use premixed gas containing N_2, O_2, and $^{13}CO_2$ with $^{13}CO_2$ at any desired concentrations (Sigma, Cambridge Isotope Laboratories).
2. Alternatively mix a blend of nitrogen and oxygen with pure $^{13}CO_2$ gas using flow meters. Each gas is controlled by separate flow meter at a fixed rate. Total flow is maintained at 2–5 L/min dependent on the amount of biomass being labeled (*see* **Note 6**).
3. Concentrations of N_2, O_2, $^{12}CO_2$, and $^{13}CO_2$ in the labeling chamber are monitored by a benchtop residual gas analyzer (RGA) which ensures that the concentration of $^{13}CO_2$ is maintained at 400 or 800 ppm or the level determined by research goals.

3.9 Determine the Labeling Time Course

1. Preliminary labeling experiments indicated that in C_3 plants, the average ^{13}C labeling of most metabolites increased rapidly over time. Therefore short time points that cover 0–15 min are generally desirable. Samples are collected more frequently near the beginning of the labeling experiment. In our hands, time points less than 10 s are hard to reliably obtain without an alternative labeling configuration. Target four time points within the first 2 min.
2. Additional long time points (i.e., 1 h) are helpful to access the active or inactive state of some pools; however labeling for long durations is cost prohibitive at some point.

3.10 $^{13}CO_2$ Labeling of Whole Arabidopsis Rosettes

1. Individual plants are moved into the labeling chamber at 3 weeks of age and grown in the chamber for an additional week before the labeling experiment.
2. The labeling experiment is performed during the late morning, consistent with the steady state for starch and sucrose metabolism.
3. Feed $^{13}CO_2$ enriched air into the labeling chamber at a rate of 4–5 L/min to initiate the ^{13}C incorporation.
4. Gas in the headspace in each chamber is turned over rapidly to replace $^{12}CO_2$ with $^{13}CO_2$. RGA measurements suggest that the concentration of CO_2 will be 85% ^{13}C or greater within 5 s

using this approach (Fig. 3c). Then the flow rate can be reduced substantially.

3.11 Quenching Metabolism

1. To quench metabolism at a given time point, a large amount of liquid N_2 is directly dumped on rosettes to rapidly freeze the rosettes.
2. Alternatively, custom-made freeze clamps (aluminum or copper construction are pre-chilled in liquid nitrogen and used to clamp individual leaves or entire plants (Fig. 3d). Other labeling devices are also constructed to label plants with different canopy structures (i.e., broad leaves).
3. Avoid shading or change of light intensity during this process.
4. Frozen rosettes from each plant are collected and kept at −80 °C until analysis.

3.12 Extraction of Metabolites from Labeled Rosettes

1. Frozen rosettes from $^{13}CO_2$ labeling experiments are pulverized with the aid of liquid nitrogen.
2. Weigh out 200 mg (FW) frozen leaf powder in a precooled 2 mL centrifuge tube containing a metal bead.
3. Add 1.8 mL methanol:chloroform extraction buffer (v/v, 7:3; −20 °C), 12 pmol of PIPES (internal standard), and 5 μg norvaline (internal standard) to each tube.
4. After a 5 min homogenization with a bead mill at frequency of 30 repetitions/s, extracts are transferred into ice-cold 8 mL glass vials. Add 1.2 mL extraction buffer to rinse the centrifuge tube and combine with the previous 1.8 mL extracts. Pooled extracts (3 mL) are incubated at −20 °C for 2 h with occasional vortexing (*see* **Note** 7).
5. Water-soluble metabolites are extracted with 2.4 mL ice-cold ddH_2O that is added into the 3 mL methanol:chloroform mixture, and after vigorous vortexing, centrifuged for 10 min at 2200 × *g* at 4 °C.
6. Transfer the upper methanol-H_2O phase to a new glass vial and repeat **step 5**; then combine the second upper phase with the first and dry the extracts with speed vacuum centrifugation.
7. Dried extracts are kept at −80 °C until analysis. A flow chart describing sample preparation for MS analysis is indicated in Fig. 4a.

1. Add 200 μL ddH_2O to dissolve the dried extracts from **step** 7 in Subheading 3.12.
2. Insoluble materials are filtered by 0.45 μm cellulose acetate centrifuge filters.

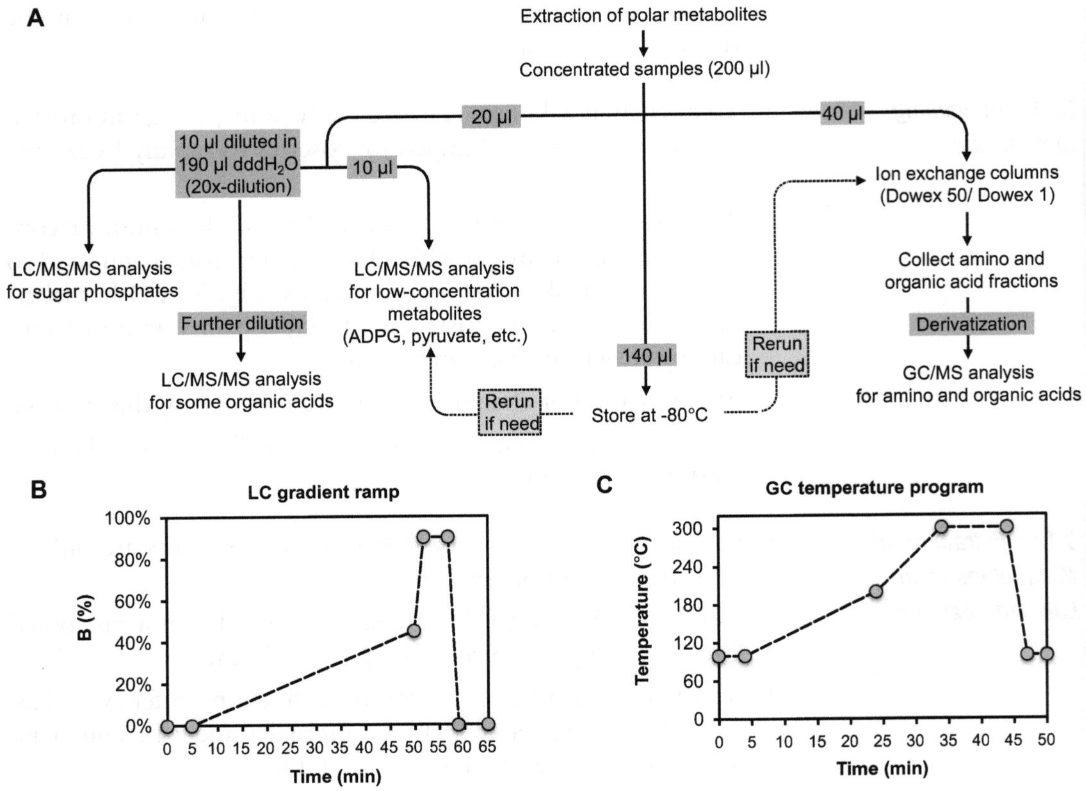

Fig. 4 Sample preparation for MS analysis. (**a**) Details about sample handling for LC/MS/MS and GC/MS analysis. (**b**) The gradient ramp profile for LC. (**c**) The temperature program for GC

3.13 Analyze Isotope Distributions of Low-Abundant Metabolic Intermediates (i.e., Sugar Phosphates) Via LC-MS/MS

3. Filtrate (~200 μL), is collected in LC injection vials and kept at −80 °C until analysis. Several low abundance metabolites such as ADPG, pyruvate, and sucrose-6-phosphate may require this concentration; however injection of 10 μL from this extract will generally not give good chromatographic peak quality for most metabolites (*see* **Note 8**).
4. Dilute 5 μL extract with 195 μL ddH$_2$O to make 20-fold diluted samples. 20 μL diluted extracts are injected onto LC-MS/MS for most sugar phosphates. Known amount of standards are loaded to HPLC vials and processed in the same way.
5. For some organic acids (malate, fumarate, succinate, etc.), 50-fold dilution may be necessary dependent on the biological tissue being analyzed.
6. Setup of ion-pairing LC-MS/MS.
   (a) Instruments: AB Sciex QTRAP4000 linked to Shimadzu HPLC (or equivalent).
   (b) Column: Synergi 4-μm Hydro-RP column (150 mm × 2 mm; Phenomenex) or equivalent.

(c) Eluent A (the ion-pairing reagent): 10 mM tributylamine + 11 mM acetic acid.

(d) Eluent B: methanol.

(e) Flow rate: 0.3 mL/min.

(f) Gradient ramp profile: 0–5 min, 0% B; 5–50 min, 0–45% B; 50–52 min, 60–90% B; 52–57 min, 90% B; 57-59 min, 90–0% B; 60–65 min, 0% B (Fig. 4b).

(g) MS settings: negative-mode electrospray ionization, multiple reaction monitoring (MRM) mode (check Table 1 for the LC-MS/MS ion transitions and method parameters; *see* **Note 9**).

3.14 Analyze Isotope Distributions of Organic and Amino Acids Via GC-MS

1. Transfer 20% (40 μL) from concentrated extracts in **step 3** of Subheading 3.13 to a new 1.5 mL centrifuge tube and add 960 μL ddH_2O, vortex, load this 1 mL sample to ion exchange columns (Dowex 50 and Dowex 1, Sigma), separate organic acids from amino acids, and collect different fractions in glass vials for subsequent analysis.
2. Aspirate the organic and amino acid fractions with gaseous N_2. Organic and amino acid standards are prepared in parallel.
3. Add 100 μL acetonitrile (fresh) and 100 μL MTBSTFA + 1% TBDMCS, vortex, and incubate at room temperature for 30 min.
4. Incubate at 100 °C on a block heater for 1 h to covert organic and amino acids to tert-butyldimethylsilyl (TBDMS) derivatives.
5. Cool for 10 min and transfer to GC vials.
6. Setup of GC-MS analysis.

   (a) Instrument: Thermo Scientific ISQ single quadrupole GC-MS (or equivalent).

   (b) Column: DB5 silica column (30 m length, 0.25 mm diameter, 0.25 μm film thickness) or equivalent.

   (c) Sample injection volume: 1 μL. A 10- to 20-fold dilution may be necessary for some organic and amino acids. Dilute in acetonitrile: MTBSTFA + 1% TBDMCS (v/v, 20:1).

   (d) Carrier gas flow rate: helium, 1.2 mL/min.

   (e) Ion source temperature: 250 °C.

   (f) Ionization mode: EI.

   (g) Scan range: 50–650 amu.

   (h) Temperature program: an initial temperature of 100 °C for 4 min, followed by a 5 °C/min ramp to 200 °C, then a

Table 1
LC-MS/MS ion transitions and method parameters

Metabolites	Parent ion ([M−H]$^-$) Formula	Mass	Product ion Formula	Mass	DP (v)	EP (v)	CE (v)	CXP (v)
PGA	$C_3H_6O_7P^-$	185	$[PO_3]^-$	79	−35	−10	−44	−3
DHAP	$C_3H_6O_6P^-$	169	$[H_2PO_4]^-$	97	−40	−5	−16	−23
FBP	$C_6H_{13}O_{12}P_2^-$	339	$[H_2PO_4]^-$	97	−37	−10	−25	−5
F6P	$C_6H_{12}O_9P^-$	259	$[H_2PO_4]^-$	97	−25	−11	−21	−10
G6P	$C_6H_{12}O_9P^-$	259	$[H_2PO_4]^-$	97	−25	−11	−21	−10
S7P	$C_7H_{14}O_{10}P^-$	289	$[H_2PO_4]^-$	97	−26	−3	−30	−6
P5P	$C_5H_{10}O_8P^-$	229	$[H_2PO_4]^-$	97	−28	−7	−19	−5
RUBP	$C_5H_{11}O_{11}P_2^-$	309	$[H_2PO_4]^-$	97	−28	−10	−31	−16
G1P	$C_6H_{12}O_9P^-$	259	$[PO_3]^-$	79	−27	−9	−58	−5
ADPG	$C_{16}H_{24}N_5O_{15}P_2^-$	588	$C_6H_{11}O_8P^-$	241	−50	−10	−44	−7
UDPG	$C_{15}H_{23}N_2O_{17}P_2^-$	565	$C_9H_{12}N_2O_9P^-$	323	−45	−8	−36	−7
PEP	$C_3H_4O_6^-$	167	$[PO_3]^-$	79	−37	−8	−22	−6
Pyruvate	$C_3H_3O_3^-$	87	$C_2H_3O^-$	43	−24	−7	−13	−3
Aconitate	$C_6H_5O_6^-$	173	$-2CO_2$	85	−35	−10	−17	−5
αKG	$C_5H_5O_5^-$	145	$-CO_2$	101	−29	−5	−14	−7
Succinate	$C_4H_5O_4^-$	117	$-CO_2$	73	−41	−10	−18.5	−2
Fumarate	$C_2HO_3^-$	115	$-CO_2$	71	−32	−8	−13	−3
Malate	$C_4H_5O_5^-$	133	$-H_2O$	115	−30	−10	−18	−4
[a]2PG	$C_2H_4O_6P^-$	155	$[PO_3]^-$	79	−35	−10	−20	−5
[a]Glycerate	$C_3H_5O_4^-$	105	$C_2H_3O_3^-$	75	−40	−3.8	−14	−3.8
[a]Glycolate	$C_2H_3O_3^-$	75	-CO	47	−40	−3.8	−14	−3.8
[a]Glyoxylate	$C_2HO_3^-$	73	-CO	45	−50	−9	−11.2	−5

[a]Important photorespiratory intermediates

10 °C/min ramp to 300 °C, and a final heating at 300 °C for 10 min (Fig. 4c).

(i) Scan range: 50–650 amu.

(j) Scan mode: full and selective ion monitoring (SIM) scan.

3.15 Quantify Vascular Sucrose and Amino Acid Contents

1. Cut fully expanded rosettes from 4-week-old Arabidopsis plants and place the rosettes on a glass plate.

2. Press the leaf vein with glass pipette from the leaf tip to the cut end and collect vascular exudates with a 5 μL capillary pipet.
3. Add ribitol and norvaline to the solution as internal standards.
4. Process vascular exudates through Dowex 50 ion exchange column to separate sugars from amino acids.
5. Sucrose concentration in the sugar fraction is quantified with GC-MS as described above in Subheading 3.4.
6. Amino acid fractions are dried down, converted to their TBDMS derivatives, and quantified with GC-MS (*see* Subheading 3.14). Standard curves for each amino acid are plotted using amino acid standards.

3.16 Analyze Isotopic Labeling Data Via LC-MS/MS

1. Identify the chromatographic peaks based on the retention time (RT) of standards and MS/MS fragmentation spectra. The parent ion mass and the RT of peaks are often used to associate the chromatographic peaks to metabolites of interest. The identification of peaks can be further confirmed by searching databases containing ESI-MS/MS information.
2. Identify ions for mass isotopomer profiling. In negative-ESI mode, the mass of parent ion is generally the base mass of any metabolite for short-term labeling experiments. For most sugar phosphates (except ADPG and UDPG), the daughter ion is the phosphate group (97 or 79). Therefore the carbon backbones of molecules are not broken. For ADPG, UDPG, and many organic acids, daughter ions are formed with breaking of carbon-carbon bonds and the daughter ion may or may not contain the labeled carbon atoms. A list of potential parent and daughter ions for metabolites of interest can be found in Tables 2 and 3. For photorespiration, it is important to get mass isotopomer distribution (MID) measurements of 2PG, glycolate, and glyoxylate. The abundance of the latter two can be low.
3. Integrate peak area for mass isotopomers using software that is: custom-made, freeware, or available with the instrument (*see* **Note 10**).
4. Calculate MIDs and correct for natural abundance of isotopes. Some MFA software can perform mass corrections internally; therefore in this case only raw, uncorrected MIDs data are required as a model input.
5. Calculate the mean and standard error of MIDs for each metabolite at each time point.

3.17 Analyze Isotopic Labeling Data by GC-MS

1. Identify the chromatographic peaks based on the RT of standard peaks and the MS fingerprint of peaks. The TBDMS derivatives produce characteristic $[M-57]^+$, $[M-85]^+$, and

Table 2
Specific compound dependent MS parameters used in MRM with LC-MS/MS

Metabolites	C atoms	Isotopomers	$Q_1[Q_3]$ (*m/z*) of mass isotopomers
PGA	3	[M] – [M+3]	185[79], 186[79], 187[79], 188[79]
DHAP	3	[M] – [M+3]	169[97], 170[97], 171[97], 172[97]
FBP	6	[M] – [M+6]	339[97], 340[97], 341[97], 342[97], 343[97], 344[97], 345[97]
G6P&F6P	6	[M] – [M+6]	259[97], 260[97], 261[97], 262[97], 263[97], 264[97], 265[97]
P5P	5	[M] – [M+5]	229[97], 230[97], 231[97], 232[97], 233[97], 234[97]
S7P	7	[M] – [M+7]	289[97], 290[97], 291[97], 292[97], 293[97], 294[97], 295[97], 296[97]
RUBP	5	[M] – [M+5]	309[97], 310[97], 311[97], 312[97], 313[97], 314[97]
G1P	6	[M] – [M+6]	259[79], 260[79], 261[79], 262[79], 263[79], 264[79], 265[79]
[a]2PG	2	[M] – [M+2]	155[79], 156[79], 157[79]
[a]Glycolate	2	[M] – [M+2]	75[47], 76[47], 76[48], 77[48]
[a]Glyoxylate	2	[M] – [M+2]	73[45], 74[45], 74[46], 75[46]
[a]Glycerate	3	[M] – [M+3]	105[75], 106[75], 106[76], 107[76], 107[77], 108[77]
PEP	3	[M] – [M+3]	167[79], 168[79], 169[79], 170[79]
Aconitate	5	[M] – [M+5]	173[85], 174[85], 174[86], 175[85], 175[86], 175[87], 176[86], 176[87], 176[88], 177[87], 177[88], 177[89], 178[88], 178[89], 179[89]
αKG	5	[M] – [M+5]	145[101], 146[101], 146[102], 147[102], 147[103], 148[103], 148[104], 149[104], 149[105], 150[105]
Succinate	4	[M] – [M+4]	117[73], 118[73], 118[74], 119[74], 119[75], 120[75], 120[76], 121[76]
Fumarate	4	[M] – [M+4]	115[71], 116[71], 116[72], 117[72], 117[73], 118[73], 118[74], 119[74]
Malate	4	[M] – [M+4]	133[115], 134[116], 135[117], 136[118], 137[119]

[a]Important photorespiratory intermediates

Table 3
Isotopomer-dependent MS parameters for UDPG and ADPG used in MRM with LC-MS/MS

	Mass isotopomers	Q1[Q3] (*m/z*) of mass isotopomers
	[M]	565[323]
	[M+1]	566[323], 566[324]
	[M+2]	567[323], 567[324], 567[325]
	[M+3]	568[323], 568[324], 568[325], 568[326]
	[M+4]	569[323], 569[324], 569[325], 569[326], 569[327]
	[M+5]	570[323], 570[324], 570[325], 570[326], 570[327], 570[328]
	[M+6]	571[323], 571[324], 571[325], 571[326], 571[327], 571[328], 571[329]
UDPG	[M+7]	572[324], 572[325], 572[326], 572[327], 572[328], 572[329], 572[330]
	[M+8]	573[325], 573[326], 573[327], 573[328], 573[329], 573[330], 573[331]
	[M+9]	574[326], 574[327], 574[328], 574[329], 574[330], 574[331], 574[332]
	[M+10]	575[327], 575[328], 575[329], 575[330], 575[331], 575[332]
	[M+11]	576[328], 576[329], 576[330], 576[331], 576[332]
	[M+12]	577[329], 577[330], 577[331], 577[332]
	[M+13]	578[330], 578[331], 578[332]
	[M+14]	579[331], 579[332]
	[M+15]	580[332]
	[M]	588[241]
	[M+1]	589[241], 589[242]
	[M+2]	590[241], 590[242], 590[243]
	[M+3]	591[241], 591[242], 591[243], 591[244]
	[M+4]	592[241], 592[242], 592[243], 592[244], 592[245]
	[M+5]	593[241], 593[242], 593[243], 593[244], 593[245], 593[246]
	[M+6]	594[241], 594[242], 594[243], 594[244], 594[245], 594[246], 594[247]
	[M+7]	595[241], 595[242], 595[243], 595[244], 595[245], 595[246], 595[247]
ADPG	[M+8]	596[241], 596[242], 596[243], 596[244], 596[245], 596[246], 596[247]
	[M+9]	597[241], 597[242], 597[243], 597[244], 597[245], 597[246], 596[247]
	[M+10]	598[241], 598[242], 598[243], 598[244], 598[245], 598[246], 598[247]
	[M+11]	599[242], 599[243], 599[244], 599[245], 599[246], 599[247]
	[M+12]	600[243], 600[244], 600[245], 600[246], 600[247]
	[M+13]	601[244], 601[245], 601[246], 601[247]
	[M+14]	602[245], 602[246], 602[247]
	[M+15]	603[246], 603[247]
	[M+16]	604[247]

$[M-157]^+$ fragment ions that can help to confirm the identified peaks by manually searching the available MS library.

2. Identify ions for mass isotopomer analysis. Highly abundant fragment ions chosen for amino and organic acids can be found in Table 4. For photorespiration, it is vital to get MID data for glycine and serine.
3. Integrate mass isotopomer peaks.
4. Calculate relative MIDs and correct for natural isotope abundance.
5. Assess the precision and accuracy of MIDs prior to inclusion in modeling by comparison to unlabeled control samples or standards of known isotope composition.

3.18 Build the Flux Model

As summarized in the introduction [14, 15], current techniques for quantifying photorespiration all have limitations. As an alternative strategy, we use INST-MFA to quantitatively describe the integrated photorespiratory pathways and avoid other measurements and assumptions inherent to these methods. As indicated in Fig. 2, INST-MFA is concerned with solving an "inverse problem" where fluxes and pool sizes are estimated from measured labeling patterns and extracellular rates through the means of an iterative least-squares fitting procedure. At each iteration, a "forward problem" is solved where an isotopomer model is used to simulate labeling measurements for a given metabolic network and a given set of parameter estimates. The discrepancy between the simulated and measured labeling patterns is then assessed, and the parameter estimates are updated to achieve an improved fit. Once convergence to the best-fit solution is obtained, the procedure terminates and the optimal flux and pool size estimates are returned. INST-MFA can consider a combination of "active" and "inactive" pools such as glycine and serine that contribute to photorespiration but also other aspects of central metabolism (e.g., amino acid and folate metabolism) that may have significantly different labeling dynamics. Furthermore, INST-MFA does not require the direct measurement of pool sizes. The contributions of pool size and isotopic labeling measurements to estimated parameters as evaluated through sensitivity analysis [28] confirm that the flux determination via INST-MFA is not strongly correlated to direct pool size information.

1. Network specification. Reconstruct a metabolic network based upon the interactions between photorespiration, photosynthesis, and other central metabolic pathways.
   (a) Build the metabolic network based upon biochemical literature and annotated plant genomes (Fig. 5). The metabolic network must describe both the stoichiometry of all enzymatic reactions under consideration and atom

Table 4
Specific compound dependent MS parameters used in SIM with GC-MS

Amino and organic acids	Mass range	Carbon atoms	Fragmentation
[a]Serine 23	288–291	2, 3	$M-C_7H_{15}O_2Si$
[a]Serine 12	302–305	1, 2	$M-C_7H_{17}O_1Si$
[a]Serine 123	390–394	1, 2, 3	$M-C_4H_9$
[a]Glycine 2	218–220	2	$M-C_5H_9O$
[a]Glycine 12	246–249	1, 2	$M-C_4H_9$
Alanine 23	232–235	2, 3	$M-C_5H_9O$
Alanine 123	260–264	1, 2, 3	$M-C_4H_9$
Aspartate234	316–320	2, 3, 4	$M-C_7H_{15}O_2Si$
Aspartate 234	390–394	2, 3, 4	$M-C_5H_9O$
Aspartate 1234	418–423	1, 2, 3, 4	$M-C_4H_9$
Threonine 234	376–379	2, 3, 4	$M-C_5H_9O$
Threonine 1234	404–408	1, 2, 3, 4	$M-C_4H_9$
Glutamate 2345	330–334	2, 3, 4, 5	$M-C_7H_{15}O_2Si$
Glutamate 2345	404–408	2, 3, 4, 5	$M-C_5H_9O$
Proline 2345	258–262	2, 3, 4, 5	$M-C_5H_9O$
Asparagine 1234	417–421	1, 2, 3, 4	$M-C_4H_9$
Asparagine 12	302–304	1, 2	$M-C_8H_{18}NOSi$
Glutamine12345	431–436	1, 2, 3, 4, 5	$M-C_4H_9$
Citrate 123,456	591–598	1, 2, 3, 4, 5, 6	$M-C_4H_9$
Succinate 1234	289–294	1, 2, 3, 4	$M-C_4H_9$
Succinate 1234	331–336	1, 2, 3, 4	$M-CH_3$
Fumarate 1234	287–292	1, 2, 3, 4	$M-C_4H_9$
Malate 1234	419–424	1, 2, 3, 4	$M-C_4H_9$
Malate 1234	461–466	1, 2, 3, 4	$M-CH_3$

[a]Important photorespiratory intermediates

transitions for each reaction. Each metabolite is classified as balanced or unbalanced depending on whether it can accumulate, deplete, or remains at a constant level. For steady state operation the intermediates of metabolism are balanced and input substrates and output products are unbalanced. Reactions are then classified as either reversible or irreversible. The mass balance is supplied in a simple text format: e.g., "A (abc) → B (ab) + C (c)."

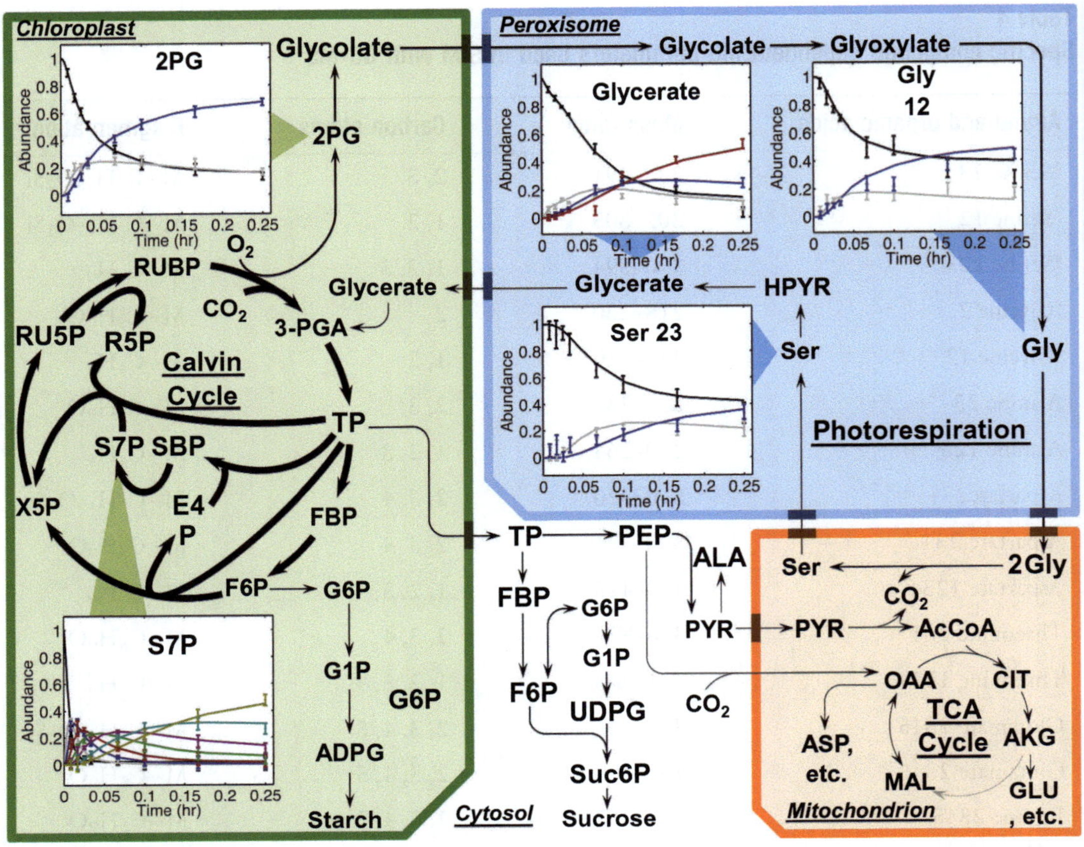

Fig. 5 Metabolic network of central metabolism in leaves. Thickness of arrows does not indicate actual fluxes. Measured (*data points*) and simulated (*lines*) MIDs of 2PG, S7P, glycerate, serine, and glycine are *highlighted*

In this example the uppercase letters are metabolite names and lower case letters indicate individual atoms that are subject to labeling and rearrangement through metabolism. The modeling process should attempt to reconcile all available experimental measurements (in this case, including isotopic labeling data, net CO_2 assimilation rate, starch production rate, and the steady state sucrose and amino acid concentrations in vascular exudates) without adding undue complexity that would lead to parameter overfitting and poor sensitivity.

(b) When dictated by poor fitting, or possibly the availability of organelle-specific measurements, modeled networks are designed to incorporate subcellular compartmentation. Different organelles can independently maintain the same biochemical reactions; therefore compartmentalized pools for each intermediate should be treated as separate nodes in isotopomer models. In the case of leaf photosynthetic metabolism, accessing subcellular compartmentation is facilitated by investigation of MIDs of ADPG and

UDPG, which are the respective precursors of starch and sucrose synthesis. Precursors that originate within distinct subcellular locations (plastid and cytosol, respectively) can be labeled to different extents reflecting differences in metabolism; however sometimes the differences can be small and may not justify compartmentalization. Other subcellular readouts from isotopically labeled proteins [30–33]; lipids [34, 35], or cell wall [34, 36] have helped describe metabolic steady states in other parts of plants.

(c) Describe the cellular heterogeneity of leaves. Both the multicellular and heterogeneous anatomy of leaves and the mixing of compartmentalized pools during cell lysis contribute to photosynthetically less active or "inactive" pools of metabolic intermediates. Inactive pools are not significantly enriched within 15 min which consequently dilute ^{13}C labeling. To reflect this aspect of central metabolism, dilution parameters to accommodate inactive pools are required for statistically acceptable fitting of labeling data. Inactive pools are built into the metabolic network through the use of dilution reactions, for example, the merging of two dilution fluxes to form the measured ADPG pool

0*ADPG.x (abcdef) → ADPG (abcdef)
0*ADPG.cp (abcdef) → ADPG (abcdef)

in combination with other reactions that generate ADPG. cp as part of metabolism
G1P.cp (abcdeg) → ADPG.cp (abcdef)
and a "sink" reaction that is assigned a value of 100
ADPG → sink
ADPG.cp is the chloroplast precursor for starch and ADPG.x is an unlabeled pool of ADPG that includes its presence in other cells potentially or places where its labeling and turnover are substantially slower, thus diluting the readout of measured ADPG relative to ADPG.cp (i.e., the active pool). The zero stoichiometric coefficients for ADPG.x and ADPG.cp in the dilution reactions ensure that these pseudo-reactions do not perturb the mass balance on ADPG.cp, rather they just contribute to the measured ADPG labeling. The value of "sink" is assigned to 100 so that each estimated dilution flux represents its percent contribution to the measured ADPG pool. The pool size of the mixed ADPG pool should be assigned a value of 0 to model instantaneous mixing of the segregated ADPG pools upon cell lysis.

(d) Longer labeling experiments (i.e., 1 h) can help to assess the sizes of inactive pools. The relative abundances of M0 isotopomer at 60 min can either be directly applied to modeling or be used as an independent validation for the model-derived dilution parameters based on shorter labeling experiments (i.e., 15 min).

(e) Comparison between the model-derived dilution parameters (a fraction of the total intracellular pool) and measured inactive pools at 60 min. For a number of CBB intermediates, the inactive pool sizes are relatively small and in qualitative agreement with the M0 abundances measured at 60 min. For amino acids, especially those involved in photorespiration, some qualitative discrepancies exist. These amino acids exist in multiple subcellular pools, some of which turn over at time scales that differ from those more rapidly labeled CBB intermediates leading to the complex labeling dynamics observed. Furthermore, these discrepancies change when environmental conditions change (i.e., low light to high light), suggesting photorespiration plays different roles in amino acid metabolism.

2. Mathematical aspects of the network that are accounted for in a computational model.

(a) The stoichiometric matrix *S*. Fluxes must satisfy the stoichiometric constraint $S.\nu = 0$, where *S* is the stoichiometric matrix and ν is the flux vector. *S* is a $k \times j$ matrix, with *k* metabolites and *j* fluxes. Reversible reactions are specified as forward and backward fluxes, respectively.

(b) The null space is evaluated to define the choices for free flux parameters. Free fluxes are determined by $\nu = N.u$ where *N* is the null space matrix of *S* and *u* is the vector of free fluxes. The size of the null space matrix and the number of independent flux variables are determined by the rank of the stoichiometric matrix.

(c) The minimal set of elementary metabolic units (EMUs) required to simulate the MIDs of the measured metabolites is established. An EMU is defined as a distinct subset of a metabolite's atoms and can exist in a variety of mass states depending on its isotopic composition. Decomposing the isotopomer network into EMUs reduces the computational demand and results in isotopomer distributions of EMUs that contribute to the available measurements. The number of ordinary differential equations (ODEs) that are needed is minimized enabling much faster convergence. Consequently, this strategy can increase the efficiency of solving the inverse problem of INST-MFA because each iteration of the parameter estimation procedure can be accomplished in minimal time.

The application of EMUs to INST-MFA makes use of Dulmage-Mendelsohn-decomposition [37, 38] and more details are described in [23].

3. Perform simulations to optimize flux and pool size parameters. Flux measurements, isotope labeling measurements, and isotope tracer substrates are specified along with the atom transition-based metabolic network and computational simulations are performed to best fit the data and establish flux parameters. The parameter estimation is repeated starting with random combinations of fluxes that are constrained to be within the solution space (i.e., within the flux polygon, or null space solution). The convergence to the same solution repeatedly that also has the best chi-square fit indicates the best estimates for fluxes globally.
4. Evaluate the goodness-of-fit. The convergence to a final solution will continue until no further improvements are made to the sum of squared residuals (SSR) of the objective function. Testing the goodness-of-fit will determine whether the optimal solution is statistically acceptable based on the minimized SSR. The minimized variance-weighted SSR is compared to a chi-square distribution with $n-p$ degrees of freedom (DOF), where n is the number of independent measurements and p is the number of estimated parameters. To achieve an acceptable fit, the calculated SSR must fall within the expected range of $[\chi^2_{\alpha/2}, \chi^2_{1-\alpha/2}]$, where α is a chosen threshold value corresponding to the desired confidence level (i.e., 0.05 for 95% confidence or 0.01 for 99% confidence). Inspecting individual residual error values for measurements or plotting the simulated and measured MIDs of each MS fragment ion is informative as it can indicate gross errors in the network, erroneous measurements, or areas of the network that are not accurately represented and require further consideration.
5. Calculate the 95% confidence intervals using either parameter continuation methods [39] or Monte Carlo analysis. Large confidence intervals indicate poor flux determination and small confidence intervals indicate well-determined fluxes for the given network description.
6. Report the flux values and flux uncertainties through generation of a flux map. After the flux determination is completed and an optimal solution is obtained, a flux map will provide the best visual depiction of metabolite flow through the network (*see* **Note 11**).
7. Computational assessment of the sensitivity of flux estimates (Fig. 6). Sensitivity expressions are generated by taking the first derivative of isotopomer balances with respect to each free flux. Solve the sensitivity equations to calculate the first derivatives

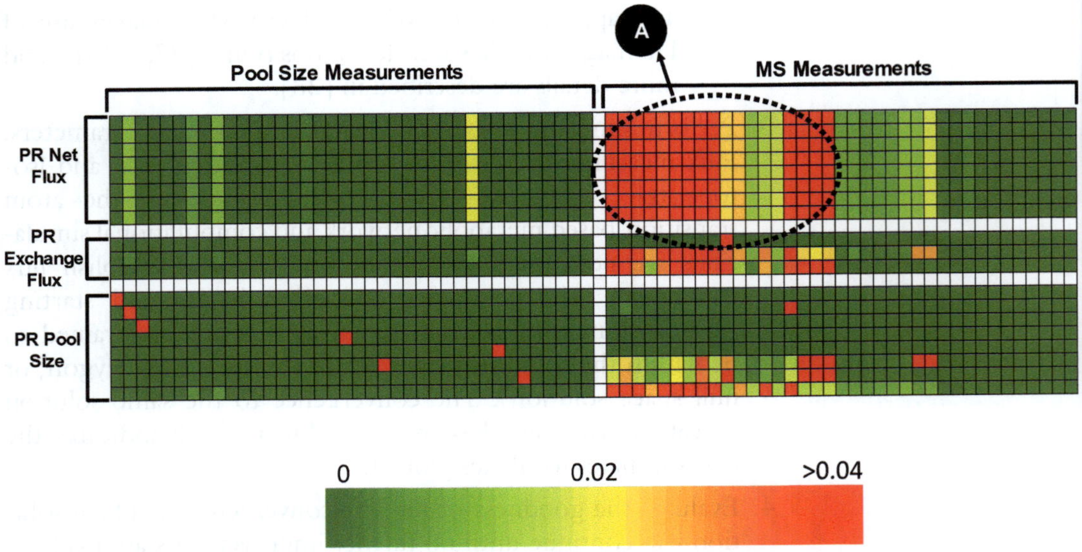

Fig. 6 Contribution Matrix. This heat map shows the contribution of pool sizes and MS labeling measurements to estimated parameters in photorespiration (PR). The effects of the pool size measurements on estimated fluxes were investigated by treating the pool sizes as measurements and computing the contribution matrix based on local parameter sensitivities [39]. Each element of this contribution matrix represents the fractional contribution of each measurement (*columns*) to the local variance of each estimated parameter (*rows*). This heat map shows that the MS labeling measurements are more important than pool size measurements in determining the estimated net flux, exchange flux, and pool size in photorespiration. Specifically, the MS measurements that had the greatest impact on the PR net flux estimates were the measurements included in the Calvin cycle and photorespiration: RUBP, PGA, DHAP, FBP, F6P, P5P, R5P, S7P, 2PG, PEP, and serine (as noted by the cells in *red* in "A")

of the measured MIDs with respect to the free parameters. Examining the sensitivity through heat maps (e.g., [28]) can indicate which measurements were most crucial in establishing a given flux value and can potentially inspire further studies or better explain results.

4 Notes

1. To enhance the efficiency of derivatization, ensure acetonitrile and other components have not taken up moisture.
2. Arabidopsis rosettes are not big enough to cover the standard 2 × 3 cm leaf chamber. The actual leaf area of each rosette must be individually measured, recorded, and calculated via ImageJ.
3. Since the purpose of measuring RuBisCO protein content is to establish a rough estimate of when RuBisCO is maximally present more sensitive and time-consuming methods such as western blots are generally not necessary.
4. Plants photosynthetic metabolism is closely linked with their developmental stages. Choices of appropriate developmental

stages for isotopic labeling investigations are determined by various research goals.

5. Plant metabolism must be maintained in a fixed state during the transient labeling. Leaves usually exhibit limited metabolic steady state. To determine the appropriate time window for transient labeling, one must take the diurnal changes of metabolic activities into consideration. Sucrose and starch are produced by photosynthetic metabolism in leaves and therefore are the most direct indicators for evaluating if metabolism stays steadily or not.
6. The minimum total flow rate is determined based on the volume of headspace in each labeling chamber. For the Arabidopsis labeling chambers (Fig. 3b) with a 60 mL headspace, a total flow rate in 2 L/min can turn over the gas in headspace for more than 30 times within the first minute. This design makes sure the concentration of $^{13}CO_2$ can achieve the desired level within seconds (Fig. 3c).
7. The 2 h cold incubation step is vital for efficient extraction of polar metabolites, which cannot be skipped.
8. To test the chromatographic peak quality, the mass isotopomer distribution data of each metabolite in unlabeled samples are mass corrected to see if they match the natural isotope abundances. Poor mass correction results suggest that samples are either too concentrated or too diluted.
9. Abbreviations: *PGA* phosphoglycerate, *DHAP* dihydroxyacetone phosphate, *FBP* fructose 1,6-bisphosphate, *F6P* fructose-6-phosphate, *G6P* glucose-6-phosphate, *S7P* sedoheptulose-7-phosphate, *P5P* pentose-5-phosphates, *R5P* ribose-5-phosphate, *RUBP* ribulose-1,5-bisphosphate, *G1P* glucose-1-phosphate, *ADPG* ADP-glucose, *UDPG* UDP-glucose, *PEP* phosphoenolpyruvate, *αKG* alpha-ketoglutarate, *2PG* 2-phosphoglycolate.
10. Available LC-MS/MS freeware include MZmine (http://mzmine.github.io) and XCMS (http://metlin.scripps.edu/xcms), with the latter running in the R statistical programming environment. Both programs require the conversion of raw data files into a nonproprietary format, such as mzXML, or NetCDF.
11. Available tools for flux map visualization include: FluxMap [40], Omix [41], FluxViz [42], OptFlux [43], and Escher [44].

Acknowledgements

Any product or trademark mentioned here does not imply a warranty, guarantee, or endorsement by the authors or their affiliations over other suitable alternatives. Mass spectrometry methods were developed using instrumentation in part supported by National Science Foundation grants (DBI-1427621 and DBI-0521250 for acquisition of QTRAP LC-MS/MS instruments). We acknowledge support from the Department of Energy (DE-AR0000202) to F. M. and D.K.A.

References

1. Bowes G, Ogren WL, Hageman RH (1971) Phosphoglycolate production catalyzed by ribulose diphosphate carboxylase. Biochem Biophys Res Commun 45(3):716–722. 0006-291X(71)90475-X [pii]
2. Ogren WL, Bowes G (1971) Ribulose diphosphate carboxylase regulates soybean photorespiration. Nat New Biol 230(13):159–160
3. Ogren WL (1984) Photorespiration: pathways, regulation, and modification. Annu Rev Plant Physiol 35(1):415–442. doi:10.1146/annurev.pp.35.060184.002215
4. Sage RF, Sage TL, Kocacinar F (2012) Photorespiration and the evolution of C4 photosynthesis. Annu Rev Plant Biol 63:19–47. doi:10.1146/annurev-arplant-042811-105511
5. Tcherkez GG, Farquhar GD, Andrews TJ (2006) Despite slow catalysis and confused substrate specificity, all ribulose bisphosphate carboxylases may be nearly perfectly optimized. Proc Natl Acad Sci U S A 103(19):7246–7251. doi:10.1073/pnas.0600605103. 0600605103 [pii]
6. Parry MAJ, Madgwick PJ, Carvalho JFC, Andralojc PJ (2007) Paper presented at international workshop on increasing wheat yield potential, CIMMYT, Obregon, Mexico, 20–24 Mar 2006. Prospects for increasing photosynthesis by overcoming the limitations of Rubisco. J Agric Sci 145(01):31–43. doi:10.1017/S0021859606006666
7. Bauwe H, Hagemann M, Fernie AR (2010) Photorespiration: players, partners and origin. Trends Plant Sci 15(6):330–336. doi:10.1016/j.tplants.2010.03.006. S1360-1385(10)00062-2 [pii]
8. Bauwe H, Hagemann M, Kern R, Timm S (2012) Photorespiration has a dual origin and manifold links to central metabolism. Curr Opin Plant Biol 15(3):269–275. doi:10.1016/j.pbi.2012.01.008. S1369-5266(12)00009-X [pii]
9. Servaites JC (1977) Chemical inhibition of the glycolate pathway in soybean leaf cells. Plant Physiol 60(4):461–466
10. Wingler A, Lea PJ, Leegood RC (1997) Control of photosynthesis in barley plants with reduced activities of glycine decarboxylase. Planta 202(2):171–178
11. Heineke D, Bykova N, Gardestrom P, Bauwe H (2001) Metabolic response of potato plants to an antisense reduction of the P-protein of glycine decarboxylase. Planta 212(5–6):880–887
12. Rachmilevitch S, Cousins AB, Bloom AJ (2004) Nitrate assimilation in plant shoots depends on photorespiration. Proc Natl Acad Sci U S A 101(31):11506–11510. doi:10.1073/pnas.0404388101. 0404388101 [pii]
13. Tcherkez G, Bligny R, Gout E, Mahe A, Hodges M, Cornic G (2008) Respiratory metabolism of illuminated leaves depends on CO_2 and O_2 conditions. Proc Natl Acad Sci U S A 105(2):797–802. doi:10.1073/pnas.0708947105. 0708947105 [pii]
14. Sharkey TD (1988) Estimating the rate of photorespiration in leaves. Physiol Plant 73(1):147–152. doi:10.1111/j.1399-3054.1988.tb09205.x
15. Busch FA (2013) Current methods for estimating the rate of photorespiration in leaves. Plant Biol (Stuttg) 15(4):648–655. doi:10.1111/j.1438-8677.2012.00694.x
16. Dandekar T, Fieselmann A, Majeed S, Ahmed Z (2014) Software applications toward quantitative metabolic flux analysis and modeling. Brief Bioinform 15(1):91–107. doi:10.1093/bib/bbs065. bbs065 [pii]
17. Young JD (2014) INCA: a computational platform for isotopically non-stationary metabolic flux analysis. Bioinformatics 30(9):1333–1335. doi:10.1093/bioinformatics/btu015. btu015 [pii]

18. Kajihata S, Furusawa C, Matsuda F, Shimizu H (2014) OpenMebius: an open source software for isotopically nonstationary ^{13}C-based metabolic flux analysis. Biomed Res Int 2014:627014. doi:10.1155/2014/627014
19. Yuan J, Fowler WU, Kimball E, Lu W, Rabinowitz JD (2006) Kinetic flux profiling of nitrogen assimilation in *Escherichia coli*. Nat Chem Biol 2(10):529–530. doi:10.1038/nchembio816. nchembio816 [pii]
20. Yuan J, Bennett BD, Rabinowitz JD (2008) Kinetic flux profiling for quantitation of cellular metabolic fluxes. Nat Protoc 3(8):1328–1340. doi:10.1038/nprot.2008.131. nprot.2008.131 [pii]
21. Szecowka M, Heise R, Tohge T, Nunes-Nesi A, Vosloh D, Huege J, Feil R, Lunn J, Nikoloski Z, Stitt M, Fernie AR, Arrivault S (2013) Metabolic fluxes in an illuminated Arabidopsis rosette. Plant Cell 25(2):694–714. doi:10.1105/tpc.112.106989. tpc.112.106989 [pii]
22. Antoniewicz MR, Kelleher JK, Stephanopoulos G (2007) Elementary metabolite units (EMU): a novel framework for modeling isotopic distributions. Metab Eng 9(1):68–86. doi:10.1016/j.ymben.2006.09.001. S1096-7176(06)00084-X [pii]
23. Young JD, Walther JL, Antoniewicz MR, Yoo H, Stephanopoulos G (2008) An elementary metabolite unit (EMU) based method of isotopically nonstationary flux analysis. Biotechnol Bioeng 99(3):686–699. doi:10.1002/bit.21632
24. Young JD, Shastri AA, Stephanopoulos G, Morgan JA (2011) Mapping photoautotrophic metabolism with isotopically nonstationary (13)C flux analysis. Metab Eng 13(6):656–665. doi:10.1016/j.ymben.2011.08.002. S1096-7176(11)00088-7 [pii]
25. Adebiyi AO, Jazmin LJ, Young JD (2015) ^{13}C flux analysis of cyanobacterial metabolism. Photosynth Res 126(1):19–32. doi:10.1007/s11120-014-0045-1
26. Xiong W, Lee TC, Rommelfanger S, Gjersing E, Cano M, Maness PC, Ghirardi M, Yu J (2015) Phosphoketolase pathway contributes to carbon metabolism in cyanobacteria. Nat Plants 2:15187. doi:10.1038/nplants.2015.187. nplants2015187 [pii]
27. Wu C, Xiong W, Dai J, Wu Q (2015) Genome-based metabolic mapping and ^{13}C flux analysis reveal systematic properties of an oleaginous microalga *Chlorella protothecoides*. Plant Physiol 167(2):586–599. doi:10.1104/pp.114.250688. pp.114.250688 [pii]
28. Ma F, Jazmin LJ, Young JD, Allen DK (2014) Isotopically nonstationary ^{13}C flux analysis of changes in *Arabidopsis thaliana* leaf metabolism due to high light acclimation. Proc Natl Acad Sci U S A 111(47):16967–16972. doi:10.1073/pnas.1319485111. 1319485111 [pii]
29. Arnon DI (1949) Copper enzymes in isolated chloroplasts. Polyphenoloxidase in *Beta vulgaris*. Plant Physiol 24(1):1–15
30. Allen DK, Laclair RW, Ohlrogge JB, Shachar-Hill Y (2012) Isotope labelling of Rubisco subunits provides in vivo information on subcellular biosynthesis and exchange of amino acids between compartments. Plant Cell Environ 35(7):1232–1244. doi:10.1111/j.1365-3040.2012.02485.x
31. Mandy DE, Goldford JE, Yang H, Allen DK, Libourel IG (2014) Metabolic flux analysis using (1)(3)C peptide label measurements. Plant J 77(3):476–486. doi:10.1111/tpj.12390
32. Allen DK, Goldford J, Gierse JK, Mandy D, Diepenbrock C, Libourel IG (2014) Quantification of peptide m/z distributions from ^{13}C-labeled cultures with high-resolution mass spectrometry. Anal Chem 86(3):1894–1901. doi:10.1021/ac403985w
33. Allen DK, Evans BS, Libourel IG (2014) Analysis of isotopic labeling in peptide fragments by tandem mass spectrometry. PLoS One 9(3):e91537. doi:10.1371/journal.pone.0091537. PONE-D-13-05119 [pii]
34. Allen DK, Shachar-Hill Y, Ohlrogge JB (2007) Compartment-specific labeling information in ^{13}C metabolic flux analysis of plants. Phytochemistry 68(16–18):2197–2210. doi:10.1016/j.phytochem.2007.04.010. S0031-9422(07)00246-4 [pii]
35. Lonien J, Schwender J (2009) Analysis of metabolic flux phenotypes for two Arabidopsis mutants with severe impairment in seed storage lipid synthesis. Plant Physiol 151(3):1617–1634. doi:10.1104/pp.109.144121. pp.109.144121 [pii]
36. Sriram G, Iyer VV, Bruce Fulton D, Shanks JV (2007) Identification of hexose hydrolysis products in metabolic flux analytes: a case study of levulinic acid in plant protein hydrolysate. Metab Eng 9(5–6):442–451. doi:10.1016/j.ymben.2007.07.003. S1096-7176(07)00045-6 [pii]
37. Dulmage AL, Mendelsohn NS (1958) Coverings of bipartite graphs. Can J Math 10:517–534
38. Pothen A, Fan C-J (1990) Computing the block triangular form of a sparse matrix. ACM Trans Math Softw 16(4):303–324. doi:10.1145/98267.98287

39. Antoniewicz MR, Kelleher JK, Stephanopoulos G (2006) Determination of confidence intervals of metabolic fluxes estimated from stable isotope measurements. Metab Eng 8 (4):324–337. doi:10.1016/j.ymben.2006.01.004. S1096-7176(06)00006-1 [pii]
40. Rohn H, Hartmann A, Junker A, Junker BH, Schreiber F (2012) FluxMap: a VANTED add-on for the visual exploration of flux distributions in biological networks. BMC Syst Biol 6:33. doi:10.1186/1752-0509-6-33. 1752-0509-6-33 [pii]
41. Droste P, Miebach S, Niedenführ S, Wiechert W, Nöh K (2011) Visualizing multi-omics data in metabolic networks with the software Omix—a case study. Biosystems 105 (2):154–161. doi:10.1016/j.biosystems.2011.04.003
42. Konig M, Holzhutter HG (2010) Fluxviz – Cytoscape plug-in for visualization of flux distributions in networks. Genome Inform 24:96–103. 9781848166585_0008 [pii]
43. Rocha I, Maia P, Evangelista P, Vilaca P, Soares S, Pinto JP, Nielsen J, Patil KR, Ferreira EC, Rocha M (2010) OptFlux: an open-source software platform for in silico metabolic engineering. BMC Syst Biol 4:45. doi:10.1186/1752-0509-4-45. 1752-0509-4-45 [pii]
44. King ZA, Drager A, Ebrahim A, Sonnenschein N, Lewis NE, Palsson BO (2015) Escher: a web application for building, sharing, and embedding data-rich visualizations of biological pathways. PLoS Comput Biol 11 (8):e1004321. doi:10.1371/journal.pcbi.1004321. PCOMPBIOL-D-15-00439 [pii]

Chapter 13

Genome-Scale Modeling of Photorespiratory Pathway Manipulation

Anika Küken and Zoran Nikoloski

Abstract

Quantifying the redistribution of metabolic reaction fluxes under experimental scenarios that affect the photorespiratory pathway can provide insights about the coupling of this pathway with other parts of metabolism. However, differential flux profiling on a genome-scale level remains the biggest challenge in modern systems biology. Here we present a protocol for applying a constraint-based approach, termed iReMet-Flux, that integrates data about relative metabolite levels in a stoichiometric metabolic model to predict differential fluxes at a genome-scale level under mild modeling assumptions. We demonstrate how iReMet-Flux can be employed to investigate the interplay between photorespiration and other pathways at a genome-scale level, and complements flux profiling methods based on radioactive tracer labeling.

Key words Constraint-based modelling, Differential flux profiling, Data integration, Metabolomics, Genome-scale metabolic networks

1 Introduction

The role and interplay of the photorespiratory pathway with the rest of plant metabolism has long been in the focus of photosynthetic research and computational biology. For instance, it is now known that the role of the photorespiratory pathway goes beyond recycling of organic carbon to the Calvin-Benson cycle (CBC), and mounting evidence points to its important role in shaping the energetics of photosynthesis, compartmental reductant exchange, nitrate assimilation, one-carbon metabolism, and redox signal transduction [1–3]. Nevertheless, thorough understanding of the implications of tempering with the photorespiratory pathway (e.g., gene knockout or overexpression) for the whole plant metabolism is still lacking.

One way to understand the role of photorespiration is to quantify the effect of modulating different enzymatic steps on redistribution of fluxes not only in photorespiration but also in other pathways. Since enzyme-catalyzed reactions and their fluxes are

Alisdair R. Fernie et al. (eds.), *Photorespiration: Methods and Protocols*, Methods in Molecular Biology, vol. 1653,
DOI 10.1007/978-1-4939-7225-8_13, © Springer Science+Business Media LLC 2017

mutually related and depend on the involved metabolite pools, changes in metabolite levels may be related to changes in fluxes, although a direct relationship is not always obvious [4–6]. The determination of metabolic fluxes is a laborious task, as it is based on radioactive tracer labeling in combination with modeling [7]. To this end, computational methods used in ^{13}C metabolic flux analysis (^{13}C-MFA) have facilitated the estimation of steady-state fluxes on the level of small-scale networks [8]. These methods are well established in elucidating fluxomes of central carbon metabolism of prokaryotes and, recently, of more complex eukaryotes, including plants [9–12]. In addition, methods which use only the dynamics of the wash-out of the unlabeled fraction in metabolic pools have also been applied to provide flux estimates in canonical plant pathways, including starch and sucrose syntheses [11].

In contrast, constraint-based approaches, like flux balance analysis (FBA) [13], allow predicting flues on a genome-scale level without usage of data, but under the assumption that the system optimizes an objective (e.g., biomass yield [14]) at steady state. However, predictions from general optimization programs often result in multiple optimal steady-state flux distributions. To reduce the solution space of feasible flux phenotypes, high-throughput data obtained from various cellular levels, i.e., the transcriptome, proteome and metabolome, have been used [15, 16].

To date, there exist only a few methods allowing for integration of metabolite levels into large-scale constraint-based modeling approaches able to predict flux phenotypes. A recent review investigated the applicability of these approaches to large-scale networks [17] and identified some shortcomings including: (1) parameter-dependence (e.g., Integrative Omics-Metabolic Analysis [18]), as well as the need for data about metabolite concentrations, (2) consideration of presence/absence of the metabolite only (e.g., Gene Inactivation Moderated by Metabolism, Metabolomics and Expression GIM3E [19]), (3) the need for time-series data about absolute metabolite level alongside with transcriptomics and/or proteomics read-outs (e.g., TREM-Flux [20]), or (4) integration of thermodynamics constraints which leads to nonlinear optimization, and therefore may not predict the global optimum (e.g., thermodynamics-based metabolic flux analysis (TMFA) [21]). However, none of these methods allow for differential flux profiling.

In this chapter, we describe a detailed protocol for the usage of a recently devised constraint-based method, termed integration of Relative Metabolite Levels for Flux prediction, iReMet-Flux [22]. This approach allows the inclusion of data about relative metabolite levels that are closer to the physiological state, while maintaining the simplicity of a constraint-based framework and being (largely) parameter-independent. The approach only relies on the assumptions that reaction kinetics follows the law of mass action and that organisms tend to minimize change in flux between two different environments. With the aid of iReMet-Flux the investigation of changes in activity of pathways which are beyond the point of the natural tracer ($^{13}CO_2$) become possible. The approach has already been used to provide predictions about changes in fluxes in all reactions of a genome-scale metabolic network of *Arabidopsis thaliana* in which relative metabolite levels between Col0 and four knockout mutants for key steps of the photorespiratory pathway were integrated. Therefore, iReMet-Flux paves the wave for mechanistic dissection of the interplay between pathways in variety of organisms and experimental scenarios.

2 Materials

2.1 Software Packages

1. Standard personal computer that can run MATLAB R2007b or above (http://mathworks.com).
2. COBRA Toolbox version 1.3.4 or above (http://systemsbiology.ucsd.edu/downloads/COBRAToolbox).
3. SBML Toolbox for MATLAB to read models in SBML format (http://sbml.org/Software/SBMLToolbox).
4. TOMLAB Optimization Toolbox (http://tomopt.com) with the Sparse Non-linear OPTimizer (SNOPT).

2.2 Metabolic Model Reconstruction and Metabolomics Data

1. Large-scale metabolic models that have been previously published in peer-reviewed journals can be downloaded in SBML format from databases, like BIGG (http://bigg.ucsd.edu/) or Biomodels (http://www.ebi.ac.uk/biomodels-main/), or the model can be accessed directly from their original publication (*see* **Note 1**).
2. Metabolomics data under two different experimental scenarios for a set of metabolites included in the large-scale metabolic network saved in a text file (e.g., CSV format).

3 Methods

3.1 Prerequisites

1. Install MATLAB and the required Toolboxes (SBML Toolbox and COBRA Toolbox, TOMLAB Optimization Toolbox).
2. Change into directory where COBRA Toolbox is located and initiate it by entering the command *initCobraToolbox* (*see* **Note 2**).
3. Initiate TOMLAB Optimization Toolbox by changing into TOMLAB located directory and entering command *startup*.
4. Change to the working directory including SBML file of interest and metabolomics data file(s).

3.2 Importing and Adapting Data

1. Load metabolic reconstruction from SBML file into MATLAB using the *readCbModel* function of the COBRA Toolbox (for a protocol about using functions of the COBRA Toolbox *see* [23]).
2. Among others, the obtained structured variable *model* includes fields important in setting up the optimization problem:

   model.S—the stoichiometric matrix S with n rows corresponding to metabolites and p columns corresponding to reactions, is a mathematical model representation consisting of stoichiometric coefficients s_{ij} indicating the molarity with which a metabolite x_i enters reaction r_j.

   model.rev—vector of size $p \times 1$ indicating reaction reversibility, includes 1 for reversible reactions and 0 otherwise.

   model.lb. and *model.ub*—vectors of size $p \times 1$ including values for lower (*lb*) and upper bound (*ub*) on flux through a particular reaction; by default *ub* is set to 1000, *lb*. is set to 0 for irreversible reactions and −1000 for reversible reactions.

   model.c—vector of size $p \times 1$ containing objective coefficients for each reaction in the system.
3. Change model metabolite import based on experimental growth conditions.
4. To model gene knockout mutants constrain the flux through the reactions related to the respective gene to zero.
5. Metabolomics data for experimental scenarios A and B can be imported from text files using the preferred MATLAB function for reading text files into a matrix (e.g., *importdata*, *dlmread*, *readtable*).

3.3 Setting Up the Optimization Problem of iReMet-Flux

This section describes notations and assumptions used in formulating iReMet-Flux shown in Eq. 1. The resulting quadratic problem with linear constraints can be solved by nonlinear programming solvers like SNOPT provided by the TOMLAB Optimization Toolbox. A description of how to amplify the quadratic problem for investigating the variability of selected reaction fluxes in optimal solutions can be found in the original publication [22].

1. The flux v through an irreversible reaction i, modeled by mass action kinetics, is given by $v_i = k_i E_i \prod_{j=1}^{n} x_j^{s_{ji}}$, where k_i is the rate constant, E_i denotes the concentration of the enzyme, x_j, the concentration of metabolite j, $1 \leq j \leq n$ and s_{ji} denotes the stoichiometric coefficient of metabolite j entering as a substrate in reaction i (*see* **Note 3**).
2. Taking the ratio over experimental condition A and B (e.g., wild type and photorespiratory mutants) we obtain expression

$\frac{v_i^A}{v_i^B} = \frac{E_i^A}{E_i^B} \prod_{j=1}^{n} \left(\frac{x_j^A}{x_j^B} \right)^{s_{ji}}$ which is independent of k_i.

3. From this expression we obtain the notation of variables used in the problem formulation of iReMet-Flux shown in Eq. 1: $r_j = \frac{x_j^A}{x_j^B}$, $q_i = \frac{E_i^A}{E_i^B}$ and, therefore, $\frac{v_i^A}{v_i^B} = q_i \prod_{j=1}^{n} r_j^{s_{ji}}$, or equivalently, $v_i^A = \left(q_i \prod_{j=1}^{n} r_j^{s_{ji}} \right) v_i^B$.
4. If ratio r_j has been experimentally quantified, then $\chi(r_j) = 1$, and $\chi(r_j) = 0$, otherwise.
5. For metabolites j with a measured ratio, $\widehat{r}_j^{\min} \leq \widehat{r}_j \leq \widehat{r}_j^{\max}$, the lower bound $r_j^{\min} = \widehat{r}_j^{\min}$ and upper bound $r_j^{\max} = \widehat{r}_j^{\max}$, respectively (*see* **Note 4**).
6. For metabolites which cannot be measured use extremal values found in a specific experiment.
7. The same strategy can be employed for ratios on enzymes, where measured ratios of enzyme activities can be used to set boundaries.
8. The set of substrates of reaction i is denoted by $E(i)$ and the set of irreversible reactions with at least one measured substrate ratio is denoted by $\mathfrak{I} = \left\{ i | \sum_{j \in E(i)} \chi(r_j) > 0 \right\}$ (*see* **Note 5**).
9. Additional assumptions: (1) the system operates in a (quasi) steady-state, and therefore $Sv^A = Sv^B = 0$, (2) the system optimizes for a certain task, here, biomass yield (*see* **Note 6**), thus, biomass yield in scenario B is a fraction f of the optimal biomass yield in scenario A, and (3) the distance between the two flux distributions for scenarios A and B is minimal, specifically $\|v^A - v^B\|_2$ is minimized (*see* **Note 7**).
10. To obtain v_{biomass}^A and v_{biomass}^B apply standard FBA with objective function maximize v_{biomass} under steady-state constraint and $v_{\min} \leq v \leq v_{\max}$.

$$z = \min_{v^A, v^B} \left\| v^A - v^B \right\|_2$$

s.t.

$$Sv^A = Sv^B = 0$$

$$v_{\text{biomass}}^B \geq f v_{\text{biomass}}^A$$

$$v_{\min}^A \leq v^A \leq v_{\max}^A$$

$$v_{\min}^B \leq v^B \leq v_{\max}^B$$

$$\forall i \in \mathfrak{I} : v_i^B \prod_{j \in E(i)} \left(r_j^{\min} \right)^{S_{ij}} \leq v_i^A \leq v_i^B \prod_{j \in E(i)} \left(r_j^{\max} \right)^{S_{ij}}$$

$$\forall j \in \{l | \chi(r_l) = 1\} : r_j^{\min} = \hat{r}_j^{\min}, r_j^{\max} = \hat{r}_j^{\max}$$

$$\forall j \in \{l | \chi(r_l) = 0\} : r_j^{\min} = \min_l \hat{r}_l^{\min}, r_j^{\min} = \max_l \hat{r}_l^{\max} \quad (1)$$

3.4 Possible Applications

1. Determine the extent of flux redistribution for each of the scenarios compared by investigating the predicted values of the optimal Euclidean distance between steady-state flux distributions for scenarios *A* and *B*.
2. Find reactions with the largest sum of flux changes over all analyzed scenarios *B* with respect to scenario *A* by ranking the observed alteration in flux.
3. Find reactions with the largest deviation from the assumed mass action kinetics, as assessed by the value of the introduced relaxation (*see* **Note 7**).

4 Notes

1. The SBML file describing the model has to include at least the following information: stoichiometry of each reaction, upper/lower bounds of each reaction, and objective function coefficients for each reaction.
2. In case *initCobraToolbox* leads to an error try adding the COBRA path to you MATLAB search path by command *addpath(genpath('path to COBRA folder'))*.
3. Incorporation of ratio constraints depends on a reaction's reversibility; in the case of reversible reactions a mixed-integer quadratic program (QP) with nonlinear constraints has to be solved which is computationally demanding.
4. In applying the approach with a compartmentalized model it is assumed that the measured relative levels were uniform across all compartments.
5. The presented version of iReMet-Flux is restricted to the subset of irreversible reactions. For the formulation of reversible reactions *see* Supplementary information of the original publication [22].
6. In contrast to MOMA which also minimizes the Euclidean distance between flux distributions from two scenarios, iReMet-Flux minimizes the Euclidean distance between any two flux distributions which are compatible with the constraints, and, hence, iReMet-Flux does not depend on a (precomputed) reference flux distribution. Model reaction(s) related to the biomass yield or task optimized by the system are indicated by nonzero objective function coefficients in variable *model.c*.

7. Mass action kinetics assumes that an enzyme is not substrate saturated; however, this may not be the case for all enzymes considered in a large-scale model. In this case, the metabolite ratio constraints are expected to lead to infeasibilities. To resolve this issue, iReMet-Flux can be updated to account for deviation from the mass-action-like form of the fluxes by inclusion of slack variables, ε_i, that relax the flux constraints based on the metabolite ratios:

$$\forall i \in \mathfrak{I} : v_i^B \prod_{j \in E(i)} \left(r_j^{\min}\right)^{S_{ij}} \leq v_i^A + \varepsilon_i \leq v_i^B \prod_{j \in E(i)} \left(r_j^{\max}\right)^{S_{ij}}.$$

In addition, the objective function is then updated to: $\min_{v^A, v^B} \left\| v^A - v^B \right\|_2 + \epsilon \left\| \varepsilon \right\|_1$, where ϵ denotes a weighting factor for the deviations away from the assumed mass action kinetics.

Acknowledgements

A.K. and Z.N. acknowledge financial support by the Max Planck Society.

References

1. Foyer CH, Bloom AJ, Queval G et al (2009) Photorespiratory metabolism: genes, mutants, energetics, and redox signaling. Annu Rev Plant Biol 60:455–484. doi:10.1146/annurev.arplant.043008.091948
2. Bauwe H, Hagemann M, Kern R et al (2012) Photorespiration has a dual origin and manifold links to central metabolism. Curr Opin Plant Biol 15(3):269–275. doi:10.1016/j.pbi.2012.01.008
3. Fernie AR, Bauwe H, Eisenhut M et al (2013) Perspectives on plant photorespiratory metabolism. Plant Biol (Stuttg) 15(4):748–753. doi:10 1111/j.1438-8677.2012.00693.x
4. Fernie AR, Roessner U, Geigenberger P (2001) The sucrose analog palatinose leads to a stimulation of sucrose degradation and starch synthesis when supplied to discs of growing potato tubers. Plant Physiol 125 (4):1967–1977
5. Arrivault S, Guenther M, Ivakov A et al (2009) Use of reverse-phase liquid chromatography, linked to tandem mass spectrometry, to profile the Calvin cycle and other metabolic intermediates in Arabidopsis rosettes at different carbon dioxide concentrations. Plant J 59 (5):826–839. doi:10.1111/j.1365-313X.2009.03902.x
6. Stitt M (1996) Metabolic regulation of photosynthesis. In: Photosynthesis and the environment, vol 5. Springer, Netherlands
7. Kruger NJ, Ratcliffe RG (2015) Fluxes through plant metabolic networks: measurements, predictions, insights and challenges. Biochem J 465(1):27–38. doi:10.1042/BJ20140984
8. Zamboni N (2011) ^{13}C metabolic flux analysis in complex systems. Curr Opin Biotechnol 22 (1):103–108. doi:10.1016/j.copbio.2010.08.009
9. Heise R, Arrivault S, Szecowka M et al (2014) Flux profiling of photosynthetic carbon metabolism in intact plants. Nat Protoc 9 (8):1803–1824. doi:10.1038/nprot.2014.115
10. Ma F, Jazmin LJ, Young JD et al (2014) Isotopically nonstationary 13C flux analysis of changes in *Arabidopsis thaliana* leaf metabolism due to high light acclimation. Proc Natl Acad Sci U S A 111(47):16967–16972. doi:10.1073/pnas.1319485111
11. Szecowka M, Heise R, Tohge T et al (2013) Metabolic fluxes in an illuminated Arabidopsis rosette. Plant Cell 25(2):694–714. doi:10.1105/tpc.112.106989

12. Young JD, Shastri AA, Stephanopoulos G et al (2011) Mapping photoautotrophic metabolism with isotopically nonstationary (13)C flux analysis. Metab Eng 13(6):656–665. doi:10.1016/j.ymben.2011.08.002
13. Sweetlove LJ, Ratcliffe RG (2011) Flux-balance modeling of plant metabolism. Front Plant Sci 2:38. doi:10.3389/fpls.2011.00038
14. Orth JD, Thiele I, Palsson BO (2010) What is flux balance analysis? Nat Biotechnol 28 (3):245–248. doi:10.1038/nbt.1614
15. Bordbar A, Monk JM, King ZA et al (2014) Constraint-based models predict metabolic and associated cellular functions. Nat Rev Genet 15(2):107–120. doi:10.1038/nrg3643
16. Robaina Estevez S, Nikoloski Z (2014) Generalized framework for context-specific metabolic model extraction methods. Front Plant Sci 5:491. doi:10.3389/fpls.2014.00491
17. Topfer N, Kleessen S, Nikoloski Z (2015) Integration of metabolomics data into metabolic networks. Front Plant Sci 6:49. doi:10.3389/fpls.2015.00049
18. Yizhak K, Benyamini T, Liebermeister W et al (2010) Integrating quantitative proteomics and metabolomics with a genome-scale metabolic network model. Bioinformatics 26(12): i255–i260. doi:10.1093/bioinformatics/btq183
19. Schmidt BJ, Ebrahim A, TO M et al (2013) GIM3E: condition-specific models of cellular metabolism developed from metabolomics and expression data. Bioinformatics 29 (22):2900–2908. doi:10.1093/bioinformatics/btt493
20. Kleessen S, Irgang S, Klie S et al (2015) Integration of transcriptomics and metabolomics data specifies the metabolic response of Chlamydomonas to rapamycin treatment. Plant J 81 (5):822–835. doi:10.1111/tpj.12763
21. Henry CS, Broadbelt LJ, Hatzimanikatis V (2007) Thermodynamics-based metabolic flux analysis. Biophys J 92(5):1792–1805. doi:10.1529/biophysj.106.093138
22. Sajitz-Hermstein M, Topfer N, Kleessen S et al (2016) iReMet-flux: constraint-based approach for integrating relative metabolite levels into a stoichiometric metabolic models. Bioinformatics 32(17):i755–i762. doi:10.1093/bioinformatics/btw465
23. Schellenberger J, Que R, Fleming RM et al (2011) Quantitative prediction of cellular metabolism with constraint-based models: the COBRA Toolbox v2.0. Nat Protoc 6 (9):1290–1307. doi:10.1038/nprot.2011.308

Chapter 14

Kinetic Modeling of Photorespiration

Honglong Zhao, Yi Xiao, and Xin-Guang Zhu

Abstract

Dynamic systems modeling is a method to study systematic properties of a complex system. The basic principles, procedures, and tools available to develop a dynamic systems model of complex metabolic processes are detailed. Here, a photosynthetic carbon metabolism model, which includes the Calvin Benson cycle, photorespiration, and starch and sucrose synthesis pathways, is used as an example to illustrate the whole process of model development.

Key words Dynamic systems modeling, Modularized design, Ordinary differential equations, Model validation, Parameter estimation, Optimization, Sensitivity analysis

1 Introduction

About 30% of assimilated CO_2 is further released in mitochondria by photorespiratory pathway; therefore dramatic improvement in photosynthetic efficiency and hence crop yields can be achieved if photorespiratory CO_2 loss can be decreased [1]. Various options have been tested to decrease the photorespiratory CO_2 loss by engineering in plants with exogenous photorespiratory bypass pathways [2–4]. Some mixed successes have been achieved in this endeavor, *see* review by [5, 6]. Furthermore, an increasing complex picture of interactions between photorespiration and the Calvin-Benson cycle is emerging with recent progresses showing promoting effects of photorespiration on the CO_2 uptake rate. For example, certain photorespiratory mutants have lower photosynthetic rate and biomass [7, 8]; overexpression of H subunit of glycine decarboxylase leads to concurrent increase in photorespiratory flux, photosynthesis, and biomass [8, 9]. Understanding the complex interactions between photorespiratory pathway with the overall plant primary metabolism is becoming a major theme of photorespiratory research. In addition to the biochemical and molecular biology methods currently used to study photorespiration, dynamic systems modeling can be used to explore the system

Alisdair R. Fernie et al. (eds.), *Photorespiration: Methods and Protocols*, Methods in Molecular Biology, vol. 1653,
DOI 10.1007/978-1-4939-7225-8_14, © Springer Science+Business Media LLC 2017

properties of photorespiration, such as its performance under different conditions, interactions with other parts of the plant primary metabolism, and engineering new bypass pathways for higher photosynthesis [10].

In addition to study the impacts of different environmental and genetic perturbations on a metabolism, dynamic systems models have a number of additional applications. Firstly, it can be used to estimate the control coefficient of individual enzyme on metabolic fluxes by calculating ratio between the percentage changes in reaction fluxes to percentage changes in the activity of a particular enzyme. Secondly, it can be used to identify the optimal resource distribution to gain desired property, such as protein-nitrogen redistribution to gain an increased photosynthetic efficiency [11]. Thirdly, the systems model can also be used as a basis for further theoretical studies, such as analysis of model structure [12, 13] and parameter estimation [14].

In this chapter, we will describe a detailed protocol for establishing a dynamic systems model of the photosynthetic carbon metabolism, which includes photorespiration, and its close interacting Calvin Benson cycle, starch and sucrose synthesis pathways.

2 Materials

1. A common computer. For a typical dynamic systems model, the current desk-top computers with a RAM larger than 4G suffices.
2. Any ordinary differential equation solver. MATLAB (the Mathworks, Incorp., Natick, MA, USA) software is commonly used in developing dynamic systems models.

3 Methods

3.1 Information Collection to Set up a Model Diagram

1. Compiling all the reactions and transporters in photosynthetic carbon metabolism to develop a complete metabolic network. The information can be collected from the following databases (Table 1).
2. Dividing the photosynthetic carbon metabolism into three modules. A module is a collection of the metabolic reactions forming a close-related unit of network (*see* **Note 1**). The three modules are module of the Calvin Benson cycle, module of the photorespiratory pathway (*see* **Note 1**), and module of the sucrose synthesis pathway.
3. Collecting the basic parameters for the enzymes and transporters involved in the model. The parameters related to the rates of enzymatic reactions and transport processes are collected.

Table 1
Metabolic pathway databases that are freely available online

Database	Link of the website
KEGG	http://www.genome.jp/kegg/pathway.html
Gramene	http://pathway.iplantcollaborative.org/
MetaCrop	http://metacrop.ipk-gatersleben.de
PMN	http://www.plantcyc.org/
BioModels	https://www.ebi.ac.uk/biomodels-main/
Expasy	https://www.expasy.org/
Plant Reactome	http://plantreactome.gramene.org/
Solcyc	http://solcyc.solgenomics.net/
Wiki Pathway	http://www.wikipathways.org/index.php/WikiPathways
Biocyc	http://www.biocyc.org/

Table 2
Basic parameters used to describe enzyme characteristics

Parameters	The meanings of the parameters
EC number	Enzyme commission number for enzymes
K_e	Equilibrium constant (if the reaction is reversible)
K_{cat}	Catalytic number of enzyme
M	Molecular weight of enzyme
V_{max}	The maximum velocity of reaction
A	Activators (activating type)
K_a	Their corresponding activation constants
I	Inhibitors (inhibiting type)
K_i	Their corresponding inhibition constants
K_m	Michaelis–Menten constants
$[A]$	The concentration of substrate A

Most of the reactions follow Michaelis–Menten kinetics and the related parameters are listed in Table 2. The values of these parameters are either collected from peer-reviewed publications (*see* **Notes 2, 3**, and **4**), databases (Table 3), or experimental measurement directly.

Table 3
Enzyme parameter databases that are freely available online

Database	Link of the website
Brenda	http://www.brenda-enzymes.org/
BioModels	https://www.ebi.ac.uk/biomodels-main/
BioNumbers	http://bionumbers.hms.harvard.edu/

3.2 Developing Equations for Modules

1. Developing rate equations for each biochemical reaction or transportation

For most enzymatic reactions, we develop their rate equations in the form of Michaelis–Menten kinetic equations. Generalized rate equations for different reaction types are used [11, 15].
Type I: Single substrate and reversible reactions

$$\mathbf{A} \leftrightarrow \mathbf{B}$$

$$v = \frac{V_{\max}{}^{*}\left([A] - \frac{[B]}{K_e}\right)}{K_{mA}{}^{*}\left(1 + \frac{[A]}{K_{mA}} + \frac{[B]}{K_{mB}}\right)} \tag{1}$$

where v represents the rate of this reaction, $V_{\max}$ is the maximum velocity of the enzyme; $[A]$ and $[B]$ represent the concentrations of substrate A and B in corresponding cellular compartment, and K_e represents the equilibrium constant of the reaction. For reactions which are in equilibrium, their reaction rates are described as:

$$[A] = \frac{K_e{}^{*}\,[T]}{1 + K_e} \tag{2}$$

$$[B] = \frac{[T]}{1 + K_e} \tag{3}$$

$$[T] = [A] + [B] \tag{4}$$

where $[T]$ represents the total concentration of A and B, K_e represents the equilibrium constant of the reaction. $[A]$ and $[B]$ represent the concentrations of substrate A and B in corresponding cellular compartment, respectively.

Type II: Single substrate and irreversible reactions

$$\mathbf{A} \rightarrow \mathbf{B}$$

$$v = \frac{V_{\max}{}^{*}[A]}{K_{mA} + [A]} \tag{5}$$

where $[A]$ is the concentration of A; V_{max} is the maximum velocity of the reaction; K_{mA} is the Michaelis–Menten constant for A.

Type III: Double substrates and reversible reactions

$$\mathbf{A} + \mathbf{B} \leftrightarrow \mathbf{P} + \mathbf{Q}$$

$$v = \frac{V_{max}{}^{*}\left([A]^{*}[B] - \frac{[P]^{*}[Q]}{K_e}\right)}{K_{mA}{}^{*}K_{mB}{}^{*}\left(1 + \frac{[A]}{K_{mA}} + \frac{[B]}{K_{mB}} + \frac{[P]}{K_{mP}} + \frac{[Q]}{K_{mQ}} + \frac{[A]}{K_{mA}}{}^{*}\frac{[B]}{K_{mB}} + \frac{[P]}{K_{mP}}{}^{*}\frac{[Q]}{K_{mQ}}\right)} \quad (6)$$

$[A]$, $[B]$, $[P]$, and $[Q]$ represent the concentrations of substrate A, B, P, and Q in corresponding cellular compartment; K_{mA}, K_{mB}, K_{mP}, and K_{mQ} represent the Michaelis–Menten constants of A, B, P, and Q, respectively. V_{max} means the maximum velocity of the enzyme catalyzing positive reaction.

Type IV: Double substrates and irreversible reactions

$$\mathbf{A} + \mathbf{B} \rightarrow \mathbf{P} + \mathbf{Q}.$$

$$v = \frac{V_{max}{}^{*}[A]^{*}[B]}{([A] + K_{mA})^{*}([B] + K_{mB})} \quad (7)$$

$[A]$ and $[B]$ represent the concentrations of substrate A and B in corresponding cellular compartment; K_{mA} and K_{mB} represent the Michaelis–Menten constants of A and B, respectively. V_{max} means the maximum velocity of the enzyme.

Type V: Competitive regulatory reactions

This type of inhibition changes the apparent Michaelis–Menten constant (K_m) of corresponding substrate instead of the maximum velocity (V_{max}) of enzyme. For example, an inhibitor X competes with substrate A, then we replace K_{mA} with K'_{mA} in Eqs. 5–7:

$$K'_{mA} = K_{mA}{}^{*}\left(1 + \frac{[X]}{K_{iX}}\right) \quad (8)$$

where $[X]$ is the concentration of competitive inhibitor X; K'_{mA} is the modified apparent Michaelis–Menten constant with the inhibitor X; K_{iX} is the inhibition constant of X.

Type VI: Noncompetitive regulatory reactions

The noncompetitive inhibitor changes the maximum velocity of the enzyme (V_{max}). To model this, we replace V_{max} with V'_{max} in Eqs. 5–7:

$$V'_{max} = \frac{V_{max}}{\left(1 + \frac{[X]}{K_{iX}}\right)} \quad (9)$$

where $[X]$ is the concentration of noncompetitive inhibitor X; V'_{max} is the modified apparent maximum velocity with the inhibitor X; K_{iX} is the inhibition constant of X.

Type VII: Reactions that do not follow Michaelis–Menten kinetics

Some reactions such as reactions catalyzed by Rubisco in Calvin-Benson Cycle or by ADPGGP in starch synthesis, and transporters on organelles' membrane, do not follow the Michaelis–Menten kinetics. We use rate equations from publications which have described the measured data accurately. For example, the concentration of active sites of Rubisco in the chloroplast stroma is of the same order of magnitude as the concentration of its substrate, RuBP, which makes Rubisco catalyzed reaction does not follow Michaelis–Menten kinetics. The following equation is used to calculate the rate of carboxylation and oxygenation to total RuBP concentration (R_t) catalyzed by Rubisco [16]:

$$v_c = W_c \times \min\left\{1, \frac{[R_t]}{[E_t]}\right\} \tag{10}$$

where W_c (the rate of RuBP carboxylation) is described as the following equation [16]:

$$W_c = \frac{V_{Cmax}{}^{*}[CO_2]}{[CO_2] + K_{MCO_2}{}^{*}\left(1 + \frac{[O_2]}{K_{MO_2}}\right)} \tag{11}$$

In Eq. 11, V_{Cmax} represents the maximum rate of RuBP carboxylation; K_{MCO_2} and K_{MO_2} represent the Michaelis–Menten constants of CO_2 and O_2, respectively. $[CO_2]$ and $[O_2]$ represent CO_2 and O_2 concentrations around Rubisco catalytic site. The active site of Rubisco can also be bound by substrates of Calvin cycle other than RuBP leading to competitive inhibition of RuBP carboxylation [17], which leads to modification of the rate equation as following [10]:

$$v_c = \frac{[RuBP]^{*}\,W_c{}^{*}\min}{\left([RuBP] + K_{RuBP}{}^{*}\left(1 + \frac{[PGA]}{K_{iPGA}} + \frac{[FBP]}{K_{iFBP}} + \frac{[SBP]}{K_{iSBP}} + \frac{[Pi]}{K_{iPi}} + \frac{[NADPH]}{K_{iNADPH}}\right)\right)} \tag{12}$$

where the K_{RuBP} is the Michaelis–Menten constant for RuBP; $[RuBP]$, $[PGA]$, $[FBP]$, $[SBP]$, $[Pi]$, and $[NADPH]$ represent the concentration of substrates RuBP, PGA, FBP, SBP, Pi, and NADPH in chloroplast; K_{iPGA}, K_{iFBP}, K_{iSBP}, K_{iPi}, and K_{iNADPH} represent inhibition constants for PGA, FBP, SBP, Pi, and NADPH, respectively.

The reaction catalyzed by ADPGPP in starch synthesis pathway follows equation developed by Pettersson and Ryde-Pettersson [18].

$$v=\frac{V_{\max}{}^{*}[G1P]^{*}[ATP]}{([G1P]+K_{\mathrm{mG1P}})^{*}\left(\left(1+\frac{[ADP]}{K_{\mathrm{iADP}}}\right)^{*}([ATP]+K_{\mathrm{mATP}})+\frac{K_{\mathrm{mATP}}{}^{*}[Pi]}{K_{\mathrm{aPGA}}{}^{*}[PGA]+K_{\mathrm{aF6P}}{}^{*}[F6P]+K_{\mathrm{aFBP}}{}^{*}[FBP]}\right)} \tag{13}$$

where $[G1P]$, $[ATP]$, $[ADP]$, $[Pi]$, $[PGA]$, $[F6P]$, and $[FBP]$ represent the concentration of G1P, ATP, ADP, Pi, PGA, F6P, and FBP; K_{mG1P} and K_{mATP} represent the Michaelis–Menten constants for G1P and ATP, respectively; K_{aPGA}, K_{aF6P}, and K_{aFBP} represent the activation constant for PGA, F6P, and FBP, respectively.

Type VIII: Transportation of metabolite between compartments

The following generalized equation is used to calculate the rate of transportation of metabolites between intercellular compartments [19]:

$$v_{\mathrm{t}}=\frac{V_{\max}{}^{*}\left([A]_{\mathrm{a}}-\frac{[A]_{\mathrm{b}}}{K_{\mathrm{e}}}\right)}{K_{\mathrm{mA}}+[A]_{\mathrm{a}}} \tag{14}$$

where $[A]_{\mathrm{a}}$ and $[A]_{\mathrm{b}}$ represent the concentrations of substrate A in compartment a and compartment b, respectively. If special rate equations have been established in literature for particular transporters, those can be used directly. Pettersson and Ryde-Pettersson established a model to describe the rate of phosphate translocators on chloroplastic membrane as the following equation set [18]:

$$N=1+\left(1+\frac{K_{\mathrm{mPiout}}}{[Pi]_{\mathrm{out}}}\right)^{*}\left(\frac{[Pi]}{K_{\mathrm{mPi}}}+\frac{[PGA]}{K_{\mathrm{mPGA}}}+\frac{[GAP]}{K_{\mathrm{mGAP}}}+\frac{[DHAP]}{K_{\mathrm{mDHAP}}}\right) \tag{15}$$

$$v_{\mathrm{PGA}}=\frac{V_{\mathrm{ex}}{}^{*}[PGA]}{N^{*}K_{\mathrm{mPGA}}} \tag{16}$$

$$v_{\mathrm{GAP}}=\frac{V_{\mathrm{ex}}{}^{*}[GAP]}{N^{*}K_{\mathrm{mGAP}}} \tag{17}$$

$$v_{\mathrm{DHAP}}=\frac{V_{\mathrm{ex}}{}^{*}[DHAP]}{N^{*}K_{\mathrm{mDHAP}}} \tag{18}$$

where $[Pi]_{\mathrm{out}}$ and K_{mPiout} represent phosphate concentration in cytosol and Michaelis–Menten constant for phosphate in cytosol, respectively; V_{ex} means the maximum velocity of phosphate translocator of exporting photosynthate; $[Pi]$, $[PGA]$,

$[GAP]$, and $[DPGA]$ represent the concentration of Pi, PGA, GAP, and DPGA in chloroplast, respectively; K_{mPi}, K_{mPGA}, K_{mGAP}, and K_{mDHAP} represent the Michaelis–Menten constants for Pi,PGA,GAP, and DPGA in chloroplast, respectively.

2. Developing algebraic equations for conserved quantities

   In the photosynthetic carbon metabolism, the total concentration of ATP and ADP ($[CA]$), and total concentration of NADP and NADPH ($[CN]$), and total concentration of phosphate ($[CP]$) are assumed to be constant [11] and all of them are represented as algebraic equations 19–21 .

$$[CA] = [ATP] + [ADP] \tag{19}$$

$$[CN] = [NADP] + [NADPH] \tag{20}$$

$$\begin{aligned}[CP] = {} & 2^*[BPGA] + 2^*[FBP] + 2^*[SBP] + 2^*[RuBP] \\ & + [Pi] + [PGA] + [GAP] + [DHAP] + [F6P] \\ & + [E4P] + [S7P] + [Xu5P] + [Ri5P] + [Ru5P] \\ & + [G6P] + [G1P] + [ATP] + [PGCA]\end{aligned} \tag{21}$$

   where [compound] represents the concentration of involved compound.

3. Developing ordinary differential equations (ODEs) for compounds in the model

   Each ODE represents the rate of concentration change of a particular compound, which equals the difference between the total rate of metabolite production and total rate of metabolite consumption. For example, RuBP is produced from the phosphorylation of Ribulose 5-P, catalyzed by the enzyme Ru5P kinase, with a reaction rate v_{13} and it is consumed by oxygenation (v_{111}) and carboxylation (v_1) of RuBP (Fig. 1). Thus, the rate of RuBP concentration change is represented by the following ODE:

$$\frac{d[\mathrm{RuBP}]}{dt} = v_{13} - v_1 - v_{111} \tag{22}$$

   The volume of the chloroplast stroma is different from that of the cytosol in a typical higher plant cell [20–22], with the ratio between the volume of the stroma and that of the cytosol varying among species and under different conditions [20–22]. As a simplification, a ratio of 1:1 is assumed in calculating concentrations in cytosol and stroma. ODEs for each individual compound in the model are developed. All values for constants used in this model, rate equations, ODEs, and algebraic equations can be accessed at “http://www.plantphysiol.org/content/145/2/513.long.”

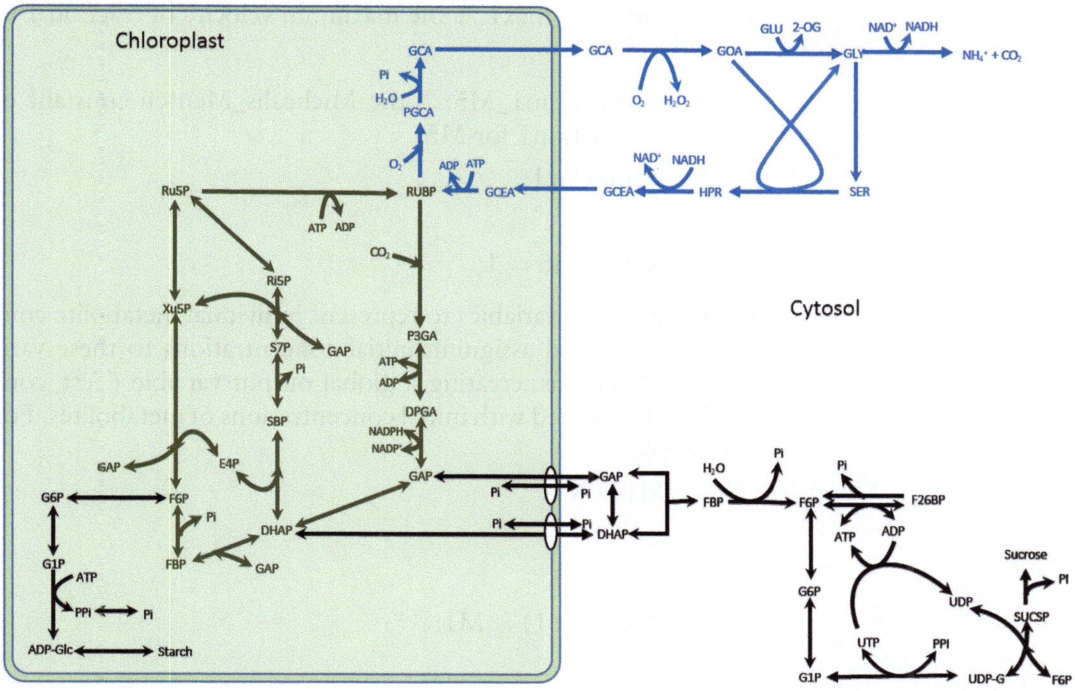

Fig. 1 The diagram of photosynthetic carbon metabolic model. The C_3 photosynthetic or Calvin Benson cycle is indicated n *green*; the photorespiratory pathway is indicated as *blue*; the sucrose and starch synthesis pathways are indicated as *black*. Abbreviations: *ADPG* ADP-glucose, *DHAP* Dihydroxyacetone-phosphate, *DPGA* 1,3-bisphosphoglycerate, *E4P* Erythrose 4-phosphate, *F6P* Fructose 6-phosphate, *FBP* Fructose 1,6-bisphosphate, *F26BP* Fructose 2,6-bisphosphate, *G1P* Glucose 1-phosphate, *G6P* Glucose 6-phosphate, *GAP* Glyceraldehyde 3-phosphate, *GCA* Glycollate, *GCEA* Glycerate, *GLY* Glycine, *GOA* Glyoxylate, *HPR* Hydroxypyruvate, *OPOP* Pyrophosphate, *PGA* 3-Phosphoglycerate, *PGCA* 3-Phosphoglycollate, *Ri5P* Ribose 5-phosphate, *Ru5P* Ribulose 5-phosphate, *RuBP* Ribulose 1,5-biphosphate, *S7P* Sedoheptulose 7-phosphate, *SBP* Sedoheptulose 1,7-bisphosphate, *SER* Serine, *UDPGlu* Uridine Diphosphate Glucose, *SUC* Sucrose, *SUCP* Sucrose phosphate, *UDP* Uridine 50-diphosphate, *UTP* Uridine-50-triphosphate, *Xu5P* Xylulose 5-phosphate, *GLU* Glutamate, *2-OG* a-Ketoglutarate, *Pi* phosphate. This diagram is reproduced with permission from [11]

3.3 Computational Implementation of the Model

Computational implementation of a dynamic systems model includes four categories of functions: a function for parameter initialization, a function for calculating reactions rate, a function for calculating the ODEs, and a function for organizing other functions to solve the systems and display the model results.

3.3.1 Creating a Function for Initialization of Model Parameter

The values of kinetic parameters used in rate equations, the initial concentration of metabolites, and the total concentration of the conserved quantities are initialized in this function (*see* **Note 5**).

1. Creating a function for parameter initialization (Test_Ini.m).

   function Test_con = Test_Ini(Begin)

2. Defining global variables for kinetic parameters and assigning values based on collected parameter values from step 3.1. For example,

```
global V_max1; % the maximum velocity of reaction 1
...
global K_m1_M5; % the Michealis_Menten constant of
    reaction1 for M5
V_max1=1;
...
K_m1_M5 = 1;
```

3. Creating local variables to represent individual metabolite concentrations and assigning initial concentrations to these variables. After this, creating a global output variable (Test_con) which is assigned with initial concentrations of metabolites. For example,

```
M1 = 3
...
Mn=0.35
Test_con(1) = M1;
...
Test_con(n) = Mn;
```

4. Defining global variables (CA, CP, CN) for the conserved quantities and assigning values compatible to the initial concentrations of their components. For example,

```
global [CA];
global [CP];
[CA] = [ATP] + [ADP];
[CP] = [Pi] + [ATP] + [M3] + 2*[PPi] + 2*[M4] + ...;
```

3.3.2 Create a Function for Calculating Reaction Rates (Test_Rate.m)

1. Create a function which calculates rates of different reactions. function Test_Velocity = Test_Rate(t, Test_con) where *t* is the parameters to describe the time label; Test_con is the output argument of Test_ini.m function.

2. Obtaining the kinetic parameters and other constants by declaring global parameters which are initialized in the Test_ini.m file. For example,

```
global V_max1;
...
global K_m1_M5;
global CP;
global CA;
...
```

3. Assigning the Test_con set to corresponding metabolite concentrations. For example,

   M1 = Test_con(1);

   ...

   Mn = Test_con(n);

4. Writing out the algebraic equations representing the conserved quantifies. For example,

   global [CA];

   global [CP];

   [ADP] = [CA] - [ATP];

   [Pi] = [CP] – ([ATP] + [M3] +2*[PPi] + 2*[M4] + ...);

5. Calculating the reaction rates using the rate equations for all reactions. For example,

   v_1=(V_max1*(M1-M2]/K_e1))/(K_m1_M1*(1 + M1/K_m1_M1+M2/K_m1_M2));

   ...

   v_7=V_max5*M3/(K_m5_M3 + M3);

6. Assigning the calculated reaction rates to an output variable Test_Velocity. For example,

   Test_Velocity(1) = v_1;

   ...

   Test_Velocity(7) = v_7;

3.3.3 Creating a Function for Representing the Mass Balance Equations (Test_mb.m)

1. Creating a new function which includes all the ODEs (Test_mb.m).

   function Test_MB = Test_mb(t, Test_con, Param)

2. Calling the rate function and obtaining the calculated rates for each reaction. For example,

   Test_Velocity = Test_Rate(t, Test_con, Param)

   v_1 = Test_Velocity(1);

   ...

   v_7 = Test_Velocity(7);

3. Calculating the rate of concentration change for each individual metabolite and outputting the value to an output variable (Test_MB). For example:

   Test_MB(1) = v_1-v_2-v_3;

   ...

   Test_MB(n) = v_4-v_2;

3.3.4 Creating a Function (Test_Drive.m) to Solve the Model and Display the Simulation Results

1. Calling the initial function and call the *ode15s* function to solve the model;

   Begin = 1;

   Test_con = Test_Ini(Begin);

   [Tt,d] = ode15s(@Test_mb,[0,t], Test_con);

2. Displaying the result of the simulation by plotting the time variable (*t*) as *X* axes and concentrations of different metabolites (*d*) in the *Y* axes.

   plot(Tt,d(:,index)); % index represents the individual metabolite.

3.3.5 Calling the Test_Drive to run the model

Test_Drive;

3.3.6 Validation of the model

Model validation is done through comparison of the model prediction with experimental observations. Three criteria are taken to validate a model. First, check whether the model can reach a steady state solution; secondly, check whether the predicted rates of different reactions and concentrations of involved metabolites are within the range of experimentally observed values; thirdly, check whether the model prediction can regain a steady state when the external conditions (e.g., CO_2 levels, light levels) are perturbed. Here we use the model to predict the response of photosynthetic CO_2 uptake rate under different CO_2 levels and compare the predicted response curve to the observed one. If the model fails to reach any of the above listed criteria, the model needs to be adjusted starting from step 3.1.

3.3.7 Integrating Sub-Models to Form a Complete Model

After completing individual sub-models representing individual modules, the sub-models can be combined together to form the integrated model. The same procedures above will be used except that the rate equations and ODEs are developed already and only those equations (rate equations, ODEs, and algebraic equations) closely related to the reactions of sub-modules should be modified. For example, in the sub-module of the Calvin-Benson cycle, it does not consider the influence of photorespiratory products on the rate equation of Rubisco, which can be incorporated into the integrated model of the photosynthetic carbon metabolism model. The complete package for the photosynthetic carbon metabolism model is accessible at "http://www.plantphysiol.org/content/145/2/513.long."

4 Notes

1. Although many database contain information about different metabolic pathways and related enzymes, authors suggest readers to check each reference to confirm the completeness and accuracy of the information before using them in the model development.
2. Please pay attention to units of parameters used in the model parameters.
3. This dynamic systems model assumes that all substrates and enzymes are uniformly distributed inside the correspondingly compartment.
4. Parameters used in the model are classified as either private variable or global variables. The private variable can only be accessed in the function it is defined; while the global variable, once declared, can be accessed in different functions.
5. An updated model of the photorespiratory pathway can be found at [10], which includes not only the basic photorespiratory pathways as discussed here, but also the different photorespiratory bypass pathways.

References

1. Long SP, Marshall-Colon A, Zhu XG (2015) Meeting the global food demand of the future by engineering crop photosynthesis and yield potential. Cell 161:56–66
2. Kebeish R, Niessen M, Thiruveedhi K et al (2007) Chloroplastic photorespiratory bypass increases photosynthesis and biomass production in *Arabidopsis thaliana*. Nat Biotechnol 25:593–599
3. Maier A, Fahnenstich H, Von Caemmerer S et al (2012) Transgenic introduction of a glycolate oxidative cycle into A-thaliana chloroplasts leads to growth improvement. Front Plant Sci 3:38
4. Nolke G, Houdelet M, Kreuzaler F et al (2014) The expression of a recombinant glycolate dehydrogenase polyprotein in potato (*Solanum tuberosum*) plastids strongly enhances photosynthesis and tuber yield. Plant Biotechnol J 12:734–742
5. Hodges M, Dellero Y, Keech O et al (2016) Perspectives for a better understanding of the metabolic integration of photorespiration within a complex plant primary metabolism network. J Exp Bot 67:3015–3026
6. Obata T, Florian A, Timm S et al (2016) On the metabolic interactions of (photo)respiration. J Exp Bot 67:3003–3014
7. Dellero Y, Lamothe-Sibold M, Jossier M et al (2015) *Arabidopsis thaliana* ggt1 photorespiratory mutants maintain leaf carbon/nitrogen balance by reducing RuBisCO content and plant growth. Plant J 83:1005–1018
8. Timm S, Florian A, Arrivault S et al (2012) Glycine decarboxylase controls photosynthesis and plant growth. FEBS Lett 586:3692–3697
9. Timm S, Wittmiss M, Gamlien S et al (2015) Mitochondrial dihydrolipoyl dehydrogenase activity shapes photosynthesis and photorespiration of *Arabidopsis thaliana*. Plant Cell 27:1968–1984
10. Xin CP, Tholen D, Devloo V et al (2015) The benefits of photorespiratory bypasses: how can they work? Plant Physiol 167:574–585
11. Zhu XG, De Sturler E, Long SP (2007) Optimizing the distribution of resources between enzymes of carbon metabolism can dramatically increase photosynthetic rate: a numerical simulation using an evolutionary algorithm. Plant Physiol 145:513–526

12. Gutenkunst RN, Waterfall JJ, Casey FP et al (2007) Universally sloppy parameter sensitivities in systems biology models. PLoS Comput Biol 3:e189
13. Transtrum MK, Qiu P (2014) Model reduction by manifold boundaries. Phys Rev Lett 113:098701
14. Transtrum MK, Machta BB, Brown KS et al (2015) Perspective: sloppiness and emergent theories in physics, biology, and beyond. J Chem Phys 143:010901
15. Rogers A, Gibon Y (2009) Enzyme kinetics: theory and practice. In: Schwender J (ed) Plant metabolic networks. Springer, New York, pp 71–103
16. Von Caemmerer S (2000) Biochemical models of leaf photosynthesis. Csiro Publishing, Collingwood, VIC
17. Badger MR, Lorimer GH (1981) Interaction of sugar phosphates with the catalytic site of ribulose-1,5-bisphosphate carboxylase. Biochemist 20:2219–2225
18. Pettersson G, Rydepettersson U (1988) A mathematical-model of the Calvin photosynthesis cycle. Eur J Biochem 175:661–672
19. Wang Y, Long SP, Zhu XG (2014) Elements required for an efficient NADP-malic enzyme type C4 photosynthesis. Plant Physiol 164:2231–2246
20. Leidreiter K, Kruse A, Heineke D et al (1995) Subcellular volumes and metabolite concentrations in potato (*Solanum Tuberosum* cv *Desiree*) leaves. Bot Acta 108:439–444
21. Winter H, Robinson DG, Heldt HW (1993) Subcellular volumes and metabolite concentrations in barley leaves. Planta 191:180–190
22. Winter H, Robinson DG, Heldt HW (1994) Subcellular volumes and metabolite concentrations in spinach leaves. Planta 193:530–535

Chapter 15

Investigating the Role of the Photorespiratory Pathway in Non-photosynthetic Tissues

Alisdair R. Fernie, Hermann Bauwe, and Lee J. Sweetlove

Abstract

Whilst photorespiration represents one of the dominant pathway fluxes in photosynthetic tissues there are hints from publically available gene expression data such as that housed in the bioarray resource (BAR; www.bar.utoronto.ca) that several of the constituent enzymes are present in roots and other heterotrophic tissues. Here we describe a protocol based on modification of the gaseous environment surrounding individual tissues of mutant and wild type Arabidopsis and evaluation of the consequences. This method could additionally easily be used for larger plants.

Key words Photorespiratory mutants, Metabolite analysis, Split-plant gas experiment

1 Introduction

Photorespiration is an essential metabolic process in leaves which facilitates recovery of carbon lost by the oxygenase reaction of Rubisco and avoids the accumulation of the noxious product of this reaction, 2-phosphoglycolate (2PG) [1]. However, there is also considerable evidence to suggest that photorespiration has a far more complex role during normal growth than the mere detoxification of 2PG and corresponding recovery of 3-phosphoglycerate (3-PGA) [1]. Thus the current view is that photorespiration is an essential mechanism to maintain optimal photosynthesis, normal growth, and development of oxygenic photosynthetic organisms in an oxygen-containing environment [1]. However, beyond this, evaluation of the bioarray resource (http://www.bar.utoronto.ca; [2]) reveals that several transcripts associated with enzymes associated with the photorespiratory process are highly expressed in roots and other heterotrophic with some transcripts even exclusively being expressed in heterotrophic tissues [3]. Comparison of the pep2pro proteome database (http://fgcz-pep2pro.uzh.ch; [4]) revealed considerable abundance of several of the enzymes in roots (for complete details *see* [5]). Similarly, these proteins were found to

Alisdair R. Fernie et al. (eds.), *Photorespiration: Methods and Protocols*, Methods in Molecular Biology, vol. 1653,
DOI 10.1007/978-1-4939-7225-8_15, © Springer Science+Business Media LLC 2017

be highly abundant in heterotrophic tissues following exposure to stress [6]. In spite of the knowledge concerning the presence of several enzymes of, or at least associated with, photorespiration and their consequent inclusion in the recently published root-specific genome-scale model of Arabidopsis [7], surprisingly little is known concerning the functionality of the pathway in this tissue. That being said two studies carried out in the 1990s demonstrated the presence of activities of photorespiratory enzymes in heterotrophic tissues [8, 9] with the activity of, for example, glycine decarboxylase of heterotrophic tissues of cereals reaching 20% of that observed in photosynthetic tissues. Driven by a desire to better understand the role of the photorespiratory pathway in heterotrophic tissues we recently established an approach by which the specific role of this pathway could be evaluated by using mutants of Arabidopsis deficient in the expression of enzymes of the photorespiratory pathway under conditions in which the gaseous environment of the aerial part of the plant but not the root were experimentally altered prior to determining alterations in the metabolite content using the protocol described in Chapter 7.

2 Materials

2.1 Plant Growth

Wild type (Col0) or T-DNA insertional knockout mutants such as, for example, those with clear but tolerable reductions of photorespiratory flow (e.g., *hpr1-1* and *gox3-1*), or in which essential reactions of photorespiration are blocked (e.g., *pglp1-1, glyk1-1* and *shm1-2*) [10] can be obtained from resource centers such as the Nottingham Arabidopsis Stock Center (NASC) or the SENDAI Arabidopsis Seed Stock Center (SASSC) and validated as detailed in Chapter 8. They can be subsequently grown in a custom-built apparatus within a controlled growth cabinet. For this purpose plants should be grown hydroponically with the root solution being bubbled with normal air whilst the rosette tissues are simultaneously exposed to a high CO_2 containing atmosphere (1%) which is necessary to allow normal growth of such mutants.

1. Take care to avoid microbial infections throughout the work as these can ruin your hydroponic culture.
2. Germinate sterilized (e.g., by treatment with about 5% sodium hypochlorite, which is liquid household bleach) seeds on sterile 50% Murashige-Skoog (MS) medium including vitamins plus 1% sucrose, solidified with 0.8% Agar, in Petri dishes. Appropriate selection agents/antibiotics such as kanamycin may be added if desired. Keep at 4 °C for at least 2 days to break dormancy and then under illumination at room temperature.

3. Remove lids from 1.5 ml Eppendorf microtubes and fill microtubes completely with sterile 50% solidified MS medium as above but without sucrose to minimize microbial growth.
4. Assemble hydroponic growth tray outside the growth cabinet.
5. Transfer seedlings to the prepared microtubes, one per microtube but a few more than actually needed, and incubate in the growth cabinet under environmental conditions (CO_2, photoperiod, temperature) to be used for hydroponics for at least 1 week. Examine for proper and uniform root growth.
6. Cut off the lower 5–7 mm by using a suitable cutter and place microtubes with seedlings into the openings in the lid of the tray. Use standard microtubes to close all unused openings. Place tray into a controlled environment growth cabinet with regulated air CO_2 content (1%).
7. Fill tray with 50% MS medium including vitamins (or a liquid fertilizer such as 0.2% Wuxal in tap water, the shown tray requires about 8 L) to about 15–20 mm below upper edge to allow the microtubes to immerse for a few millimeters. Connect the tray's air inlet to an oil-free air pump of adequate capacity, for example, Laboport N 811.KN.18 (11.5 L/min, operating pressure 2 bar). The air outlet goes outside the growth cabinet via a tube though a dual-channel air pump can serve both the inlet and the outlet channel.
8. Nutrient solution should be completely exchanged once per week.

2.2 Metabolite Profiling

After about 4 weeks growth, root material was harvested and flash-frozen in liquid nitrogen and stored at −80 °C prior to measurement of metabolites using an established GC-MS metabolite profiling method (*see* below and [11]).

2.3 Reagents

1. Argon 5.0.
2. Chloroform for liquid chromatography (*see* **Note 1**).
3. Helium 5.0 carrier gas.
4. Methanol, gradient grade for liquid chromatography.
5. *N*-Methyl-*N*-(trimethylsilyl)trifluoroacetamide (MSTFA reagent); prepared in 1-ml vials and stored at 4 °C (*see* **Note 2**).
6. Orange silica gel.
7. Ribitol.

2.4 Equipment

1. Autosampler and software (CTC Combi PAL and PAL cycle composer software version 1.5.0; CTC Analytics); the configuration comprises an agitator-incubator oven, a 98-sample tray for 2.0-ml vials, a 32-sample tray for 10–20-ml vials, three 100-ml solvent reservoirs (i.e., a syringe wash station and a liquid version 25-µl syringe kit mounted on the robotic autosampler arm).

2. Conical single taper split/splitless liner.
3. Gas chromatograph, 6890N, split/splitless injector with electronic pressure control up to 150 psi.
4. GL14 glass vials.
5. MDN-35 capillary column, 30-m length, 0.32-mm inner diameter, 0.25-μm film thickness.
6. Micro-vials: 1.5 ml, safe-lock, tapered bottom, and 2.0 ml, screw cap, round bottom.
7. Oscillating ball mill, MM200.
8. Time-of-flight mass spectrometer.
9. Steel balls, VA5mm.
10. Screw caps for GL14 glass vials.
11. Teflon adaptor for 1.5–2.0 ml micro-vials.
12. Automated mass spectral deconvolution and identification system (AMDIS; National Institute of Standards and Technology).
13. ChromaTOF chromatography processing and mass spectral deconvolution software, version 1.00 or higher, driver 1.61 or higher (LECO Instrumente), running on a state-of-the-art computer with a minimum of 512-MB RAM and an 1.0G-Hz Pentium IV processor or equivalent.
14. R: a Language and Environment for Statistical Computing (R Foundation for Statistical Computing).

3 Methods

3.1 Reagent Setup

1. Dissolve the derivatization reagent methoxyamine hydrochloride at 20 mg/ml in pure pyridine at 20–25 °C. This reagent needs to be prepared freshly before the experiment (*see* **Note 3**).
2. Preparation of the retention time index standard mixture. Dissolve FAMEs in chloroform at a final concentration of 0.4 ml/ml or 0.8 mg/ml for liquid or solid standards. Reagent can be stored at −4 °C. Esters included are methylcaprylate, methyl pelargonate, methylcaprate, methyllaurate, methylmyristate, methylpalmitate, methylstearate, methyleicosanoate, methyldocosanoate, lignoceric acid methylester, methylhexacosanoate, methyloctacosanoate, and triacontanoic acid methylester [12]. Alternatively, alkanes [13] or fatty acids [14] can be used.
3. Sampling and extraction. Sample root material in 2-ml, screw cap, round bottom tubes. Define the exact mass of plant sample (~100 mg of fresh weight) and rapidly freeze the sample-containing vial using liquid nitrogen or an equivalent low-temperature liquid.

4. To homogenize the tissue, place steel balls into the sample tubes and insert samples into precooled Teflon adaptors. Homogenize in ball mill for 2 min at 20 Hz (*see* **Note 4**).
5. Add 1400 μl of 100% methanol (precooled at −20 °C) and vortex for 10 s.
6. Add 60 μl of Ribitol (0.2 mg/ml stock in dH_2O) as an internal quantitative standard and vortex for 10 s.
7. Shake for 10 min at 70 °C in a thermomixer at 950 rpm.
8. Centrifuge for 10 min at 11,000 × *g*.
9. Transfer supernatant to a glass vial.
10. Add 750 μl of chloroform (−20 °C).
11. Add 1500 μl dH_2O (4 °C) and vortex for 10 s.
12. Centrifuge 15 min at 2200 × *g*.
13. Transfer 150 μl from the upper phase (polar phase) into a fresh 1.5-ml tube.
14. Dry in a vacuum concentrator without heating.
15. Before freezing the aliquots at −80 °C, fill the tubes with argon gas and place them inside a plastic bag containing silica bead desiccant. Argon-filled sample vials prevent the extract from oxidization and degradation by reactions through components of atmospheric air (*see* **Notes 5** and **6**).
16. Samples should be stored at −80 °C in a vacuum concentrator for 30 min before derivatization.
17. Add 40 μl of methoxyamination reagent (*see* **step 1**) to the aliquots (*see* **Notes** 7 and **8**).
18. Also prepare one derivatization reaction using an empty reaction tube as a control.
19. Shake for 2 h at 37 °C.
20. Prepare MSTFA reagent with 20 μl/ml of retention time index standard mixture (*see* **step 2**).
21. Add 70 μl of the solution prepared in **step 19** to the sample aliquots.
22. Shake for 30 min at 37 °C.
23. Transfer into glass vials suitable for GC-MS analysis (*see* **Note 9**).
24. GC-TOF-MS injection parameters: Inject 1 μl of sample at 230 °C in splitless mode with helium carrier gas flow set to 2 ml/min by using the autosampler setup (*see* Subheading 2.4) The flow rate is kept constant with electronic pressure control enabled (*see* **Note 10**).
25. GC-TOF-MS chromatography parameters. Perform chromatography with a 30-m MDN-35 capillary column. The temperature program should be isothermal for 2 min at 80 °C,

followed by a 15 °C per min ramp to 330 °C, and holding at this temperature for 6 min. Cooling should be as rapid as instrument specifications allow. The transfer line temperature should be set to 250 °C and match ion source conditions [15].

26. GC-TOF-MS mass spectrometer parameters. Set the ion source to maximum instrument specifications, 250 °C. The recorded mass range should be *m/z* 70 to *m/z* 600 at 20 scans/s. Proceed the remaining monitored chromatography time with a 170-s solvent delay with filaments turned off. Manual mass defect should be set to 0, filament bias current should be −70 V, and detector voltage should be ~1700–1850 V. The instrument should be automatically tuned according to the manufacturer's instructions.
27. Data evaluation: Deconvolution. Metabolite profiling leads to complex chromatograms characterized by coeluting compounds and vast differences in the relative abundance of the different compounds. Although problematic, these issues can be partially resolved by deconvolution of the chromatograms. The machine supplier's software, e.g., ChromaTOF, offers a built-in deconvolution algorithm. Deconvoluted spectra can be exported as plain text files for further processing (*see* **Note 11**).
28. Data evaluation: Peak assignment following retention time index (RI). The retention time index (RI) is arguably the most important parameter for peak assignment. In our experience it is absolutely crucial that each chromatogram is corrected for retention times separately, as even within a day absolute retention times show variance that, combined with the fact that the complex mixtures apparent in plant extracts result in highly complex chromatograms, can lead to false peak annotations (*see* **Note 12**). A reference list—containing name, expected RI, allowed RI variation, and unique mass for a number of metabolites—can be initially prepared by evaluating the GC-MS spectra resulting from the evaluation of a mixed sample pooled from aliquots of the whole measurement set in conjunction with available GC-MS library sets that contain the detectable photorespiratory intermediates [16–18].

3.2 Data Interpretation

Once the metabolite data have been collected a number of comparative analyses can be carried out. First, comparison of the wild type and mutant metabolite content provides information concerning whether the mutation provokes a metabolic change within the tissue under study. If so detailed analysis will reveal which other pathways of primary metabolism are effected and consequently provide first hints as to the role of the enzyme under study within heterotrophic metabolism. Finally, comparing the root mutant metabolite profile with those previously determined in actively photosynthesizing tissues [10] allows one to contrast the role of the constituent enzymes of the photorespiratory cycle between the different tissue types.

4 Notes

1. *Caution*: Chloroform is toxic and should be handled under a fume hood.
2. *Caution*: Reagent is extremely toxic and should be handled under the fume hood.
3. *Caution*: Reagents are extremely toxic and should be handled under the fume hood.
4. Pause Point Frozen homogenate can be stored at −80 °C for up to 3 months.
5. *Caution*: Halogenic reagents and solutions should be disposed with halogenic waste.
6. Pause Point Samples can be stored at −80 °C for up to 3 months.
7. *Caution*: Derivatization reagents are extremely toxic. Handle with absolute care. Work with gloves and under the fume hood.
8. Critical step: In the process of derivatization, condensation of reagents appears on the wall and lid of the reaction tubes; therefore, centrifugation of the reaction mixture is essential after every incubation step.
9. Critical step **steps 15–22** have been shown to be very critical. In this protocol we use derivatization reagent in supersaturated concentrations to ensure completion of derivatization.
10. Optionally, but especially recommended in cases of high metabolite concentrations, injection can be done in split mode with the split ratio adjusted to 1:25.
11. We suggest the following parameters for the deconvolution process (with acceptable range in parentheses). In all instances we used the machine manufacturer's recommended approaches, which we have found to be highly appropriate: Baseline offset = 1 (0.5–1); Smoothing = 5 data points (3–7); Peak width = 3 s (3–4 s); *S/N* (signal-to-noise ratio) = 10 (2–15).
12. In principle, it should be possible to annotate each compound based on its unique mass spectrum and RI [16–18]. In metabolite profiling, however, the presence of coeluting compounds present in high dynamic range [19] can mean that reliance on these parameters proves to be difficult. This is even more pronounced when the coeluting compounds have one or more masses in common. For this purpose, we decided to use

a combination of a very precise relative retention time as described above and one or more mass traces unique within this retention time window for the assignment of a given peak to a compound according to a prepared reference list using an algorithm that achieves the following: all peaks within the specified time window are evaluated, and the peak showing the maximum intensity for the predefined unique ion is chosen.

Acknowledgement

Work in the H.B. and A.R.F. laboratories was funded by the DFG Forschergruppe 1186 Promics Photorespiration: Origin and Metabolic Integration in Interacting Compartments. We wish to thank Mrs. Kathrin Jahnke at Rostock University for technical assistance.

References

1. Bauwe H, Hagemann M, Kern R, Timm S (2012) Photorespiration has a dual origin and manifold links to central metabolism. Curr Opin Plant Biol 15(3):269–275
2. Toufighi K, Brady SM, Austin R, Ly E, Provart NJ (2005) The botany array resource: e-Northerns, expression angling, and promoter analyses. Plant J 43(1):153–163
3. Engel N et al (2011) The presequence of Arabidopsis serine hydroxymethyltransferase SHM2 selectively prevents import into mesophyll mitochondria. Plant Physiol 157 (4):1711–1720
4. Baerenfaller K et al (2008) Genome-scale proteomics reveals *Arabidopsis thaliana* gene models and proteome dynamics. Science 320 (5878):938–941
5. Nunes-Nesi A et al (2014) Is there a metabolic requirement for photorespiratory enzyme activities in heterotrophic tissues? Mol Plant 7 (1):248–251
6. Jacoby RP, Millar AH, Taylor NL (2013) Application of selected reaction monitoring mass spectrometry to field-grown crop plants to allow dissection of the molecular mechanisms of abiotic stress tolerance. Front Plant Sci 4:20
7. Mintz-Oron S et al (2012) Reconstruction of Arabidopsis metabolic network models accounting for subcellular compartmentalization and tissue-specificity. Proc Natl Acad Sci U S A 109(1):339–344
8. Lernmark U, Henricson D, Wigge B, Gardestrom P (1991) Glycine oxidation in mitochondria isolated from light grown and etiolated plant-tissue. Physiol Plant 82 (3):339–344
9. Igamberdiev AU, Bykova NV, Gardestrom P (1997) Involvement of cyanide-resistant and rotenone-insensitive pathways of mitochondrial electron transport during oxidation of glycine in higher plants. FEBS Lett 412 (2):265–269
10. Timm S et al (2013) Serine acts as a metabolic signal for the transcriptional control of photorespiration-related genes in Arabidopsis (1 W). Plant Physiol 162(1):379–389
11. Lisec J, Schauer N, Kopka J, Willmitzer L, Fernie AR (2006) Gas chromatography mass spectrometry-based metabolite profiling in plants. Nat Protoc 1(1):387–396
12. Weckwerth W, Loureiro ME, Wenzel K, Fiehn O (2004) Differential metabolic networks unravel the effects of silent plant phenotypes. Proc Natl Acad Sci U S A 101(20):7809–7814
13. Roessner-Tunali U et al (2003) Metabolic profiling of transgenic tomato plants overexpressing hexokinase reveals that the influence of hexose phosphorylation diminishes during fruit development. Plant Physiol 133(1):84–99
14. Roessner U, Willmitzer L, Fernie AR (2001) High-resolution metabolic phenotyping of genetically and environmentally diverse potato tuber systems. Identification of phenocopies. Plant Physiol 127(3):749–764
15. Schad M, Mungur R, Fiehn O, Kehr J (2005) Metabolic profiling of laser microdissected vascular bundles of *Arabidopsis thaliana*. Plant Methods 1:2

16. Wagner C, Sefkow M, Kopka J (2003) Construction and application of a mass spectral and retention time index database generated from plant GC/EI-TOF-MS metabolite profiles. Phytochemistry 62(6):887–900
17. Kopka J et al (2005) GMD@CSB.DB: the Golm metabolome database. Bioinformatics 21(8):1635–1638
18. Schauer N et al (2005) GC-MS libraries for the rapid identification of metabolites in complex biological samples. FEBS Lett 579 (6):1332–1337
19. Sumner LW, Mendes P, Dixon RA (2003) Plant metabolomics: large-scale phytochemistry in the functional genomics era. Phytochemistry 62(6):817–836

Chapter 16

Studying the Function of the Phosphorylated Pathway of Serine Biosynthesis in *Arabidopsis thaliana*

Stephan Krueger, Ruben M. Benstein, Sabine Wulfert, Armand D. Anoman, María Flores-Tornero, and Roc Ros

Abstract

Photorespiration is an essential pathway in photosynthetic organisms and is particularly important to detoxify and recycle 2-phosphoglycolate (2-PG), a by-product of oxygenic photosynthesis. The enzymes that catalyze the reactions in the photorespiratory core cycle and closely associated pathways have been identified; however, open questions remain concerning the metabolic network in which photorespiration is embedded. The amino acid serine represents one of the major intermediates in the photorespiratory pathway and photorespiration is thought to be the major source of serine in plants. The restriction of photorespiration to autotrophic cells raises questions concerning the source of serine in heterotrophic tissues. Recently, the phosphorylated pathway of serine biosynthesis has been found to be extremely important for plant development and metabolism. In this protocol, we describe a detailed methodological workflow to analyze the generative and vegetative phenotypes of plants deficient in the phosphorylated pathway of serine biosynthesis, which together allow a better understanding of its function in plants.

Key words Autotrophic metabolism, Heterotrophic metabolism, Serine, Phosphorylated pathway, Vegetative phenotypes

1 Introduction

Sequential reactions of the photorespiratory core pathway (Fig. 1) are embedded within a complex metabolic network of closely associated metabolic reactions [1, 2]. Mutants deficient in photorespiratory genes are lethal in ambient air, but are viable when grown under elevated CO_2 conditions [3]. Oxygen sensitivity of photorespiratory mutants has been extensively used to identify major components of the photorespiratory core pathway and associated reactions such as the detoxification of peroxisomal H_2O_2, the ammonium assimilation, the C1-metabolism, and the biosynthesis of short-chain fatty acids for lipoylation of H protein subunit of the glycine decarboxylase complex [4–8]. The enhanced oxygen sensitivity of photorespiratory mutants is the consequence of 2-PG

Alisdair R. Fernie et al. (eds.), *Photorespiration: Methods and Protocols*, Methods in Molecular Biology, vol. 1653, DOI 10.1007/978-1-4939-7225-8_16, © Springer Science+Business Media LLC 2017

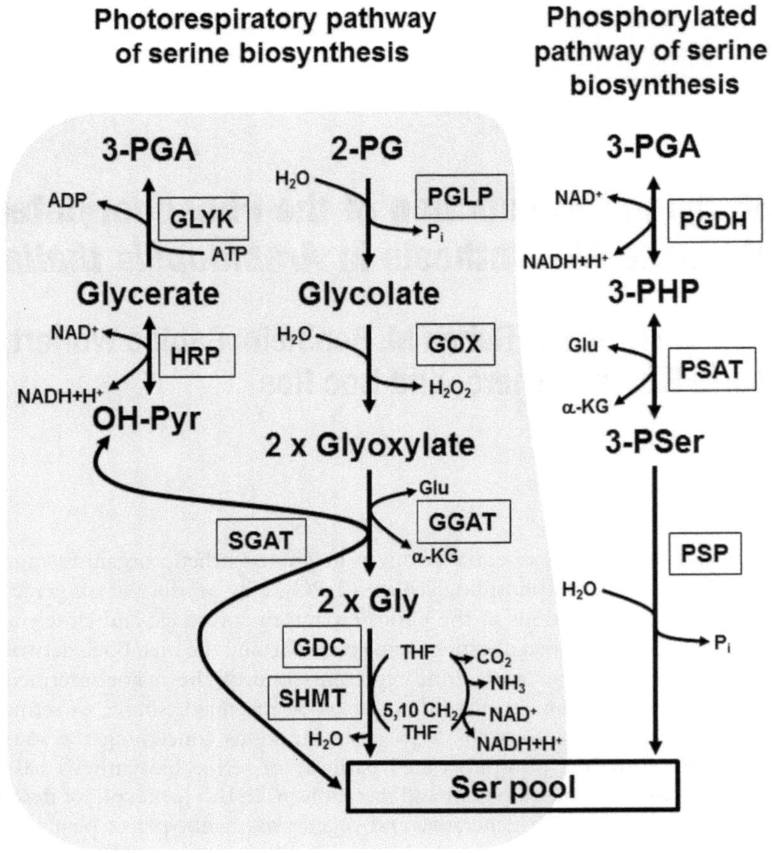

Fig. 1 Graphical overview about the photorespiratory (*left*) and the phosphorylated pathway of serine biosynthesis (*right*). The reactions within photorespiration are catalyzed by the following enzymes: PGLP, 2-phosphoglycolate phosphatase, catalyzing the dephosphorylation of 2-phosphoglycolate (2-PG); GOX, glycolate oxidase, catalyzing the oxidation of glycolate; GGAT, glutamate: glyoxylate aminotransferase catalyzing the transfer of an amino group from glutamate (Glu) to glyoxylate yielding α-ketoglutarate (α-KG) and glycine (Gly); GDC, glycine decarboxylate complex; SHMT, serine hydroxymethyltransferase, catalyzing the transfer of C_1 units from 5,10-methylene tetrahydrofolate (5,10 CH_2 THF) to Gly to form serine (Ser); SGAT, serine:glyoxylate aminotransferase, catalyzing the formation of Gly and hydroxypyruvate (OH-Pyr); HRP, OH-Pyr reductase catalyzing the formation of glycerate; GLYK, glycerate kinase, catalyzing the phosphorylation of glycerate to form 3-phosphoglycerate (3-PGA). The reactions within the phosphorylated pathway of serine biosynthesis are catalyzed by the following enzymes: PGDH, phosphoglycerate dehydrogenase, catalyzing the oxidation of 3-PGA to 3-phosphohydroxypyruvate (3-PHP); PSAT, catalyzing the transfer of an amino group from Glu to 3-PHP yielding in α-KG and phosphoserine (3-Pser); PSP, phosphoserine phosphatase, catalyzing the dephosphorylation of 3-PSer to Ser

accumulation inhibiting distinct steps in the carbon-assimilating Calvin-Benson cycle. Accumulation of 2-PG is the result of an imbalanced carbon flow through the photorespiratory core pathway, which can be also impaired by disruption of photorespiratory-associated reactions. However, not all pathways connected to photorespiration, such as the phosphorylated pathway of serine biosynthesis (PPSB), negatively influence photorespiratory carbon flow in the same manner. In plants, serine is synthesized by two major pathways, the photorespiratory pathway and the PPSB (Fig. 1) [9]. The PPSB pathway synthesizes serine from 3-phosphoglyceric acid (3-PGA) in plastids and consists of three sequential reactions that are catalyzed by 3-PGA dehydrogenase (PGDH), 3-phosphoserine aminotransferase (PSAT), and 3-phosphoserine phosphatase (PSP). The Arabidopsis genome contains genes that encode all enzymes of the PPSB (Aramemnon; http://aramemnon.botanik.uni-koeln.de): three genes have been described to encode PGDHs (*At4g34200*, *PGDH1*; *At1g17745*, *PGDH2*; *At3g19480*, *PGDH3*), two to encode PSAT (*At4g35630*, *PSAT1* and *At2g17630*, *PSAT2*) and one to encode PSP (*At1g18640*). All *PGDH*, *PSAT1,* and *PSP* genes have been cloned and in vitro activity of respective enzymes has been demonstrated [10, 11]. Previous studies on the function of the PPSB in Arabidopsis have shown that it is essential for plant growth and development. Multiple independent T-DNA insertion lines for genes of the PPSB have been isolated [9] and the analysis of these mutants has identified two essential genes of the pathway, *PGDH1* and *PSP1* [11–13]. Knockout mutants of each gene revealed a lethal phenotype due to disturbed male gametophyte and embryo development [12].

Besides its function in reproductive organs, the PPSB plays an important role during vegetative phase of plant development. Plants with reduced PPSB activity show a strong shoot and root growth retardation phenotype [11, 12], which is more severe under elevated CO_2 conditions that inhibit photorespiration [11]. Therefore, PPSB and photorespiration appear to function in concert to ensure a sufficient supply of serine at all developmental stages and to different plant organs [9].

In this chapter, we describe methods to characterize vegetative and generative phenotypes associated with PPSB deficiency in plants.

2 Materials

2.1 KEGG Database

1. The KEGG database is publicly available at "http://www.genome.jp/keg/." The version at the time this article was written was released on July 1, 2016.

2.2 Mutants and Transgenic Lines Used for Studying the PPSB Functions

1. A *PSP1* or *PGDH1* cDNA under the control of the Cauliflower mosaic virus (CAMV) 35S promoter (*35S:PGDH1* or *35S: PSP1*) complements the embryo-lethal phenotype of the *psp1* and *pgdh1* mutants [11, 12]. However, these constructs are not expressed in anther tissues and do not complement mutant's

fertility [12, 13]. Thus mutant lines expressing the 35S:PSP1 or 35S:PGDH1 constructs (*psp1 35S:PSP1* or *pgdh1 35S: PGDH1*) are an excellent tool to study male gametophyte alterations in PPSB mutants.

2. To study the vegetative alterations associated with the PPSB activity conditional mutants and silenced lines can be used. Conditional mutants express PPSB genes under the control of the heat-shock-inducible promoter of the HSP18.2 (At5g59720) gene [12, 14]. In silencing lines, the expression of PPSB genes is reduced by using either *trans*-acting small interfering RNA (tasiRNA) or artificial microRNA (amiRNA) mediated gene silencing [11, 12].

2.3 Differential Interference Contrast Microscopy of Mutant Embryos

1. Hoyer's Solution (8 g chloral hydrate in 2 ml water and 1 ml glycerol).
2. Nikon ECLIPSE E800 microscope equipped with Nomarski differential interference contrast optics and a NIKON DXM1200F digital camera with Nikon ACT-1 software.
3. Stereoscopic zoom microscope Nikon SMZ 1500 equipped with a digital sight camera DS-Fi1.

2.4 Pollen and Anther Microscopy

2.4.1 Observation of Pollen Germination

1. In vitro pollen germination medium (0.01% H_3BO_3, 5 mM KCl, 5 mM $CaCl_2$, 1 mM $MgSO_4$, 10% Sucrose, 1% Agarose, adjusted to pH 7.8 with NaOH).
2. Olympus BH-2 microscope.

2.4.2 Transmission Electron Microscopy of Anthers and Pollen

1. Karnovsky solution (2.5% paraformaldehyde, 0.5% of glutaraldehyde in sodium phosphate buffer 100 mM, pH 7.3).
2. Fixing solution (2% osmium tetroxide in water).
3. Leika EM UC6 Ultracut microtome equipped with a diamond blade (DIATOME).
4. Transmission Electron Microscope (TEM) JEOL_1010 at 60–80 kV equipped with MegaView III digital camera.

2.4.3 Nucleus Staining of Pollen with Hoechst Solution

1. Hoechst dye stock solution (10 mg/ml in water).
2. Nikon ECLIPSE E800 compound microscope equipped with an epi-fluorescence module and a NIKON DXM1200F digital camera with Nikon ACT-1 software.

2.5 Characterization of Conditional PPSB Mutants and Silencing Plants

2.5.1 Plant Growth and Heat-Shock Treatment

1. Square Petri dishes.
2. Standard ½ MS basic salt medium without sucrose.
3. Gas-permeable tape.
4. Laboratory Glass bottles (1 L).
5. Autoclave.

6. Light meter.
7. Analytical scale.
8. Digital camera Canon EOS 5D Mark II.
9. ImageJ, a public domain Java image processing program.
10. Plastic pots (200 ml volume) filled with vermiculite (ASFALTEX) and moist Kekkila substrate (enriched with macronutrients N–P–K 15–10–20) in a 1:1 rate.
11. Growth chamber with controlled lighting, photoperiod, temperature, and humidity.

2.5.2 Amino Acid Analysis

1. 1.5 and 2.0 ml safe lock tubes.
2. Laboratory Glass bottles (1 L).
3. Dissecting scissor and forceps.
4. Analytical scale.
5. Metal rack.
6. Lyophilizer.
7. Vacuum desiccator with silica gel.
8. Ball-mill tissue lyzer.
9. Temperature-controlled tabletop centrifuge.
10. Orbital tube shaker.
11. Vortexer.
12. HPLC glass vials.
13. DIONEX Ultimate 3000 HPLC System including the RF2000 fluorescence detector, the Analytical Autosampler WPS-3000-TSL for sample preparation, the TCC-3000SD column compartment, the LPG-3400SD gradient pump, and the Chromeleon 6.8 evaluation software.
14. HPLC column Hyperclone 3 μm ODS (C_{18}) 120 A; 150 × 4.6 mm.
15. Pre-column Security Guard Cartridges C_{18} 4 × 3.0 mm.
16. HPLC-grade solvents such as water, acetonitrile (ACN), methanol (MetOH), ethanol (EtOH), and tetrahydrofurane (THF).
17. 0.4 M sodium phosphate buffer (pH 7.4).
18. Derivatization solution *o*-phthalaldehyde (OPA) and borate buffer (1 M pH 10.7) (Grom Analytics).
19. Amino acid standard solution (2 mM each).

3 Methods

3.1 Identification of Serine Biosynthetic Pathways in Plants by KEGG Database Search

The Kyoto Encyclopedia of Genes and Genomes (KEGG) database provides the ability to search for metabolites and their associated metabolic pathways in many organisms.

1. To search for pathways associated with serine biosynthesis the user should select the <Pathway database> in the main menu, enter the compound names such as serine in the keyword window (*see* **Note 1**), and click the <Search> button. The <Pathway Text Search> window opens, which contains information about all pathways linked to serine.
2. To search for metabolic reactions associated with serine the user should select the <Glycine, serine and threonine metabolism> window (map00260). The window provides access to pathway details without including organismal differences.
3. To obtain organism specific information the user should select, e.g., *Arabidopsis thaliana* (thale cress) in the main menu (*see* **Note 2**). The EC number of all enzymes present in *Arabidopsis thaliana* catalyzing the biosynthesis of serine are marked in green.
4. The gene locus name of the enzymes catalyzing reactions in serine biosynthesis from glycine (photorespiration), from hydroxypyruvate (glycerate pathway), and from 3-phosphoglycerate (PPSB) can be obtained by clicking on the respective green marked EC numbers.
5. The new open window contains information about the enzyme such as enzyme class, sysname, substrates, products, and references for each isoenzyme encoded by different gene loci.
6. The gene loci name can be found in the entry field of the new open window, e.g., for EC: 1.1.1.95 (PGDH) three gene loci (At4g34200, At1g17745, and At3g19480) can be found.

3.2 Isolation of Mutants Deficient in the PPSB

In order to identify homozygous T-DNA insertional mutants of *PGDH1* and *PSP1* genes, several approaches can be followed.

1. Segregation analysis of seedling obtained from heterozygous plants based on PCR genotyping using gene-specific primers and left border primers of the T-DNA insertions (*see* **Note 3**).
2. Segregation analysis of seedling obtained from heterozygous plants based on the antibiotic resistance (or another marker) conferred by the T-DNA insertion in the mutant allele (*see* **Note 4**).
3. If the segregation ratio is different from 1:2:1 (homozygous: heterozygous:wild type) in the PCR genotyping or 1:3

(resistant:sensitive) in the antibiotic resistance approach, then a lethal phenotype might be associated with the mutant allele.

In the case of *pgdh1* and *psp1* mutants, no homozygous individuals were rescued suggesting a possible lethal phenotype. To investigate whether the lethality associated with *pgdh1* and *psp1* mutations is due to male gametophyte, female gametophyte, or embryo defects, reciprocal out-crosses of heterozygous plants as male/female parent (donor/recipient) and wild type as female/male parent (recipient/donor) should be performed. To cross the plants:

1. Wash hands and clean all material before use with ethanol (*see* **Note 5**).
2. Remove all the buds and mature flowers from the receiver flower except two or three closed buds. Make sure that these flowers are completely closed to avoid autopollination, because at this stage, anthers are immature and pollens are not viable.
3. Using fine forceps and a binocular loupe, open the flower, eliminate petals and anthers, and keep the pistil.
4. Take the mature male donor flower and pass it carefully through the pistil. Make sure that several pollen grains stuck to the stigma.
5. Check the development of the stigma several days after pollination. If the cross succeeded, the stigma should have elongated.
6. Repeat the process as many times as required to have enough seeds for the segregation analysis.
7. Collect the siliques individually in reaction tubes (*see* **Note 6**).

The segregation of the mutant alleles in the offspring from crosses should be studied by antibiotic resistance (or PCR genotyping) as described previously. Three possible scenarios could occur:

Case 1. Heterozygous mutant plant (*PGDH1pgdh1* or *PSP-1psp-1*) as male donor and wild-type plant as female receiver. In this case, a progeny resulting from this cross which presents 100% sensitivity to the antibiotic marker of the T-DNA insertion will indicate that male gametophytes carrying the mutant allele are nonviable, and thus lethal. If 50% of the seedlings are resistant, this displays a transmission efficiency [(mutant allele/wild-type allele) × 100] of 100% for the mutant male gametophyte. An antibiotic resistance lower than 50% implies a lower transmission efficiency of mutant pollen as compared to the wild type.

Case 2. Heterozygous mutant plant (*PGDH1pgdh1* or *PSP-1psp-1*) as female receiver and wild-type plant as male donor. In this case, a progeny resulting from this cross which present a 100% sensitivity to the antibiotic marker indicates that female gametophyte carrying the mutant allele are nonviable, and thus lethal. A 50% of seedling resistance will indicate a transmission efficiency [(mutant allele/wild-type allele) × 100] of 100% for the mutant female gametophyte. An antibiotic resistance lower than 50% will indicate a lower transmission efficiency of mutant pollen as compared to the wild type.

Case 3. Both offspring from crosses between heterozygous plants as male donor and wild-type plants as the female recipient and vice versa exhibit 50% kanamycin resistance. This case represents a transmission efficiency for both gametophytes carrying the mutant allele of 100%, and indicate that there are no defects which affect the transmission of the male and female mutant alleles. This result would represent a case of an embryo-lethal phenotype.

3.3 Study of Embryo Development in PPSB Mutants by Differential Interference Contrast (DIC) Microscopy

1. Dissect siliques from heterozygous *psp-1/+* or *pgdh1/+* mutant plants longitudinally at different developmental stages (usually 2, 5, 7, and 10 days after pollination). Using fine forceps and small scissors, siliques are picked up from the bottom of their pedicel to facilitate further handling.
2. Take the silique by the pedicel and place it on a slide horizontally, leaving visible to the binocular loupe the junction area of both valves to the central septum. After that, firmly holding the pedicel with the forceps (the closer to the silique, the better), perform a longitudinal dissection with a scalpel, placing the tip of the blade in the silique septum junction as close as possible to the pedicel. Then, displace the blade till the apical part of the silique is reached.
3. Gently put aside one of the silique valves, thereby exposing the seeds inside. Then release the seeds with a blade and classify them according to their morphology.
4. Repeat the procedure with the other silique valve and remove all plant material other than seeds.
5. Spread a microscope coverslip with 50 μl of clarifying solution and put the cover upside down on top of the seeds (*see* **Note 7**).
6. Seal the preparation by applying nail varnish to the four corners of the cover.
7. Incubate the preparation for a minimum of 2 h before microscope observation.

3.4 Pollen Germination Assay

1. Pour the germination medium in Petri dishes of 25 ml, divide it into four quadrants, and apply the pollen of one flower in each quadrant.

2. Conventionally each quadrant is used for a different line, for instance wild-type and mutant lines. The pollen is applied by scratching the anthers over the media.
3. The developmental stage of the pollen is crucial to succeed. Use completely and freshly opened flowers harvested just after anthesis (stage 14 according to landmark events described by [15]). A good indicator of the appropriate stage is the color of pollen, which has to be golden (*see* **Note 8**). If petals are set aside, anther should be localized above stigma. We recommend to open several flowers to see the different tones of "yellow." The brighter the pollen, the best.
4. Take images with the microscope from at least two different microscopic fields. About 200 pollen grains should be scored for each replicate.
5. As a criteria, a pollen grain is considered germinated if the pollen tube length is longer than the pollen grain diameter.

3.5 Ontogenetic Analysis of Anther and Pollen Development

3.5.1 Transmission Electron Microscopy of Anthers and Pollen

1. Collect and classify flower buds from wild-type and *pgdh1 35S:PGDH1* or *psp1 35S:PSP1* plants in different developmental groups (stages 7–12 according to [15]).
2. Take several buds from the same group and pre-fix them with Karnovsky solution for 4 h at 4 °C.
3. Wash the samples three times × 10 min in 100 mM sodium phosphate buffer.
4. Post-fix samples with 2% osmium for 2 h.
5. Wash the samples three times × 5 min in water.
6. Dehydrate tissues in increasing concentrations of ethanol (30%, 50%, 70%, 90%) for 10 min each.
7. Infiltrate tissues with increasing concentrations of LR White resin as follow:
   - 1 volume of LR White + 2 volumes of ethanol 90% for 2 h.
   - 2 volumes of LR White + 1 volume of ethanol 90% for 2 h.
   - 2 volumes of LR White + 1 volume of ethanol 100% for 2 h.
   - LR White overnight.
8. Place the sample embedded in resin in the mold and incubate at 50–60 °C till complete polymerization (*see* **Note 9**).
9. Obtain 60 nm ultrathin cuts using the ultramicrotome for subsequent visualization in the transmission electron microscope.

3.5.2 Nucleus Staining of Pollen with Hoechst Solution

1. Collect and classify flower buds from wild-type and *pgdh1 35S:PGDH1* or *psp1 35S:PSP1* plants in different developmental groups as described for transmission electron microscopy.

2. Dissect the anthers from buds using a pair of fine-tipped forceps (the same kind used for crosses) and a binocular loupe.
3. Place dissected anthers on a glass slide. Then add 10 μl of Hoechst 1:1000 from the stock solution (final concentration 10 μg/ml) directly on the stamens and immediately place a coverslip onto the sample, avoiding the formation of bubbles.
4. Next, place the slide under a microscope and once focused the anther with the 10× objective, proceed to release the pollen from the anther by applying a gentle circular pressure on the coverslip using a glass rod.
5. After confirming that the pollen has been released, remove the excess of dye that overflows from the ends of the cover and seal the preparation by applying nail varnish on all four edges of the coverslip.
6. Finally incubate the sample in the dark for at least 5 min before observation with the fluorescence microscope.
7. The preparation can be preserved for 24 h in the dark at 4 °C.

3.6 Characterization of Conditional PPSB Mutants and Silencing Lines

3.6.1 Heat-Shock Treatment of Conditional PPSB Mutants

1. Sow homozygous *psp1* conditional mutants and wild-type control seeds into vermiculite pots.
2. Place the pots into a cold room at 4 °C covered with Saran Wrap to trigger germination and to ensure high humidity.
3. After 4 days place the pots in the growth chamber and remove the Saran Wrap gradually to improve the acclimatization process. Growth conditions consist of 16 h light/8 h dark, 50% humidity, and a temperature of 22 °C during the day and 18 °C during the night.
4. To overcome lethal phenotypes and obtain offsprings, 5 days after sowing and until the end of the plant cycle, *PSP1* is induced in the conditional mutants by increasing the growth chamber temperature at 37 °C for 1 h a day.
5. For phenotypic or metabolomic characterizations, conditional mutants can be grown in Petri dishes as specified below and treated at 37 °C as described in **step 4**. Mutant characterization can be done at different time point after heat-shock induction. Non-heat-shock-induced conditional mutants as well as wild type (heat-shock and non-heat-shock treated) can be used as controls.

3.6.2 Quantification of Shoot and Root Growth

1. For phenotypical characterization, grow conditional PPSB mutants or silencing lines in square Petri dishes on ½ MS medium (*see* **Note 10**).
2. Surface sterilize seeds of conditional mutants or silencing lines by incubation in 70% EtOH for 10 min followed by 5 min in 100% EtOH.

3. Discard EtOH and air dry seeds on the clean bench.
4. Sow surface-sterilized seeds on ½ MS plates and incubate plates for 3 days at 4 °C for stratification.
5. Germinate seeds of conditional PPSB mutants or silencing lines on squared Petri dishes in horizontal orientation under long day (16 h light/8 h dark) conditions at 45% humidity and a temperature of 22 °C at day and 18 °C at night.
6. After 4 days of germination transfer seedlings to freshly prepared ½ MS plates and place plates vertically (*see* **Note 11**).
7. Set light conditions in plant height to 100 μmol photons/m^2/s and ensure uniform illumination of the plants (*see* **Note 12**).
8. Mark the position of the tip of the primary and lateral roots of each individual plant on the backside of the Petri dish every day at the same time for 10 days to calculate the root growth rate.
9. Photograph the plates from the backside for later evaluation of the root growth rate.
10. Separate shoot and root tissue of the individual plants using a dissection scissor and determine the fresh weight per plant on an analytical scale (*see* **Note 13**).
11. Evaluate the photographs to determine the root growth rate by measuring the distance between each mark of the individual roots using the "segmented line" tool of the ImageJ software (*see* **Note 14**).

3.6.3 Analysis of Amino Acid Content in Conditional PPSB Mutants and Silencing Lines

1. Harvest plant material of PPSB-silenced and wild-type plants grown on vertical ½ MS plates with a forceps and separate shoot and root tissue with a dissection scissor and determine the fresh weight of the plant material (*see* **Note 15**).
2. To avoid alterations in metabolite composition the plant material has to be immediately snap-frozen in liquid nitrogen.
3. The frozen tissue can either be directly lyophilized or stored at −80 °C for up to 3 months.
4. For lyophilization place the open Eppendorf tubes containing the frozen material in a pre-cooled (−80 °C) metal rack (*see* **Note 16**).
5. Place the tube-containing metal rack into the lyophilizer at 0.02 bar and −80 °C for at least 1 day (*see* **Note 17**). To prevent hydration of the dried material after lyophilization close the reaction tube immediately after ventilation of the lyophilizer (*see* **Note 18**). The lyophilized material should be stored in a desiccator containing silica gel protected from light and humidity or can be directly used for amino acid extraction (*see* **Note 19**).

6. A steel ball is added to the freeze-dried samples and the tissue is homogenized in a ball mill tissue lyzer for 1 min at 25 Hz (*see* **Note 20**).
7. Due to the different physicochemical properties of amino acids the homogenized plant material is extracted twice with different concentrations of EtOH. First, the dry powder is extracted with 500 μl of pre-cooled 80% EtOH in an orbital shaker at 4 °C for 20 min.
8. After centrifugation of the ethanolic extract at 20,000 rcf at 4 °C for 15 min, the supernatant (1) is transferred into a new tube and stored at 4 °C in the dark.
9. The remaining pellet is extracted a second time with 500 μl pre-cooled 50% EtOH in an orbital shaker at 4 °C for 20 min.
10. The extract is centrifuged at 20,000 rcf at 4 °C for 15 min and the obtained supernatant (2) is combined with supernatant (1) and mixed vigorously (*see* **Note 21**).
11. The combined ethanolic extract can be stored at −20 °C for several weeks or directly used for the amino acid analysis by High Performance Liquid Chromatography coupled with a Fluorescence Detector (HPLC-FD).
12. Prior HPLC analysis prepare Solvent A (2% THF and 8.8 mM sodium phosphate buffer pH 7.4) and Solvent B (20.6% ACN, 32.7% MetOH, and 18.7 mM sodium phosphate buffer pH 7.4) of the mobile phase.
13. Use the following solvent gradient for optimal separation of the amino acid mixture on a C_{18} column. $t = 0–5$ min, 0% B; $t = 5–16$ min, 0–15% B; $t = 16–26$ min, 15–50% B; $t = 26–42$ min, 60% B; $t = 42–46$ min, 100% B, $t = 46–52$ min, 100% B; $t = 52–53$ min, 100–0% B, and $t = 53–62$ min, 0% B.
14. For the quantification of amino acid concentrations by HPLC-FD, amino acid standard solutions containing all proteinogenic amino acids with increasing concentrations (1–20 μM) are analyzed in advance (*see* **Note 22**). Therefore, 1 ml of each standard solution (1, 2.5, 5, 10, 20 μM) is transferred into a HPLC glass vial and placed into the programmable HPLC autosampler.
15. To analyze the amino acid content in plants, 100 μl of the ethanolic plant extract is diluted with 900 μl HPLC-grade water and transferred into HPLC glass vials.
16. For fluorescence detection amino acids are pre-column derivatized with the fluorescent dye *o*-phthaldialdehyde (OPA) (*see* **Note 23**). Derivatization is conducted by the programmable autosampler by mixing 25 μl ethanolic extract with 5 μl borate buffer (pH 10.7; 1 M) and 25 μl 1/10 H_2O diluted OPA stock solution. After incubation for 90 s 20 μl of the

derivatization mixture is injected into the HPLC system for separation of the fluorescent amino acid derivatives onto a C_{18} column (*see* **Note 24**).

17. Fluorescence signal is monitored by a fluorescence detector (EX λ 330 nm and EM λ 450 nm) and converted into an electric signal. The presence of each amino acid derivative is detected as a peak and the size of the peak area is linear depending on the concentration of the amino acid. Based on the calibration curve the concentration of each amino acid in the plant extract is calculated and further normalized to the used fresh weight.

4 Notes

1. KEGG database allows searching for different compound names. Instead of serine, the nomenclature 2-amino-3-hydroxypropionic acid or 3-hydroxyalanine can be also used as a search item.
2. Pathway information can be obtained for many organisms of the eukaryotes, bacteria, and archea.
3. The best way to perform PCR genotyping is to use three primers in the same PCR reaction. Two primers matching the genomic sequences at both sides of the T-DNA insertion and a third one matching the left border of the T-DNA. This strategy allows to identify homozygous, heterozygous, and wild-type individuals in a single PCR reaction. Although theoretically only two fragments should be amplified under these conditions, sometimes more than two bands may appear in the electrophoresis resolving gel. This could happen when unspecific bands are amplified or when the left border primer amplify fragments with both genomic primers. This band pattern should not be a problem if the size of the specific bands can be determined, and if the band pattern of homozygous, heterozygous, and wild-type plants can be easily distinguished. Sometimes it is not possible to use three primers in the same PCR reaction because primer dimer formation or because the size of the amplified fragments overlap and therefore different genotypes cannot be distinguished easily in the resolving gel. Then two PCR reactions are necessary, one with the two genomic primers, and a second one with the left border primer and one of the genomic primers. This strategy, although necessary in some cases, have several drawbacks. If the appropriate band does not appear in the resolving gel, one cannot be sure if this is so because the amplified sequence is not present in the genome of the studied individual or simply that the PCR failed. In this

case several repeats of the same PCR reaction are required to confirm the results.

4. Many T-DNA insertional mutants loose the antibiotic resistance associated with the T-DNA insertion. In that case segregation analysis of seedling can only be done by PCR genotyping.
5. All material should be free of pollen before starting the crosses. Otherwise contamination could occur with pollen from unidentified or not desired plants.
6. Do not mix siliques just in case one of the crosses has not been done correctly.
7. In this step it is important to eliminate all bubbles in the preparation and ensuring that all seeds are in contact with the solution.
8. Do not take golden-orange pollen.
9. When placing the sample embedded in resin in the mold and before polymerization it is very important to place them in the desired position to facilitate subsequent cuts with the ultramicrotome. Otherwise the cuts will not visualize the required structures.
10. ½ MS basic salt medium without vitamins should be used as some MS variants are substituted with glycine, which influences growth of conditional PPSB mutants and silencing lines.
11. To guaranty uniform illumination plants were placed in one row approximately 3 cm below the upper edge of the plate. A total of 15 plants, five control plants and ten transgenic lines, can be placed on one plate.
12. Adjust the light intensities by using a Light Meter as slight differences in light conditions strongly influence the phenotype of conditional PPSB mutants and silencing lines.
13. Combine the root or shoot tissue of at least five plants as the fresh weight of an individual plant is not sufficient to be exactly determined by an analytical scale.
14. Photograph plotting paper as a scale to calculate the root growth rate.
15. Combine root or shoot tissue of at least five plants.
16. The metal rack is important to prevent thawing of the harvested plant material.
17. The lyophilization process should never be interrupted until the procedure is completely finished as this results in thawing of the material.
18. Hydration of the freeze-dried material leads to alteration in metabolite composition and therefore will falsify the amino acid analysis.

19. Freeze-dried material should not be stored longer than 6 months in total as metabolites lose integrity under dried conditions over time.
20. The ball mill tissue lyzer produces a very fine powder of the plant tissue; this is a prerequisite for exact determination of amino acid content in plant tissue. If no ball mill tissue lyzer is available, the tissue can also be homogenized with a conventional pistil and a mortar.
21. Mix both supernatants vigorously as they may contain different proportions of the different amino acids.
22. Most amino acids are soluble in water; however, for some, e.g., aromatic amino acids, the pH has to be adjusted to ensure complete solubilization.
23. OPA reacts in the presence of 2-mercaptoethanol only with primary amines; therefore, amino acids with a secondary amino group such as proline cannot be detected by this method.
24. OPA derivatives are unstable, which requires automatic precolumn derivatization by using a programmable autosampler.

Acknowledgement

We would like to thank Prof. Ulf-Ingo Flügge for constant support, critical comments, and fruitful discussions. We would also acknowledge the University of Cologne, the Cluster of Excellence on Plant Science (CEPLAS), the Deutsche Forschungsgemeinschaft (Grant Kr4245/1-1; Kr4245/2-1), the Spanish Government and the European Union (FEDER/BFU2012-31519, BFU2015-64204R; MINECO/FEDER), the Generalitat Valenciana (PROMETEO II/2014/052), and the University of Valencia (Atracció de Talent" fellowship to M.F.-T.) for their financial support. We thank Servei Central de Support a la Investigació Experimental (SCSIE) and Unitat Central de Investigació en Medicina (UCIM) of the Universitat de València for technical assistance.

References

1. Bauwe H, Hagemann M, Fernie AR (2010) Photorespiration: players, partners and origin. Trends Plant Sci 6:330–336
2. Hodges M, Dellero Y, Keech O, Betti M, Raghavendra AS, Sage R, Zhu XG, Allen DK, Weber AP (2016) Perspectives for a better understanding of the metabolic integration of photorespiration within a complex plant primary metabolism network. J Exp Bot 10:3015–3026
3. Somerville CR (2001) An early Arabidopsis demonstration. Resolving a few issues concerning photorespiration. Plant Physiol 1:20–24
4. Timm S, Bauwe H (2013) The variety of photorespiratory phenotypes – employing the current status for future research directions on photorespiration. Plant Biol (Stuttg) 4:737–747
5. Queval G, Issakidis-Bourguet E, Hoeberichts FA, Vandorpe M, Gakière B, Vanacker H, Miginiac-Maslow M, Van Breusegem F,

Noctor G (2007) Conditional oxidative stress responses in the Arabidopsis photorespiratory mutant cat2 demonstrate that redox state is a key modulator of daylength-dependent gene expression, and define photoperiod as a crucial factor in the regulation of H_2O_2-induced cell death. Plant J 4:640–657

6. Somerville CR, Ogren WL (1980) Inhibition of photosynthesis in Arabidopsis mutants lacking leaf glutamate synthase activity. Nature 286:257–259
7. Ewald R, Kolukisaoglu U, Bauwe U, Mikkat S, Bauwe H (2007) Mitochondrial protein lipoylation does not exclusively depend on the mtKAS pathway of de novo fatty acid synthesis in Arabidopsis. Plant Physiol 1:41–48
8. Collakova E, Goyer A, Naponelli V, Krassovskaya I, Gregory JF 3rd, Hanson AD, Shachar-Hill Y (2008) Arabidopsis 10-formyl tetrahydrofolate deformylases are essential for photorespiration. Plant Cell 7:1818–1832
9. Ros R, Muñoz-Bertomeu J, Krueger S (2014) Serine in plants: biosynthesis, metabolism, and functions. Trends Plant Sci 9:564–569
10. Ho CL, Saito K (2001) Molecular biology of the plastidic phosphorylated serine biosynthetic pathway in *Arabidopsis thaliana*. Amino Acids 3:243–259
11. Benstein RM, Ludewig K, Wulfert S, Wittek S, Gigolashvili T, Frerigmann H, Gierth M, Flügge UI, Krueger S (2013) Arabidopsis phosphoglycerate dehydrogenase1 of the phosphoserine pathway is essential for development and required for ammonium assimilation and tryptophan biosynthesis. Plant Cell 12:5011–5029
12. Cascales-Miñana B, Muñoz-Bertomeu J, Flores-Tornero M, Anoman AD, Pertusa J, Alaiz M, Osorio S, Fernie AR, Segura J, Ros R (2013) The phosphorylated pathway of serine biosynthesis is essential both for male gametophyte and embryo development and for root growth in Arabidopsis. Plant Cell 6:2084–2101
13. Toujani W, Muñoz-Bertomeu J, Flores-Tornero M, Rosa-Téllez S, Anoman AD, Alseekh S, Fernie AR, Ros R (2013) Functional characterization of the plastidial 3-phosphoglycerate dehydrogenase family in Arabidopsis. Plant Physiol 3:1164–1178
14. Matsuhara S, Jingu F, Takahashi T, Komeda Y (2000) Heat-shock tagging: a simple method for expression and isolation of plant genome DNA flanked by T-DNA insertions. Plant J 1:79–86
15. Smyth DR, Bowman JL, Meyerowitz EM (1990) Early flower development in Arabidopsis. Plant Cell 8:755–767

Chapter 17

Light Microscopy, Transmission Electron Microscopy, and Immunohistochemistry Protocols for Studying Photorespiration

Roxana Khoshravesh, Vanessa Lundsgaard-Nielsen, Stefanie Sultmanis, and Tammy L. Sage

Abstract

High-resolution images obtained from plant tissues processed for light microscopy, transmission electron microscopy, and immunohistochemistry have provided crucial links between plant subcellular structure and physiology during photorespiration as well as the impact of photorespiration on plant evolution and development. This chapter presents established protocols to guide researchers in the preparation of plant tissues for high-resolution imaging with a light and transmission electron microscope and detection of proteins using immunohistochemistry. Discussion of concepts and theory behind each step in the process from tissue preservation to staining of resin-embedded tissues is included to enhance the understanding of all steps in the procedure. We also include a brief protocol for quantification of cellular parameters from high-resolution images to help researchers rigorously test hypotheses.

Key words Light microscopy, Transmission electron microscopy, High-resolution microscopy, Immunohistochemistry, Photorespiration, Chloroplasts, Mitochondria, Peroxisomes

1 Introduction

The initial recognition of partitioning of enzymes involved in metabolism of 2-phopshogylcolate (PG) between the chloroplast, peroxisome, and mitochondrion [1, 2] was central to early studies elucidating photorespiration. A combination of methods was used to unravel this compartmentalized process and prominent amongst these procedures was high-resolution imaging of plant cells with transmission electron microscopy (TEM). The power of TEM in calibrating the initial understanding of photorespiration is underscored by the early images of photosynthetic leaf cells demonstrating a tight spatial association between the chloroplast, peroxisome, and mitochondrion [3, 4]; these snapshots provided crucial links between plant subcellular structure and physiology during

Alisdair R. Fernie et al. (eds.), *Photorespiration: Methods and Protocols*, Methods in Molecular Biology, vol. 1653,
DOI 10.1007/978-1-4939-7225-8_17, © Springer Science+Business Media LLC 2017

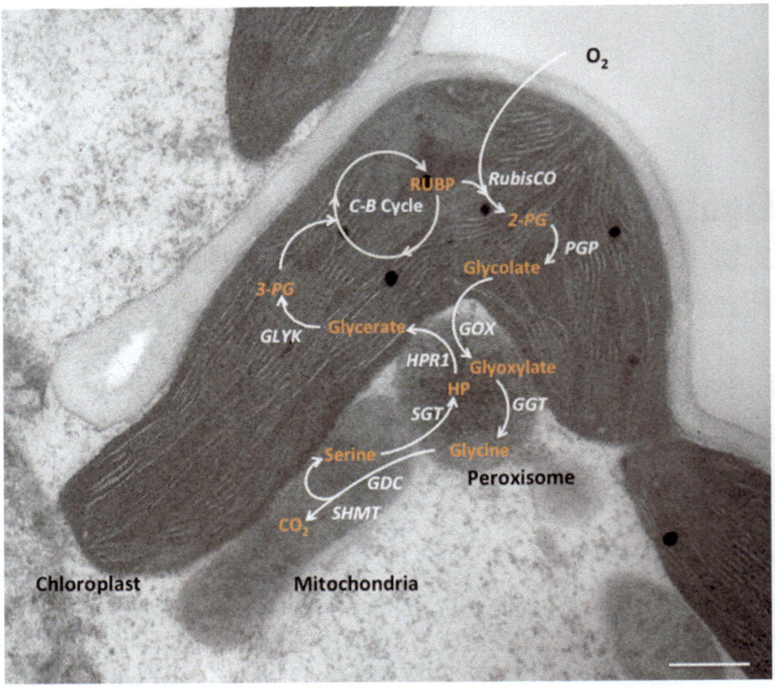

Fig. 1 Illustration of the spatial relationship between the chloroplast, peroxisome, and mitochondrion containing photorespiratory enzymes. *C-B* Calvin-Benson cycle, *GDC* glycine decarboxylase, *GGT* glutamate:glyoxylate aminotransferase, *GLYK* glycerate 3-kinase, *GOX* glycolate oxidase, *HPR1* peroxisomal hydroxypyruvate reductase, *3-PG* 3-phosphogylycerate, *2-PG* 2-phosphogylcolate, *PGP* 2PG phosphatase, *RubisCO* Ribulose 1,5-bisphosphate carboxylase/oxygenase, *SGT* serine:glyoxylate aminotransferase, *SHMT* serine hydroxymethyltransferase. Rice (*Oryza sativa*) mesophyll cell imaged by TL Sage. Bar = 500 nm

photorespiration. This important spatial relationship is illustrated in Fig. 1. Further microscopic studies detecting photorespiratory enzymes using immunohistochemical methods in combination with TEM and high-resolution light microscopy (LM) continued to solidify biochemical and physiological studies implicating chloroplasts, peroxisomes, and mitochondria in the metabolism of PG [5] and, as examples, provided data for generating and testing hypotheses regarding the role of photorespiration in the evolution of the carbon concentrating mechanism exhibited by C_4 species [5–7] and refixation of photorespired CO_2 in crop species [8]. As a final illustration, high resolution LM and TEM have also recently proven to be indispensible for assessing the influence of the subunits of the mitochondrial photorespiratory enzyme glycine decarboxylase (GDC) on photosynthesis and other metabolic pathways [9, 10] as well as plastid morphology [10]. In the latter study, the use of TEM demonstrated that reductions in the mesophyll H-subunit of GDC in rice resulted in a decrease in mesophyll cell chloroplast area and mesophyll cell wall coverage by chloroplasts. This last feature is one of the final changes in mesophyll subcellular structure during C_4 evolution [11, 12] correlating to confinement of almost all GDC in leaves to bundle sheath cells [6, 7, 13]. Reduction in chloroplast coverage is posited to enhance diffusion of intercellular CO_2 to cytoplasmic phosphoenolpyruvate decarboxylase [11, 12].

Methods used to prepare plant tissue for LM and TEM and immunohistochemical detection of proteins function to optimally preserve cellular organization and protein antigenicity for high-resolution imaging. As such, these classical techniques are powerful tools that contribute to a comprehensive understanding of gene product function, cell physiology, development, and evolution. Here we describe methods for preparation of plant tissue for high-resolution imaging of plant cells with LM and TEM. We also provide protocols for immunohistochemical detection of proteins with LM and TEM. Live-cell imaging with a confocal microscope is a valuable tool for localization of photorespiratory proteins tagged with appropriate fluorescent probes such as GFP and YFP [14, 15]. This technique has provided important data on regulation of the mitochondria localized photorespiratory enzyme serine hydroxymethyltransferase in *Arabidopsis* [16] and evolution of photorespiration through study of the peroxisomal photorespiratory enzyme glycolate oxidase in the red alga *Cyanidioschyzon merolae* [17]. However, when compared to immunohistochemical detection of proteins with TEM, live-cell imaging has a limited resolution of cellular detail and is more amenable for use with plants that are easily transformed [14, 15]. Tissues from all plant species can be easily processed for high-resolution LM and TEM imaging and immunohistochemical detection of proteins using a series of simple, well-defined procedures. These include (1) tissue fixation and dehydration; (2) infiltration and embedding with resins; and (3) sectioning and staining of resin-embedded tissue (Fig. 2). Once processed tissues have been imaged with a light or transmission electron microscope, cellular parameters can be quantified to rigorously test hypotheses.

2 Materials

2.1 Tissue Fixation

2.1.1 Primary Fixation

1. 0.1 M sodium cacodylate buffer, pH 6.9 (*see* **Note 1**).
2. EM grade glutaraldehyde aqueous solution, 25% or 50%.
3. EM grade paraformaldehyde aqueous solution, 16% or 20%.
4. Fixative: 2% glutaraldehyde in 0.1 M sodium cacodylate buffer (classical microscopy), or 1% glutaraldehyde and 1% paraformaldehyde in 0.1 M sodium cacodylate buffer [immunohistochemistry or classical microscopy for plants with cells that are difficult to fix (thick cell walls and cuticle)]. Prepare fixative immediately before use.

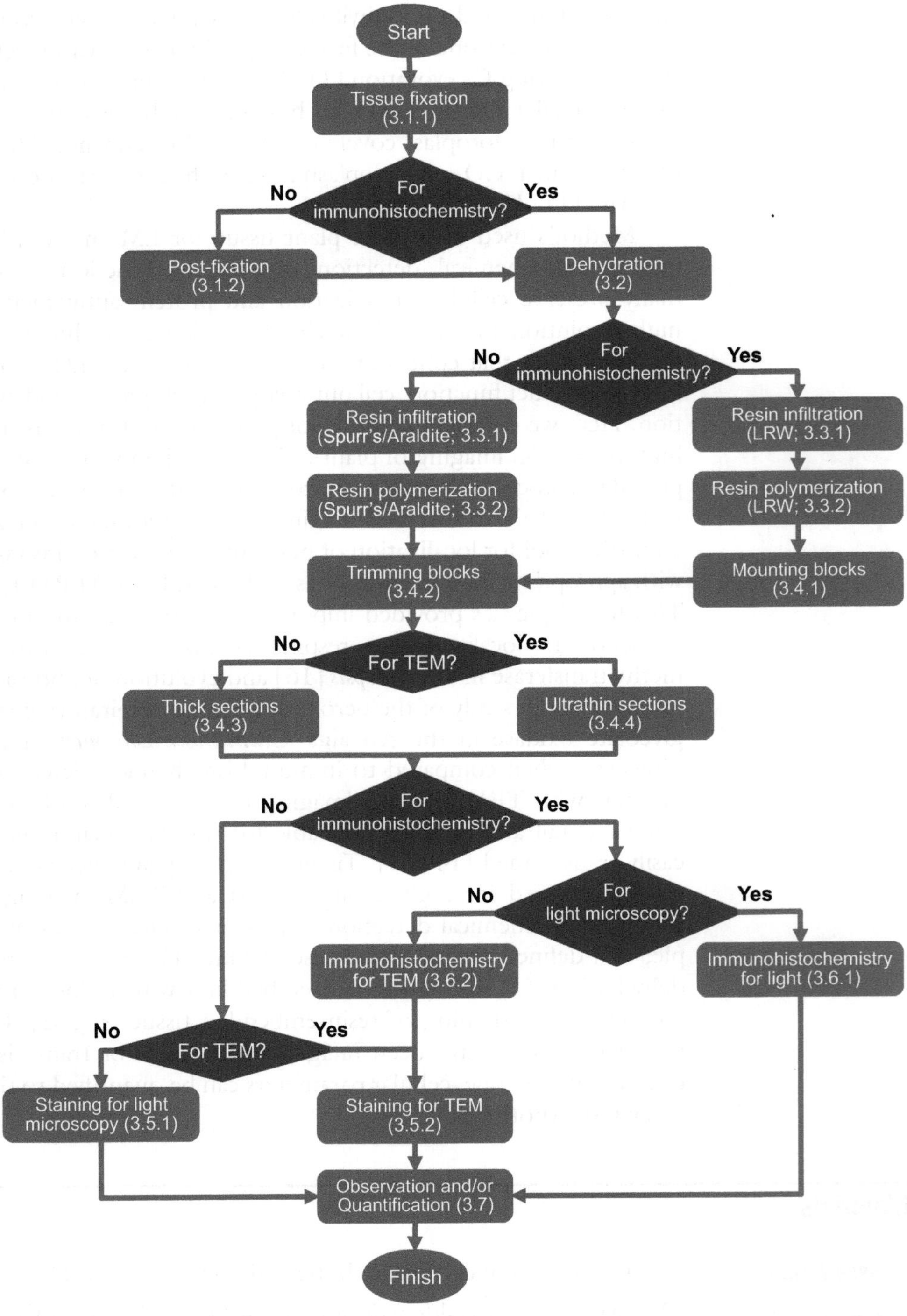

Fig. 2 Flowchart outlining successive steps for preparation of tissue for high resolution LM, TEM, and immunohistochemistry

5. 70% EtOH.
6. Double-edged razor blades (*see* **Note 2**).
7. Tweezers.
8. Dissecting microscope.
9. Borosilicate glass vials, 1 dram.
10. Transfer pipettes, 1 mL.
11. Precision wipes.

2.1.2 Post-fixation

1. 4% Osmium tetroxide (OsO_4), aqueous solution.
2. Transfer pipettes, 1 mL.
3. 0.1 M sodium cacodylate buffer, pH 6.9.

2.2 Dehydration of Fixed Tissue

1. Graded EtOH series; 10–100% in 10% increments (*see* **Note 1**).
2. Transfer pipettes, 1 mL.

2.3 Resin Infiltration and Embedding Dehydrated Tissue

1. 100% EtOH.
2. 100% acetone.
3. Flat-bed silicone embedding mold (for Spurr's and Araldite 502).
4. Medicine cups, 30 mL.
5. Transfer pipettes, 1 mL.
6. Wooden applicator stick whittled at one end with razor blade.
7. Fresh Spurr's prepared as directed by manufacturer (*see* **Note 3**).
8. Fresh London Resin White (LRW).
9. Fresh Araldite 502 prepared as directed by manufacturer (*see* **Note 4**).
10. Propylene oxide.
11. Precision wipes.
12. Oven.

2.4 Trimming Blocks of Resin-Embedded Tissue and Sectioning and Staining Resin-Embedded Tissue

2.4.1 Light Microscopy

1. Razor blades.
2. Ultramicrotomy glass strips.
3. Glass knife maker.
4. Glass knife boats (Fig. 3a).
5. Nail polish (*see* **Note 5**).
6. Double-sided tape.
7. Glass or diamond knives; Make glass knives following the instructions supplied with knife making machine. Store glass knives in a covered container. Use double-sided tape on the bottom of the knife to keep them upright. Seal the plastic boat to the knife with

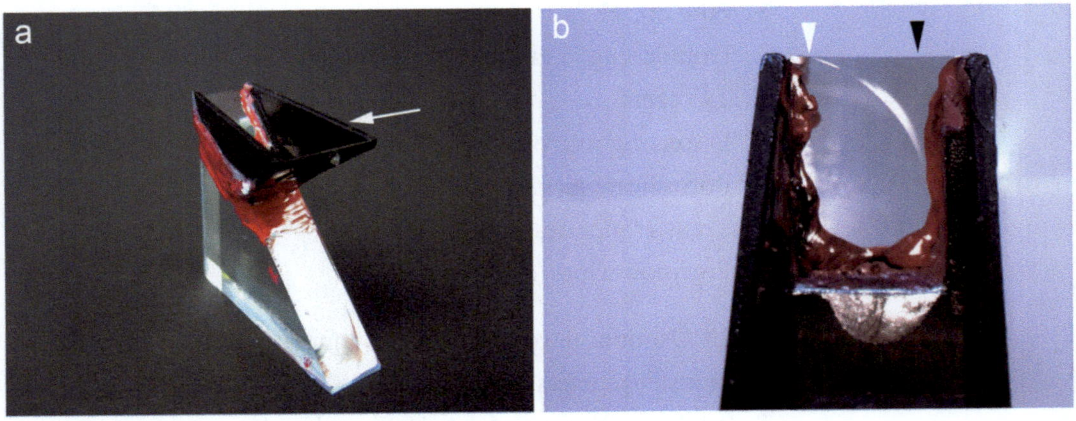

Fig. 3 A glass knife made with a knife-breaking machine illustrating (**a**) proper placement of a boat (*arrow*) attached to a glass knife with nail polish, and (**b**) the sharper (left; *white arrowhead*) and more dull (right; *black arrowhead*) sides of the knife

multiple layers of nail polish to prevent leakage of water that will be added to the boat during sectioning (*see* **Note 5**; Fig. 3a).

8. Microtome.
9. Adjustable frame jeweler's saw.
10. Diamond wire for jeweler's saw.
11. Vice grip.
12. Wooden dowel, 7 mm in length, 5 mm diameter.
13. Superglue.
14. Tweezers.
15. Adhesion microscope slides, 25 × 75 × 1 mm.
16. Glass coverslips, No. 1, 22 × 50 mm.
17. Hot plate.
18. Luer Lock disposable syringe, 50 mL.
19. Luer Lock disposable syringe filter, 0.45 μm pore size.
20. Slide mounting medium.
21. Precision analytical weights, 25, 50, or 100 g.
22. Toluidine blue [18]: 0.5% toluidine blue in 0.1% sodium carbonate, pH 11.1 (Spurr's and Araldite 502), or 0.05% toluidine blue in benzoate buffer, pH 4.4 [benzoic acid, 0.25 g; sodium benzoate, 0.29 g; 200 mL H_2O; (LRW)].
23. Glass rod (*see* **Note 6**).
24. Water- and alcohol-resistant fine tip marker.
25. 70% EtOH.
26. dH_2O.

2.4.2 Transmission Electron Microscopy

1. Double-edged razor blades.
2. Diamond or glass knife.
3. Ultramicrotome.
4. Glass microscope slide cleaned with EtOH (*see* **Note 7**).
5. Eyelash brush (*see* **Note 8**).
6. Ultrasonic bath.
7. Acetone.
8. 70% and 95% EtOH.
9. Filter paper.
10. Precision wipes.
11. 100 mL glass beaker.
12. Porcelain crucible with perforated bottom.
13. Slot, tabbed TEM grids, copper (classical microscopy) and nickel (immunohistochemistry; *see* **Note 9**). To clean new grids, add grids to the crucible and place in a glass beaker containing 40 mL acetone. Sonicate the glass beaker containing the grids in an ultrasonic bath for 1 min. Remove crucible from beaker, drain acetone. Repeat the sonication procedure after replacing the acetone with dH_2O. Rinse in 95% EtOH. Air dry grids on filter paper in Petri dish. To clean previously used grids, first sonicate for 1 min in chloroform, and proceed as with new grids.
14. Grid boxes (*see* **Note 10**).
15. Silicone staining pad (*see* **Note 11**).
16. Petri dishes.
17. Sodium hydroxide pellets.
18. Aluminum foil.
19. 2× sterile Luer Lock disposable syringes, 50 mL.
20. 2× gauge Luer Lock needles.
21. Luer Lock disposable syringe filter, 0.45 μm pore size.
22. Cotton swabs.
23. Staining dish (135 × 102 × 69 mm).
24. 1 L beaker.
25. 4% Uranyl acetate: Remove plunger from 50 mL Luer Lock syringe and seal tip with laboratory film. Add 2 g uranyl acetate to syringe. Position syringe in horizontal position, remove laboratory film, and completely depress plunger into syringe. Attach Luer Lock needle and draw in 50 mL dH_2O. Invert syringe, replace needle with 0.45 μm pore filter, shake, cover with aluminum foil, and store saturated uranyl acetate stain at 4 °C until use.

26. Lead citrate: Boil 70 mL dH_2O. Remove plunger from 50 mL Luer Lock syringe and seal tip with laboratory film. Add 1.33 g lead nitrate and 1.76 g sodium citrate to syringe. Position syringe in horizontal position, remove laboratory film, and completely depress plunger into syringe. Attach Luer Lock needle, draw in 30 mL of cooled H_2O, and seal tip with laboratory film. Shake vigorously for 2 min and gently for 30 min to produce a milky white solution. Invert syringe, replace laboratory film with the Luer Lock needle, and draw in 8 mL fresh 1 M NaOH. Solution should clear (*see* **Note 12**). Draw in dH_2O to bring volume to 50 mL. Invert syringe and replace needle with 0.45 μm syringe filter, cover with aluminum foil and store at 4 °C until use.
27. Chloroform.
28. 0.33% Formvar in fresh chloroform (*see* **Note 13**). Seal with laboratory film, stir gently, and store in desiccator.
29. Copper or nickel formvar-coated grids (Fig. 4 illustrates preparation of formvar-coated grids). Place staining dish in a shallow container and overfill staining dish with ultrapure water. Use EtOH cleaned tweezers to dip cleaned glass slide into formvar. Slowly lift slide out of coating jar and allow excess solution to run off (*see* **Note 14**; Fig. 4a). Rest coated slide against formvar jar (Fig. 4b), cover with beaker, and dry for 2 min (Fig. 4c). Etch slide using tweezers in a U-shape starting and ending at the formvar boundary (Fig. 4d). Exhale humid breath on the slide to produce a layer of condensation on the formvar and immediately immerse slide into a water-filled staining dish at a 45° angle to float the coating off onto water surface (Fig. 4e). Continue submerging the slide until the formvar sheet is released from the slide. Use tweezers to place clean copper or nickel grids shiny side up and rough side down on silver and pale gold regions of the formvar (*see* **Note 15**; Fig. 4f). Pick up sheet of grids on formvar with clean side of laboratory film and remove from water (Fig. 4g). Place in Petri dish, grid side up, and store at room temperature until use (Fig. 4h). To use, perforate formvar around grid periphery with tweezers and lift grid (Fig. 4i).

2.5 Immunohistochemical Detection of Proteins

2.5.1 Light Microscopy

1. Sections of LRW-embedded tissue on adhesion microscope slides (Subheading 3.4.3).
2. 70% EtOH.
3. Plastic container with lid (humidity chamber; *see* **Note 16**; Fig. 5a and b).
4. Precision wipes.
5. Liquid blocker/hydrophobic barrier pen.

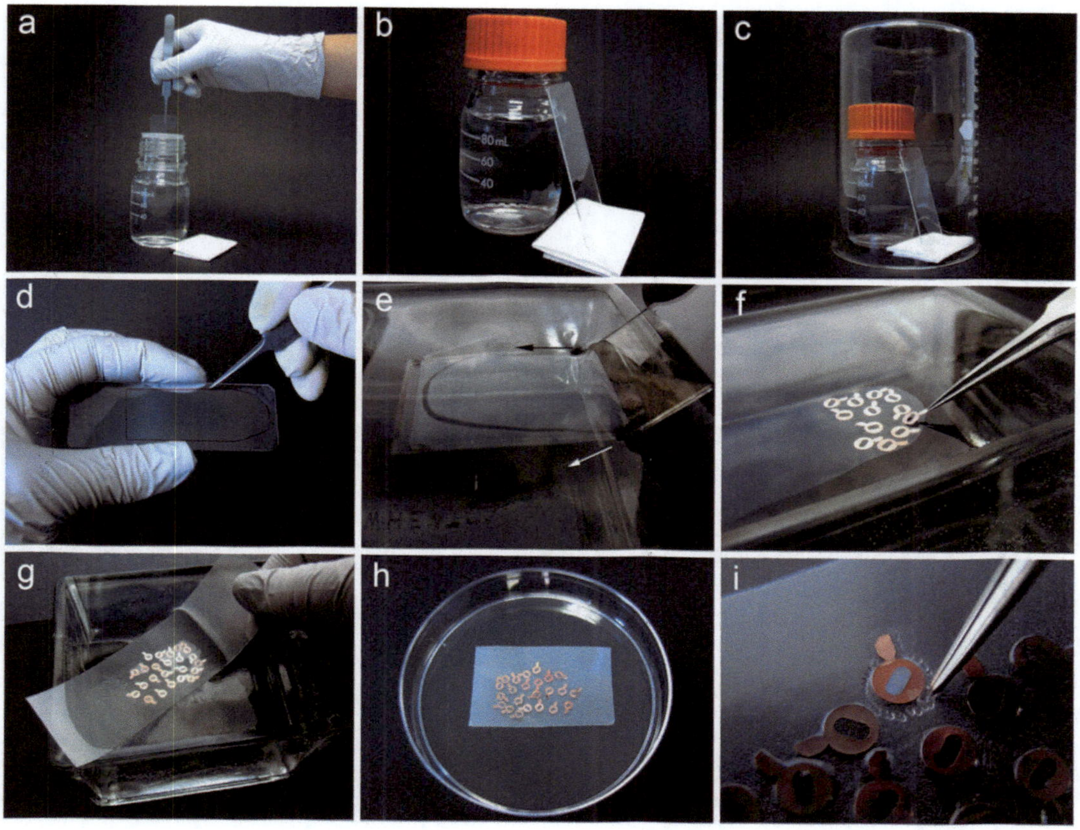

Fig. 4 Illustrations demonstrating preparation of formvar-coated grids. (**a**) EtOH-cleaned tweezers are used to dip cleaned glass slide into formvar. (**b**) The formvar-coated glass slide is placed in a vertical position. (**c**) The covered formvar-coated glass dries while resting in a vertical position. (**d**) Etching the formvar-coated glass slide with tweezers. *Solid line* shows the path that tweezers have taken to etch the formvar. (**e**) Immersing the formvar-coated slide into a water-filled staining dish at a 45° angle. *Arrows* indicate the direction of movement of the glass slide and formvar. (**f**) Placing cleaned grids rough-side down on the pale silver regions of the formvar. (**g**) Picking up a sheet of grids and formvar with laboratory film. (**h**) Grids on laboratory film placed in a Petri dish. (**i**) Perforating the formvar around the grid periphery using tweezers. The formvar-coated grid is now ready for use to pick up 70 nm thin sections for TEM

6. 50 mL conical centrifuge tube.
7. 10% BSA. Prepare in conical centrifuge tube with lid. Heat to 30 °C, stir gently to dissolve, and prevent denaturation of BSA. Store at 4 °C.
8. Tris-Buffered Saline (TBS); 10 mM Tris–HCl, 150 mM NaCl, 1% w/v BSA, pH 7.2. Store at 4 °C.
9. 1% BSA in TBS. Store at 4 °C.
10. Primary antibody with high specificity towards the protein of interest.
11. Protein A-conjugated gold, 20 nm.
12. Silver enhancement kit.

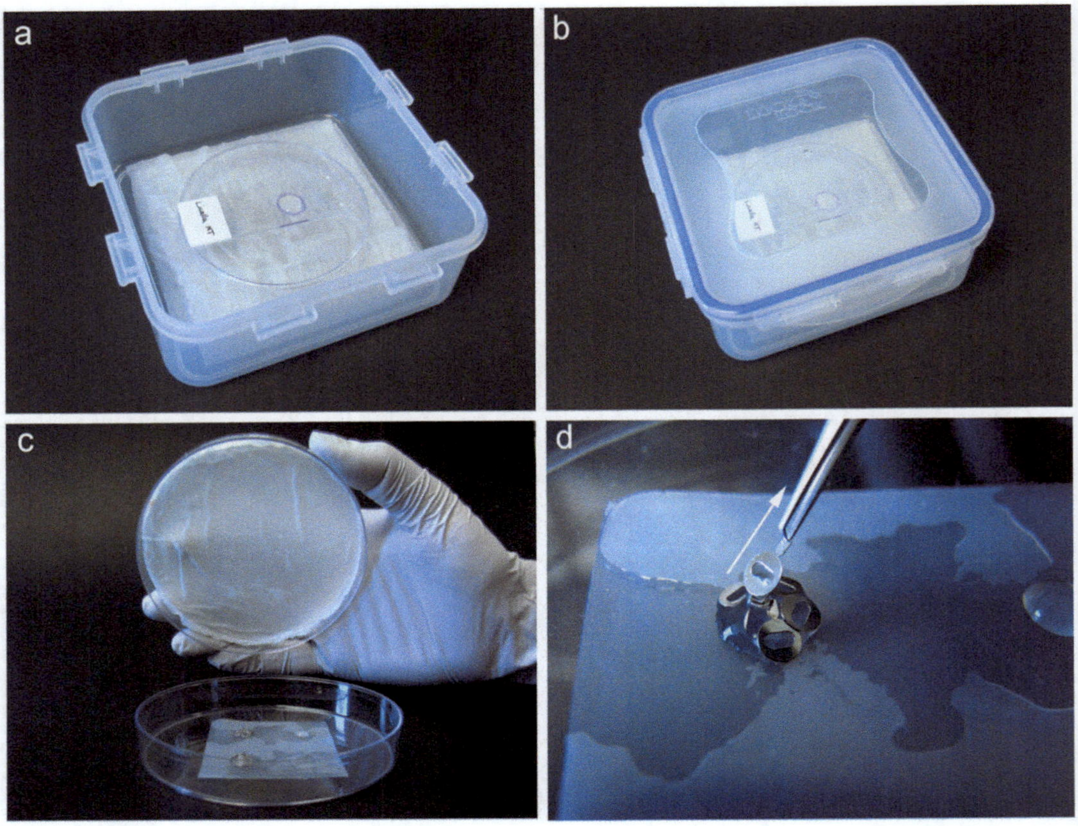

Fig. 5 Illustration of immunohistochemistry in humidity chambers designed for LM (**a**, **b**) or TEM (**c**, **d**). (**a**) A humidity chamber constructed from an airtight plastic container and precision wipes moistened with dH_2O. A Petri dish is used to elevate the slide containing sections of LRW-embedded tissue. A hydrophobic barrier pen was used to draw a boundary on the slide around the periphery of the sections (*light purple*) to prevent immunohistochemical reagents from drifting off the sections. (**b**) A closed humidity chamber during incubation steps. (**c**) A humidity chamber constructed from a Petri dish and moistened precision wipes. Laboratory film adheres to the bottom dish containing a small amount of dH_2O water. (**d**) Drops of immunohistochemical reagents are placed on the laboratory film and grids are placed in the drops. The *arrow* indicates the direction of grid removal from a drop when transferring to a new drop of solution. This direction of removal minimizes the loss of the formvar with the section

13. Glass coverslips, No. 1, 22 × 50 mm.
14. Slide mounting medium.
15. Light microscope adjusted for imaging with dark field microscopy with digital image capture capability.

2.5.2 Transmission Electron Microscopy

1. Sections of LRW-embedded tissue on formvar-coated nickel grids (Subheading 3.4.4).
2. Petri dish for humidity chamber (*see* **Note 17**; Fig. 5c).
3. Anti-capillary tweezers.
4. Laboratory film.

5. 70% EtOH.
6. Precision wipes.
7. Phosphate buffered saline (PBS); 10 mM Na_2HPO_4, 1.7 mM KH_2PO_4, 138 mM NaCl, 2.7 mM KCl, pH 7.4. Store at 4 °C (*see* **Note 18**).
8. 10% BSA (*see* Subheading 2.5.1).
9. Primary antibody with high specificity towards the protein of interest.
10. IgG-gold particle conjugated secondary antibody, 10–20 nm (*see* **Note 19**).
11. Ultrapure water.
12. Transmission electron microscope with digital image capture capability.

2.6 Quantification

1. Digital images.
2. Image J software (available for free download at http://rsbweb.nih.gov/ij/download.html).
3. Pen display tablet (*see* **Note 20**).
4. Statistical analysis software.

3 Methods

3.1 Tissue Fixation

Fixation functions to terminate cellular processes and stabilize cell architecture and chemical composition with minimal distortion to provide an accurate picture of the cell. Chemical and physical fixations are two methods used to achieve these objectives with the former being the most common and accessible and the one described in this chapter [18]. Glutaraldehyde or glutaraldehyde together with formaldehyde is the fixative of choice for standard fixation and fixation for immunohistochemistry and plants with cells that are difficult to fix (thick cell walls and cuticles), respectively [18]. Both glutaraldehyde and formaldehyde stabilize tissue primarily by cross-linking with proteins. When used in combination, the more rapidly penetrating paraformaldehyde ensures fast structural preservation of tissue and the slower moving glutaraldehyde promotes a more thorough cross-linking of proteins [18]. A second or post-fixation with osmium tetroxide (OsO_4) also forms cross-links with proteins and OsO_4 is one of the few fixatives that stabilizes lipids [19, 20]. Lipid preservation results from oxidation of unsaturated bonds of fatty acids by OsO_4. This process also results in the production of insoluble osmium that functions to add contrast to cell membranes due to the electron-dense nature of the heavy metal. In this context, OsO_4 acts as a "stain." Post-fixation with OsO_4 is essential for standard imaging of plant cells

Fig. 6 Illustration of the process of tissue fixation. (**a**) A razor blade cutting immersed tissue initially into 1–2 mm wide strips followed by (**b**) cutting strips into 1–2 mm^2 pieces. (**c**) A vial containing tissue pieces is overfilled with fixative to produce a convex meniscus. (**d**) After securing the lid, tissue pieces should immediately sink to the bottom of the vial. Note samples sinking immediately after placement of lid on vial

with transmission electron microscope except when preparing tissue for immunohistochemistry; OsO_4 reduces antigenicity and is therefore not recommended for use. Post-fixation with OsO_4 is also useful for enhancing high resolution imaging of chloroplasts and mitochondria with a compound light microscope. It is recommended that the same blocks being used for TEM be sectioned initially for LM. This will help the researcher become familiar with tissue architecture prior to viewing with a transmission electron microscope. The steps for classical fixation and fixation for immunohistochemistry are illustrated in Fig. 6. Adhere to protocols for laboratory safety during tissue fixation as the chemicals being used to preserve tissue for microscopy affect all living tissues.

3.1.1 Fixation

1. Assemble all materials for fixation in a fume hood.
2. Clean dissecting scope stage, double-edged razor blades broken in half, and tweezers with 70% EtOH and precision wipes.
3. Immerse the plant organ containing the tissue or cells of interest into a drop of fixative of choice for either classical fixation or immunohistochemistry on the dissecting scope stage plate.
4. Use razor blade to cut immersed tissue into 1–2 mm^2 pieces (*see* **Note 21**; Fig. 6a and b). Use cleaned tweezers to hold the sample.

5. Cut the tip off of a disposable, transfer pipette and draw tissue pieces into the pipette with fixative and transfer to a fixative-filled vial. Overfill the vial with fixative and quickly place the lid on the vial ensuring that no air is introduced into the vial (*see* **Note 22**; Fig. 6c). If done correctly, tissue pieces should immediately sink to the bottom of the vial (Fig. 6d).
6. Fix tissue at room temperature for a minimum of 2–3 h and a maximum of 24 h. Overnight fixation at 4 °C is recommended for immunodetection of protein antigens.

3.1.2 Post-fixation

1. Decant the fixative from the vial and replace with two 30-min washes of sodium cacodylate buffer. Ensure tissue is always covered with a small amount of liquid to prevent exposure of tissue to air.
2. Remove sodium cacodylate buffer from each sample vial and immediately add 1 mL fresh buffer. Add OsO_4 to each vial to bring final concentration to 1% or 2% OsO_4. Post-fix in OsO_4 for 2–3 h at room temperature or until tissue turns a light brown/black (*see* **Note 23**).
3. Decant OsO_4 from vial and wash with sodium cacodylate buffer (2 × 30 min).

3.2 Dehydration

Fixed plant tissue must be dehydrated before embedding in resin. Dehydration gradually removes all water from fixed tissue and replaces the tissue with a solvent that is miscible with the embedding media. Ethanol is the most frequently used dehydrating agent. Tissue can be dehydrated at room temperature and the time allotted to each successive increment within the dehydration series should allow an optimal exchange of solutions. A 30 min minimum in each successive increment should be used for plant tissues and 1 h in each increment is recommended.

1. To dehydrate, decant off the sodium cacodylate buffer with a transfer pipette and replace with 10% EtOH.
2. For each successive 10% increment increase in EtOH, continue by decanting off the old solution and replace with the new solution until completing two changes of 100% EtOH (*see* **Note 24**). Ensure tissue is always covered with a small amount of solution to prevent tissue exposure to air.

3.3 Embedding

Embedding includes the gradual removal of the dehydrating agent and infiltration of the tissue by resin and polymerization of the resin infiltrated tissue. Resins commonly used for plant tissues include glycol methacrylate, LR White acrylic resin (LRW), Spurr's, and Araldite. Glycol methacrylate and LRW are hydrophilic which enables easy penetration of stains and immunodetection of protein. Of these two resins, only LRW is stable when exposed to an electron beam and is therefore the resin of choice if imaging of tissue is done with a light or transmission electron microscope. Spurr's and Araldite are hydrophobic resins designed specifically to withstand the electron beam but they limit penetration of reagents used for histochemistry and immunodetection of proteins. Araldite versus Spurr's is the resin of choice if tissues to be fixed are difficult to infiltrate (thick cuticles, thick cell wells); Araldite requires an additional transitional solvent to be used after dehydration that enhances Araldite penetration of tissue. Spurr's, Araldite, and LRW are polymerized by heat but LRW also requires polymerization in an oxygen-free environment. In addition, if the tissue has been osmicated, the OsO_4 will add a source of oxygen which scavenges the LRW accelerant and inhibits polymerization. Thus, the use of LR White with osmicated tissue is not recommended. A variety of molds are used to eliminate oxygen from LRW during polymerization, all of which minimize the ability to control tissue orientation and therefore require the investigator to cut a small block of polymerized tissue out of the larger block for reorientation on a wooden dowel prior to sectioning. In contrast, tissue samples infiltrated with Spurr's and Araldite are easily oriented in small "ice cube tray-like" molds prior to polymerization. Here, we provide methods to infiltrate and polymerize tissue with resin for high resolution imaging with a light and transmission electron microscope (Spurr's and Araldite) and immunodetection of proteins with the light and transmission electron microscope (LRW). Adhere to protocols for laboratory safety during resin infiltration and polymerization.

3.3.1 Resin Infiltration

Prepare resins according to manufacturer's directions immediately before use (*see* **Notes 4** and **25**). To infiltrate with resin of choice, decant off the final change of 100% EtOH and replace with the below recommended increment changes of solvent:resin (*see* **Note 26**). Ensure tissue is always covered with a small amount of solution to prevent exposure of tissue to air.

LRW; Infiltrate using the following ratio of LRW to 100% EtOH for 8 h each: 1:3; 1:1; 3:1. Follow with 2 × 100% LRW.

Spurr's; Infiltrate using 10% increment increases of Spurr's in 100% EtOH for 3 h each until reaching 100% Spurr's. Follow with 2 × 100% Spurr's, 12 h each.

Araldite 502; Replace final change of 100% EtOH with 1:3 Araldite 502:100% propylene oxide. After 3 h, complete infiltration using the following ratio of Araldite to propylene oxide for 3 h each: 1:1; 3:1. Follow with 2 × 100% Araldite 502 for 3 h each (*see* **Note 27**).

3.3.2 Resin Polymerization

LRW, freshly prepared

1. Preheat oven to 60 °C.
2. Add 5 mL 100% LRW and infiltrated tissue samples to 30 mL medicine cup. Orient the tissue towards the periphery of the cup. Nest a second 30 mL medicine cup within the first. Gently press on the top medicine cup until LRW travels up the sides of the cup (*see* **Note 28**). This will remove air and create an oxygen-free environment.
3. Place double-cup mold in oven for 12 h.
4. Turn off the oven and allow samples to cool in the oven.
5. After samples have cooled, separate the cups. The resin at the base of the cup should be polymerized, while the resin that travelled up the sides of the cup will be soft. Trim soft resin from the polymerized plastic with a razor blade.

Spurr's and Araldite 502, freshly prepared

1. Preheat oven to 60 °C.
2. Clean the flat silicone embedding molds with acetone and dry.
3. Fill wells of molds 75% full with 100% fresh resin.
4. Transfer infiltrated tissue to each well with a transfer pipette, or sharpened wooden applicator stick.
5. Fully fill the tissue-containing wells with resin. Avoid introduction of air bubbles.
6. Orient tissue in the well with a sharpened wooden applicator stick to enable sectioning the plane of choice. The applicator stick can also be used to remove air bubbles.
7. Place mold in oven for 8–16 h for Spurr's resin, and 48–72 h for Araldite.
8. Turn off the oven and allow samples to cool in the oven. Samples can be removed from the mold once it has fully cooled.

3.4 Trimming and Sectioning of Resin-Embedded Tissue

Excessive amounts of resin around embedded tissue must be removed by trimming the blocks prior to sectioning. This is particularly true of LRW-embedded tissue that must also be cut out of larger disc-shaped blocks formed following embedment in medicine cups. The microscopist must experiment with the final size of the block face to determine what works best. As a general rule, the smaller the block face, the less resistance the block will meet with the knife edge resulting in better sections. However, block faces that are too small result in sections that are difficult to pick up. After blocks are trimmed of excess resin, sectioning with a microtome can be done using a glass knife with an attached knife boat or a diamond knife. Glass knives are made with a knife-making machine and must be changed frequently as they become dull after use. In contrast, diamond knives only become dull after extensive use. Although a diamond knife is more expensive in the short term, it is well worth the cost in the long term given the convenience. Diamond knives are available for both LM and TEM. The process of sectioning is challenging for even the most experienced microscopist and requires extreme patience and perseverance. Because of this, troubleshooting tips are provided.

3.4.1 Mounting Small Blocks of LRW-Embedded Tissue onto Wooden Dowels Prior to Trimming

1. Trace a trapezoid shape around one tissue sample in the LRW disc (Fig. 7a and b).
2. Mount LRW disc on the vice grip with the tissue sample upwards (Fig. 7a and b).
3. Use the saw to cut around the tissue sample in a trapezoid shape to ensure a longer, flat base (Fig. 7c).

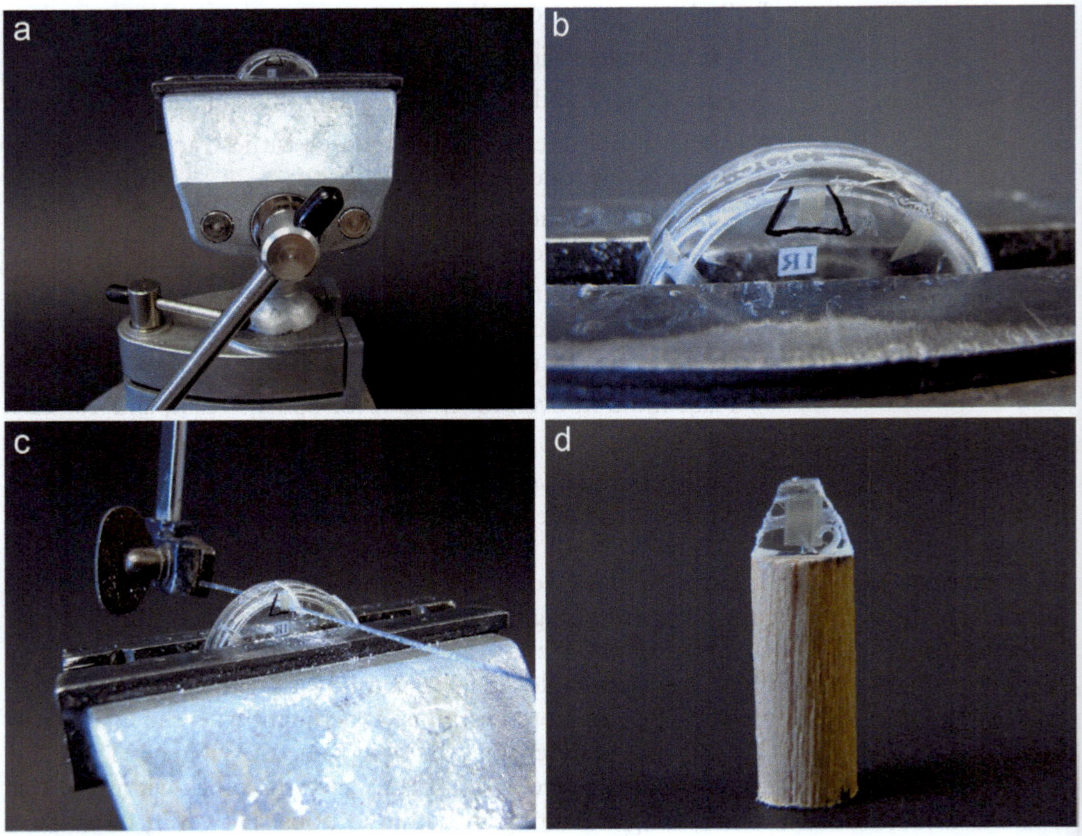

Fig. 7 Excising tissue sample from polymerized LRW discs. (**a**, **b**) Polymerized LRW disc with trapezoid-shaped black line traced around tissue sample mounted in a vice grip. (**c**) A jeweler's saw is used to cut the tissue out of the polymerized LRW disc. (**d**) The excised tissue is then mounted with superglue to an upright wooden dowel and subsequently trimmed to make an evenly shaped block face for sectioning. Note leaf tissue is clear due to a lack of post-fixation in OsO_4

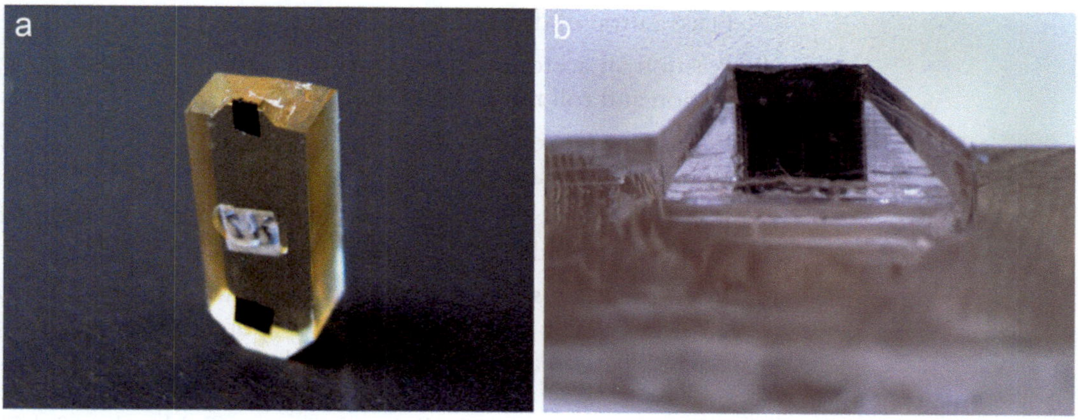

Fig. 8 A polymerized Spurr's block containing osmicated tissue. The block has been trimmed in the shape of a pyramid. This shape provides stability to the block when sectioning. Note leaf tissue is black due to post-fixation with OsO_4

4. Place a drop of superglue on a wooden dowel and use tweezers to place the LRW sample in the superglue ensuring the longer face of the trapezoid is in contact with the wooden dowel (Fig. 7d).
5. Allow the glue to dry completely.

3.4.2 Trimming Spurr's, Araldite, and LRW Blocks

1. Mount the resin block containing the tissue upright in the microtome as instructed by the manual.
2. Use one edge of a cleaned, double-edge razor blade to remove small amounts of resin around the tissue to create a pyramid shape (Fig. 8a and b). This shape provides a base of support for the block during sectioning.

3.4.3 Sectioning for Light Microscopy

1. Position the trimmed block into the microtome as instructed by the manual.
2. Position the knife in the microtome so the knife is parallel with the block face as instructed by the manual.
3. Fill the boat with distilled water such that the water level is just below the knife-edge and the leading edge of the meniscus is in contact with the knife edge.
4. Set the section thickness to 3 μm and position the duller right edge of the knife (Fig. 3b) adjacent to the embedded tissue.
5. Very slowly begin sectioning as instructed by the manual until a complete block section with tissue is coming off into the water in the knife boat. This process is known as "facing the block".
6. Once the block is faced, position the knife to use the left (sharper; Fig. 3b) side of the knife. Set the section thickness to 1.5–2 μm and continue sectioning (*see* **Notes 29–34**).

7. Place a drop of distilled water on an adhesion glass slide.
8. Position an acetone-cleaned glass rod adjacent to the floating section and roll the section onto the glass rod.
9. Release the section into the drop of water on the adhesion slide by rolling the glass rod.
10. Repeat the procedure until you have the desired number of sections on a slide.
11. Place the glass slide with the sections on a hot plate set to 45–48 °C. The water will slowly evaporate as the sections expand, flatten, and adhere to the slide (*see* **Note 35**).

3.4.4 Sectioning for Transmission Electron Microscopy

1. Proceed as sectioning for light microscopy (Subheading 3.4.3) **steps 1–4** using instructions for the ultramicrotome to obtain sections approximately 70 nm thin (dark silver to light gold; *see* **Notes 29–34**, **36** and **37**).
2. Gently wave a chloroform-soaked cotton swab 2–4 mm over the floating sections to remove compression from sections.
3. With the aid of a chloroform-cleaned eyelash brush, pick up one or more silver or light gold sections by inserting a formvar-coated grid (copper for standard imaging; nickel for immuno-histochemistry) with shiny side up into the water below the section and slowly moving the grid upwards to capture the section.
4. Place the grid with the section in a grid box until future use or immediately on a staining pad.

3.5 Staining Resin-Embedded Tissue

Characterization of plant cellular architecture following resin embedment is difficult because of the absence of contrast unless the tissue has been post-fixed with OsO_4. Numerous stains are available to intensify the visibility of cellular detail as well as histochemical detection of compounds in tissues embedded in hydrophilic and hydrophobic resins for both LM and TEM [18]. The most commonly used stain to enhance an understanding of general cellular architecture with the light microscope is toluidine blue [18]. The same blocks used for obtaining sections for LM can be used to obtain sections for TEM. It is recommended to use toluidine blue on sections for LM to become familiar with the cellular features of a sample prior to viewing sections from the same block prepared for TEM. The double contrast staining method is used on ultrathin sections for TEM. This staining technique employs the heavy metals uranium (uranyl acetate) and lead (lead citrate) [19, 20]. Uranyl acetate enhances contrast by interaction with lipids and proteins whereas lead citrate interacts with proteins and glycogens [19, 20].

3.5.1 Staining for Light Microscopy with Toluidine Blue

1. Place a dried slide containing sections on a hot plate at 45–48 °C and flood the slide with toluidine blue dispensed through a 45 μm filter fitted on a syringe.
2. Remove the slide from the hot plate after few seconds for LRW sections and 20–30 s or until a metallic halo forms around the edges of the stain for Spurr's and Araldite sections.
3. Rinse off the toluidine blue with dH_2O into a waste container. Replace the slide onto the hot plate to dry.
4. View sections with the compound light microscope. If the stain appears too light, repeat staining.
5. Dispense three small drops of slide mounting medium equidistant from each other on the slide and gently apply a coverslip to prevent air bubbles.
6. Place multiple precision weights of equal size on the coverslip of freshly mounted slide to insure a thin, even distribution of the mounting medium. Allow them to remain in place while the mounting medium fully polymerizes.
7. Mounted sections are ready for microscopic imaging in 24 h.

3.5.2 Staining for Transmission Electron Microscopy with Lead Citrate and Uranyl Acetate

1. Wash grid pad and tweezers with soap and water and rinse thoroughly. Wipe with 95–100% ethanol and rinse with dH_2O.
2. Place grids in clean, dry grid pad in a dark staining chamber (closed Petri dish covered with aluminum foil).
3. Immerse each grid in a drop of uranyl acetate dispensed through a 45 μm syringe filter. Stain for 10 min.
4. Slowly pour off the uranyl acetate, and rinse with dH_2O (3 × 5 min).
5. Place eight pellets of sodium hydroxide in a closed Petri dish to absorb CO_2. Excess CO_2 results in unwanted lead precipitate on sections.
6. Slowly pour the water off the grid pad. Place the pad with the grids in the sodium hydroxide containing staining chamber and replace the Petri dish cover.
7. Slightly open the Petri dish lid and quickly immerse each grid in a drop of lead citrate dispensed through the 45 μm syringe filter. Stain for 5 min.
8. Rinse with dH_2O (3 × 5 min).
9. Allow grids to dry in the staining pad in a covered Petri dish (*see* **Note 38**).

3.6 Immunohistochemical Detection of Proteins

The primary objective of immunohistochemistry is to define the cellular location of proteins. There are two main methods of immunohistochemistry, direct and indirect [21]. Direct employs a primary single antibody that is conjugated to a marker for microscopic

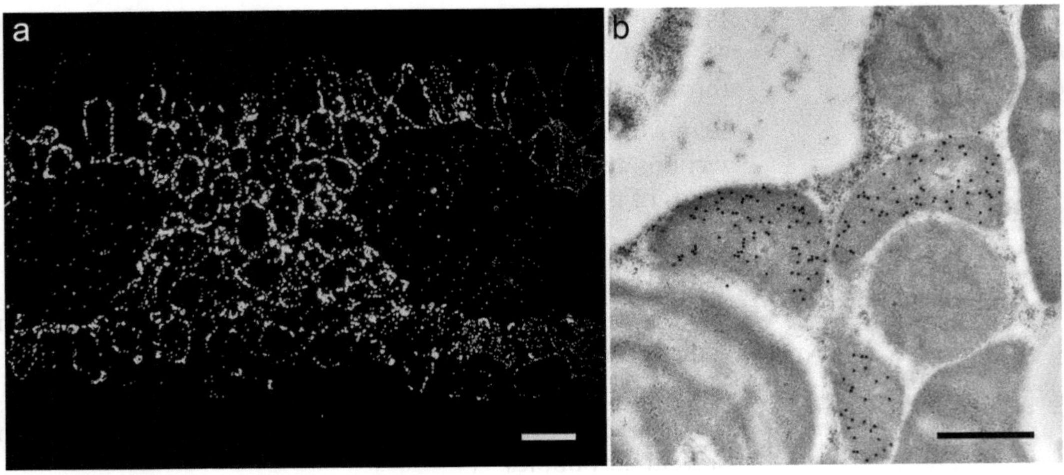

Fig. 9 Images illustrating results from immunohistochemical detection of proteins with LM (**a**) and TEM (**b**). (**a**) Silver enhancement following immunolocalization of phosphoenolpyruvate carboxylase in leaves of *Anticharis glandulosa* using methods described in Khoshravesh et al. [24]. Silver grains appear as bright spots when viewed with dark field microscopy. (**b**) Immunolocalization of the P-subunit of glycine decarboxylase in mesophyll cells of rice (*Oryza sativa*) using methods described in Khoshravesh et al. [7]. *Black dots* represent gold particles. Bars = (**a**) 30 μm; (**b**) 500 nm

detection and indirect uses a primary antibody that is unconjugated and a secondary antibody with a conjugated marker that is directed against the primary antibody for detection. The indirect method is one of the choices for TEM because it is more sensitive; several secondary antibodies will bind to the primary antibody resulting in an amplified signal. For TEM, IgG-gold particle (10–20 nm diameter) conjugated secondary antibodies are most commonly used [22, 23]. For LM, the indirect method using similar secondary antibodies or Protein A-conjugated gold combined with silver enhancement enables further signal amplification which is necessary for detection of immunolocalization events with a light microscope [23]; the gold particles nucleate silver deposition supplied in a silver salt solution in the presence of a reducing agent. Silver enhancement has also been used for TEM [23]. When viewed with a light microscope adjusted for dark field microscopy, bright spots indicate positive immunodetection of proteins (Fig. 9a; [24]). When viewed with TEM, black dots indicate a positive immunodetection event (Fig. 9b). Controls must always be performed. Controls can be run by omitting the primary antibody or using pre-immune serum. Here, we provide methods for the indirect method of immunohistochemistry for LM and TEM.

3.6.1 Immunohistochemistry for Light Microscopy

1. Collect 0.8–1 μm sections of LRW-embedded tissue of interest on adhesion slides as described in Subheading 3.4.3.
2. Use a hydrophobic barrier pen to draw a boundary on the slide around the periphery of the sections and allow to dry (*see* **Note 39**; Fig. 5a).

3. Place this slide in a humidity chamber for the remaining steps (Fig. 5a and b).
4. Dispense 10 mM TBS within the hydrophobic barrier to cover and rehydrate the tissue sections for 30 min.
5. Replace TBS with 0.1% BSA in TBS for 1 h to block nonspecific binding.
6. Wash with 10 mM TBS (3 × 20 min).
7. Incubate in a proper concentration of antibody or pre-immune/isotype specific immunoglobulin serum diluted in 10 mM TBS and 0.1% BSA for 3 h (*see* **Note 40**).
8. Wash with 10 mM TBS (3 × 20 min).
9. Incubate in protein A-gold, 10 nm (1:100) in 10 mM TBS + 0.1% BSA.
10. Rinse 10 mM TBS (3 × 20 min).
11. Expose to silver enhancement reagent according to manufacturer's instructions.
12. Rinse with tap water for 2 min, and air dry.
13. Coverslip the slides (*see* Subheading 3.5.1) and observe with dark-field microscopy.

3.6.2 Immuno-histochemistry for TEM

1. Collect 70 nm thin sections of LRW-embedded tissue on formvar-coated nickel grids as described in Subheading 3.4.4.
2. Place a clean square shape laboratory film in a humidity chamber and float grids on drops of PBS on the film for 15 min (Fig. 5c).
3. Use a pair of anti-capillary tweezers (*see* **Note 41**) to gently slide grids (Fig. 5d) from the PBS drop and transfer grids to a drop of 0.4–0.5% BSA in PBS (blocking agent) in a second humidity chamber for 15–30 min. Proceed with remaining steps by alternating between two humidity chambers.
4. Rinse grids on drops of PBS (3 × 15 min).
5. Incubate grids in primary antibody or pre-immune serum (*see* **Note 40**) diluted in 0.1% BSA in PBS for 2–3 h. A 40 μL drop is adequate for 5–6 grids.
6. Rinse grids on drops of PBS (3 × 15 min).
7. Incubate the grids in a drop of secondary antibody conjugated to gold particles (10–20 nm) in 0.1% BSA in PBS (*see* **Note 42**).
8. Rinse grids on drops of PBS (3 × 15 min).
9. Rinse grids on drops of ultrapure water (3 × 15 min).
10. Air-dry and stain with uranyl acetate and lead citrate as described in Subheading 3.5.2.
11. Observe with a transmission electron microscope. Black dots correspond to gold particles (Fig. 9b).

3.7 Quantification

Quantification of cellular parameters, to include density of immunodetection events of a given protein, is frequently essential to test hypotheses [6, 7, 11, 12]. All plants being used for quantitative assessment should be grown under the same growth conditions, and if possible, at the same time. Sampling tissue from 3–5 plants with each plant representing a replicate is recommended. 5–10 images per replicate plant most frequently provide enough data for statistical resolution. Below is a recommended procedure for quantification of various cellular parameters.

1. Make a list of the variables you intend to quantify.
2. Digitally capture enough images of the cellular features of interest for quantification with Image J (https://imagej.nih.gov/ij/docs/guide/user-guide.pdf) or other morphometric software.
3. Open a digital image in Image J and view on the pen display tablet.
4. Set the scale using Analyze menu/set scale (*see* **Note 43**).
5. Choose the proper tool. The most common tools used in cell feature quantification are free hand selection, free hand line, and multi-point tool.
6. Use the pen stylus to draw a line, trace the outline of a cell or an organelle or create points. Measure the variable using measure feature in Analyze menu or set up short cut key in your tablet.
7. Copy the values to the statistical analysis software for processing.

4 Notes

1. Use only distilled H_2O for making all solutions for TEM to include buffers, EtOH dehydration series, and stains. Ultrapure water reduces the oxidation reactions by OsO_4 and results in a peppered appearance (compare Fig. 10a and b with c) of the tissue with suboptimal cellular membrane preservation.
2. Break double-edged razor blade in half and clean with 70% EtOH to remove oil and packaging residue. Cover broken edge of the razor blade with tape to protect fingers. Resin is harder than fingers!
3. Ellis [25] provides data to modify the hardness of Spurr's resin by varying the amount of the flexibilizer, DER 736. Note that Spurr's has a shelf life of 3 days.
4. When purchasing Araldite 502, replace DMP-30 with BDMA. BDMA has a higher power of penetration in plant tissue and a longer shelf life.

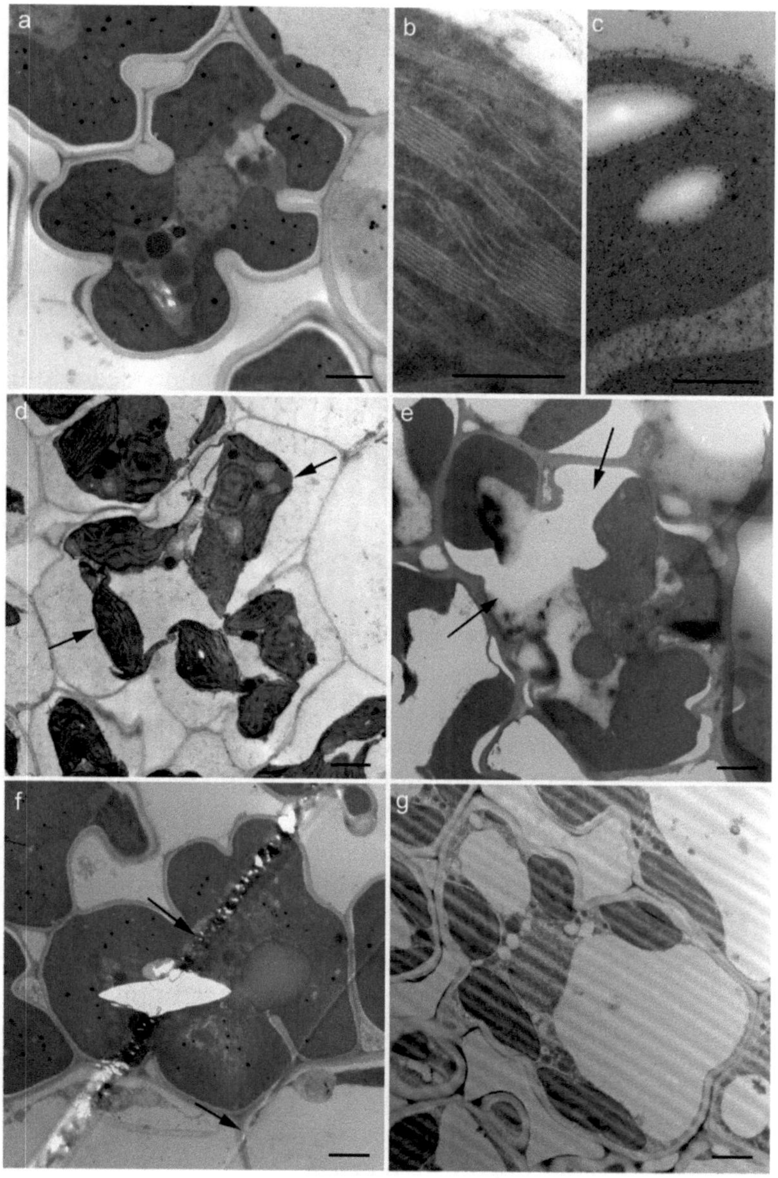

Fig. 10 Images to assist with trouble-shooting fixation, infiltration, and sectioning. (**a**) Optimally fixed rice mesophyll cell. (**b**) Optimally fixed chloroplast. (**c**) Peppered appearance of chloroplast due to the use of ultra-pure water in fixation solutions. (**d**) Suboptimal fixation of a rice mesophyll cell; note the chloroplasts and cell membrane are pulled away from the cell wall (*arrows*). (**e**) A poorly infiltrated rice mesophyll cell; note the cell components are disintegrated and parts of the cell are missing (*arrows*). (**f**) Damage (*arrow*) caused by chips in the knife. (**g**) Striations resulting from uneven thickness throughout section. Bars = (a, d, e-g) 2 μm; (b,c) 500 nm

5. Melted paraffin wax can be used in place of nail polish. Apply with a flat-ended weighing spatula.
6. Make a glass rod by sealing the open end of the tip of a glass Pasteur pipette with the flame of an alcohol burner. Then flame

the glass Pasteur pipette 3 cm behind the tip and break while flaming. Flame seals the broken ends.

7. Fisherbrand plain beveled edge pre-cleaned microscope slides (3″ × 1″ × 1.0 mm) are recommended for making formvar-coated grids because the formvar releases with ease from the slide. Do not over-clean the slides because the formvar film will not detach from the slide.
8. To make an eyelash brush, obtain an eyelash and adhere it to the end of a thin wooden applicator stick with nail polish or glue.
9. Tabbed grids are strongly recommended because they allow the microscopist to handle the grid with ease during use. Slot grids are recommended over mesh grids because a larger area of tissue can be imaged without obstruction from mesh bars.
10. Store in a grid box specifically designed for tabbed grids to accommodate the larger size of the grid.
11. Chien staining pads from Ted Pella are recommended. They provide optimal flexibility when working with tabbed grids.
12. Lead citrate is CO_2 sensitive and interactions form a white insoluble lead carbonate precipitate. All efforts must be made to prevent contact with atmospheric CO_2. If the solution does not clear following the addition of NaOH, dispose and begin again.
13. Use a freshly open bottle of chloroform to make formvar; water will compromise formvar integrity and result in the formation of holey, unstable formvar.
14. Removing the formvar-coated slide very slowly from the solution will give rise to a thin film and a quick removal results in a thick film. A thin film is desired. As well, a clean working space is vital for preparing samples for TEM. A very small invisible drop of grease may cover the entire area of interest on a tissue section. Wipe all tools and working space with ethanol and thoroughly wash hands.
15. The rough side of a grid allows more surface area for formvar adherence. The silver and gold regions of the formvar are the optimal thickness for TEM. All other colors are too thick and will compromise the ability of electrons to penetrate the samples, thereby limiting high resolution imaging.
16. To construct a humidity chamber for LM, place precision wipes on the bottom of a plastic container and wet. Place the slides on an inverted Petri dish, and seal the container.
17. To construct a humidity chamber for TEM, place precision wipes on the lid of a Petri dish and lightly wet. Add a small amount of dH_2O on the bottom of the Petri dish and place a

section of laboratory film clean side up in the water so the film lies flat in the dish. Drops for immunohistochemistry will be placed on the laboratory film.

18. Phosphate buffered saline tablets are recommended.
19. The secondary antibody raised against the IgG of the animal species in which the primary antibody has been raised should be used for TEM.
20. Wacom pen display tablets are recommended.
21. The quality of fixation and resin infiltration will be severely compromised if larger tissue pieces are processed.
22. This procedure removes the need to subject tissue to a vacuum, a commonly used practice that can lead to cellular artifacts. Do not process tissue that does not sink immediately. Shaking or inverting will ease the process of fixation. Figure 10a, b, and d illustrates optimal and sub-optimal fixation of a rice mesophyll cell, respectively.
23. A minimum post-fixation in OsO_4 of 2–3 h is sufficient for most tissues. If a longer time is needed, tissue should be left in OsO_4 solution overnight at 4 °C. Use 2% OsO_4 for tissues with lower lipid and protein content and 1% for tissues containing more lipids and protein.
24. EtOH dehydration can be completed at room temperature. Tissue samples can be stored in $\leq$70% EtOH at 4 °C overnight for LM and TEM. Samples in >70% should be processed immediately to minimize extraction of cellular materials.
25. All resins used in this protocol are toxic, so they should be used in a fume hood while wearing gloves. Read the MSDS before infiltrating samples.
26. The use of a rotating table or shaker is recommended to improve the infiltration. Figure 10e illustrates a poorly infiltrated rice mesophyll cell.
27. The shelf life of Araldite is short, and it polymerizes quickly. Thus, samples should not be left overnight in 100% Araldite.
28. To prevent air bubbles from entering between the medicine cups during LRW embedding, a thin wooden applicator or a flat-ended weighing spatula can be used to create a passageway for air to escape. Place the wooden dowel vertically between the two cups, and slowly remove the dowel as the cups are positioned together.
29. Sections may sometimes roll upon themselves. Rolling is due to either a dull knife, an excessively high water level in the knife boat, or a high speed of cutting. To eliminate rolling, change the knife, lower the water level, or reduce the speed of the microtome as the section is cut.

30. Chips in the knife cause lines and striations in the section (Fig. 10f). Chips may result from sectioning tissues that have large crystals or silica. If the tissue is causing knife imperfections, frequent changes to a new glass knife will be necessary. To avoid imperfections in the glass knife that are caused by debris external to the block, clean the knife and resin block of debris by over-filling the boat with water and gently run a precision wipe over the surface of the block, glass knife, and water. Use guidelines for cleaning the diamond knife provided by the manufacturer.
31. Sections may appear compressed or wrinkled if the knife is no longer sharp. Move to a new region of the knife and reduce sectioning speed.
32. Water should wet the knife's edge. If not, use the pipette tip (light microscopy) or eyelash brush (TEM) to guide the water meniscus to the knife's edge.
33. High water levels in the knife boat or static charge may cause water to jump onto the block. Reduce water levels in the boat with a pipette. Use a humidifier to reduce static charge in the air. Touch the block face and back of the knife with tissue to remove water from the block and knife.
34. High humidity can result in sections sticking to the block during the return stroke of sectioning. Reduce the room temperature and use a dehumidifier to eliminate this problem.
35. Sections prepared for light microscopy may wrinkle on the adhesion microscope slide during drying if the temperature of the hot plate is too high or low. A high temperature setting results in rapid drying prior to complete expansion of the section. A low temperature setting is not adequate for the section to fully expand.
36. Chatter appears on the sections as unevenly thickened lines that run parallel to the knife's edge (Fig. 10g). Chatter results from an incorrectly tightened knife, block, or microtome stage, a high cutting speed, or inappropriate knife angle. Ensure all parts are securely tightened, the tissue block (LRW blocks) is firmly attached to the wooden dowel, and the knife angle recommended by the microtome manual is employed. A fast cutting speed or a large block face may also produce chatter. Ultramicrotomes are sensitive to vibration that can be minimized by positioning the microtome on an anti-vibration table. Finally, a poorly polymerized block also causes chatter.
37. The ultramicrotome should be mounted on an anti-vibration/floating table to reduce chatter and other vibration-related imperfections in sections.

38. Gently touching the base of the grid (without touching the section) to a tissue will lessen the amount of water on the grid and reduce the drying time.
39. The hydrophobic barrier pen confines the reaction solutions to the area of sections. A black wax pencil can be used as an alternative to the hydrophobic barrier pen.
40. The concentration of primary antibody used for LM and TEM is generally 10–15× higher than used for immunoblotting. To control for primary antibody specificity, use pre-immune serum or isotype-specific immunoglobulins at the same concentration of primary antibody.
41. Anti-capillary tweezers prevents cross-contamination of the liquids between each step and prevents grids from being pulled up in the tweezers.
42. A 1:50 concentration of secondary antibody is a good starting point. To control for secondary antibody specificity, omit the primary antibody.
43. The imaging software usually contains a scale feature. Keeping the scales updated while changing the magnification is essential for consistent quantifications. To minimize quantification errors, we recommend taking all pictures with the same magnification.

References

1. Tolbert NE (1971) Microbodies – peroxisomes and glyoxysomes. Ann rev. Plant Physiol 22:45–74
2. Tolbert NE (1973) Glycolate biosynthesis. In: Horecker BL, Stadtman ER (eds) Current topics in cellular regulation, vol 7. Academic Press, New York, pp 21–50
3. Frederick SE, Newcomb EH (1969) Microbody-like organelles in leaf cells. Science 163:1353–1355
4. Frederick SE, Newcomb EH (1971) Ultrastructure and distribution of microbodies in leaves of grasses with and without CO_2-photorespiration. Planta 96:152–174
5. Hylton CM, Rawsthorne S, Smith AM, Jones DA, Woolhouse HW (1988) Glycine decarboxylase is confined to the bundle-sheath cells of leaves of C_3–C_4 intermediate species. Planta 175:452–459
6. Sage TL, Busch FA, Johnson DC, Friesen PC, Stinson CR, Stata M, Sultmanis S, Rahman BA, Rawsthorne S, Sage RF (2013) Initial events during the evolution of C_4 photosynthesis in C_3 species of *Flaveria*. Plant Physiol 163:1266–1276
7. Khoshravesh R, Stinson CR, Stata M, Busch FA, Sage RF, Ludwig M, Sage TL (2016) C_3–C_4 intermediacy in grasses: organelle enrichment and distribution, glycine decarboxylase expression, and the rise of C_2 photosynthesis. J Exp Bot 67:3065–3078
8. Busch FA, Sage TL, Cousins AB, Sage RF (2013) C_3 plants enhance rates of photosynthesis by reassimilating photorespired and respired CO_2. Plant Cell Environ 36:200–212
9. Timm S, Wittmiss M, Gamlien S, Ewald R, Florian A, Frank M, Wirtz M, Hell R, Fernie AR, Bauwe H (2015) Mitochondrial dihydrolipoyl dehydrogenase activity shapes photosynthesis and photorespiration of *Arabidopsis thaliana*. Plant Cell 27:1968–1984
10. Lin HC, Karki S, Coe RA, Bagha S, Khoshravesh R, Balahadia CP, Ver Sagun J, Tapia R, Israel WK, Montecillo F, de Luna A, Danila FR, Lazaro A, Realubit CM, Acoba MG, Sage TL, von Caemmerer S, Furbank RT, Cousins AB, Hibberd JM, Quick WP, Covshoff S (2016) Rice (*Oryza sativa* L. cv. IR64) plants with a knockdown in GDCH can survive in ambient air but exhibit a photorespiratory

deficient phenotype. Plant Cell Physiol 57:919–932

11. Stata M, Sage TL, Rennie TD, Khoshravesh R, Sultmanis S, Khaikin Y, Ludwig M, Sage RF (2014) Mesophyll cells of C_4 plants have fewer chloroplasts than those of closely related C_3 plants. Plant Cell Environ 37:2587–2600
12. Stata M, Sage TL, Hoffman N, Covshoff S, Ka-Shu Wong G, Sage RF (2016) Mesophyll chloroplast investment in C_3, C_4 and C_2 species of the genus *Flaveria*. Plant Cell Physiol 57:904–918
13. Schulze S, Mallmann J, Burscheidt J, Koczor M, Streubel M, Bauwe H, Gowik U, Westhoff P (2013) Evolution of C_4 photosynthesis in the genus *Flaveria*: establishment of a photorespiratory CO_2 pump. Plant Cell 25:2522–2535
14. Hanson MR, Köhler RH (2001) GFP imaging: methodology and application to investigate cellular compartmentation in plants. J Exp Bot 52:529–539
15. Sankaranarayanan S, Samuel MA (2015) Guiding principles for live cell imaging of plants using confocal microscopy. In: Yeung ECT, Stasolla C, Sumner MJ, Huang BQ (eds) Plant microtechniques and protocols. Springer, pp 213–224
16. Jamai A, Salomé, Schilling SH, Weber APM, McClung CR (2009) Arabidopsis photorespiratory serine hydroxymethyltransferase activity requires the mitochondrial accumulation of ferredoxin-dependent glutamate synthase. Plant Cell 21:595–606
17. Rademacher N, Kern R, Fujiwara T, Mettler-Altmann T, Miyagishima S, Hagemann M, Eisenhut M, Weber APM (2016) Photorespiratory glycolate oxidase is essential for the survival of the red alga Cyanidioschyzon merolae under ambient CO_2 conditions. J Ex Bot 67:3165–3175
18. Ruzin SE (1999) Plant microtechnique and microscopy. Oxford University Press, New York
19. Bozzola JJ, Russell LD (1992) Electron microscopy: principles and techniques for biologists. Jones and Bartlett Publishers, MA
20. Lowe J (1996) Techniques in neuropathology. In: Bancroft JD, Sevens A (eds) Theory and practice of histological techniques. Churchill Livingstone, New York
21. Kiernan JA (1990) Histological and histochemical methods: theory and practice. Pergamon Press, New York
22. Oliver C, Jamur MC (2010) Overview of antibodies for immunochemistry. Methods Mol Biol 588:3–9
23. Oliver C (2010) Use of immunogold with silver enhancement. Methods Mol Biol 588:311–316
24. Khoshravesh R, Akhani H, Sage TL, Nordenstam B, Sage RF (2012) Phylogeny and photosynthetic pathway distribution in *Anticharis* Endl. (Scrophylariaceae). J Exp Bot 63:5645–5658
25. Ellis EA (2006) Corrected formulation for Spurr low viscosity embedding medium using the replacement epoxide ERL 4221. Microsc Microanal 12(Suppl 2). doi:10.1017/S1431927606062660

Index

A

B

C

D

E

F

Alisdair R. Fernie et al. (eds.), *Photorespiration: Methods and Protocols*, Methods in Molecular Biology, vol. 1653, DOI 10.1007/978-1-4939-7225-8, © Springer Science+Business Media LLC 2017

N

O

P

Q

R

S

T

U

V

Y

MIX
Papier aus verantwortungsvollen Quellen
Paper from responsible sources
FSC® C105338

If you have any concerns about our products,
you can contact us on
ProductSafety@springernature.com

In case Publisher is established outside the EU,
the EU authorized representative is:
Springer Nature Customer Service Center GmbH
Europaplatz 3, 69115 Heidelberg, Germany

Printed by Libri Plureos GmbH
in Hamburg, Germany